W9-ACU-974

CONTROL AND DYNAMIC SYSTEMS

Advances in Theory and Applications

Volume 36

CONTRIBUTORS TO THIS VOLUME

MICHÈLE BRETON
SHAWN E. BURKE
WODEK GAWRONSKI
YACOV Y. HAIMES
ALAIN HAURIE
JAMES E. HUBBARD, JR.
JER-NAN JUANG
CHIN-PING LEE
DUAN LI
N. HARRIS MCCLAMROCH
GARY A. MCGRAW
BERC RUSTEM
MUIBI ADE. SALAMI
GEORGES ZACCOUR

CONTROL AND DYNAMIC SYSTEMS

ADVANCES IN THEORY AND APPLICATIONS

Edited by
C. T. LEONDES

School of Engineering and Applied Science
University of California, Los Angeles
Los Angeles, California

College of Engineering
University of Washington
Seattle, Washington

VOLUME 36: ADVANCES IN LARGE SCALE SYSTEMS DYNAMICS AND CONTROL

ACADEMIC PRESS, INC.
Harcourt Brace Jovanovich, Publishers
San Diego New York Boston
London Sydney Tokyo Toronto

This book is printed on acid-free paper. ∞

Academic Press, Inc.
San Diego, California 92101

United Kingdom Edition published by
Academic Press Limited
24-28 Oval Road, London NW1 7DX

Library of Congress Catalog Card Number: 64-8027

ISBN 0-12-012736-9 (alk. paper)

Printed in the United States of America
90 91 92 93 9 8 7 6 5 4 3 2 1

CONTENTS

CONTRIBUTORS

Numbers in parentheses indicate the pages on which the authors' contributions begin.

Michèle Breton (75), *GERAD and HEC, Montreal, Quebec, Canada*

Shawn E. Burke (223), *The Charles Stark Draper Laboratory Inc., Cambridge, Massachusetts 02139*

Wodek Gawronski (143), *Jet Propulsion Laboratory, California Institute of Technology, Pasadena, California 91109*

Yacov Y. Haimes (1), *Center for Risk Management of Engineering Systems, University of Virginia, Charlottesville, Virginia 22903*

Alain Haurie (75), *University of Geneva, Switzerland*

James E. Hubbard, Jr. (223), *The Charles Stark Draper Laboratory Inc., Cambridge, Massachusetts 02139*

Jer-Nan Juang (143), *NASA Langley Research Center, Hampton, Virginia 23665*

Chin-Ping Lee (107), *Department of Electrical Engineering and Computer Science, The University of Michigan, Ann Arbor, Michigan 48109*

Duan Li (1), *Center for Risk Management of Engineering Systems, University of Virginia, Charlottesville, Virginia 22903*

N. Harris McClamroch (107), *Department of Electrical Engineering and Computer Science, The University of Michigan, Ann Arbor, Michigan 48109*

Gary A. McGraw (275), *The Aerospace Corporation, El Segundo, California 90009*

Berc Rustem (17), *Department of Computing, Imperial College of Science, Technology and Medicine, London SW7 2BZ, England*

Muibi Ade. Salami (333), *Hughes Aircraft Company, Los Angeles, California 90009*

Georges Zaccour (75), *GERAD and HEC, Montreal, Quebec, Canada*

PREFACE

In the 1940s techniques for the analysis and synthesis of systems were simplistic, at best, and so were the system models. In fact, a paper was published in *Econometrica* in 1949 which represented national economies as a simple second order model primarily because the techniques available at the time would not allow otherwise. Nevertheless, this was an important paper because it put forward conceptually and somewhat analytically the important notion of feedback in quantitative economics. It was not until the latter part of the 1950s, not all that long ago, really, that a foundation for multivariable system analysis and synthesis began to be laid, and certainly Kalman played a key role in this. In any event, the rapid growth of computer power, coupled with the requirement to more effectively deal with large scale systems in such areas as engineering, industrial, and economic systems, resulted in and continues to result in advances in techniques for dealing with large scale systems problems. Indeed, these trends were recognized and manifestly treated at various times in earlier volumes in this series. For example, Volumes 14 and 15 were devoted to techniques for modeling (the beginning of the system design process) large scale systems such as jet engines, electric power systems, chemical processes, industrial processes, water resource systems, and the thyroid hormonal system of the human body. it is now time to revisit the area of large scale systems in this series, and so the theme for this volume is "Advances in Large Scale Systems Dynamics and Control."

The first contribution, "Multiobjective Decision-Tree Analysis in Industrial Systems," by Yacov Y. Haimes and Duan Li, deals with the fact that many large scale systems involve multiobjectives, and thus multiobjective analysis needs to be performed to provide the decisionmaker(s) with a set of policies with a resultant set of trade-offs between the different objectives. Two substantive examples are presented in this contribution, and by these examples the essential significance of multiobjective decision-tree analysis in applied situations is clearly demonstrated. Professor Haimes is probably the single most visible contributor on the international scene in this area of major applied significance, and so this contribution by him and a colleague constitutes a most appropriate beginning for this volume. The next contribution, "Methods for Optimal Economic Policy Design," by Berc Rustem, presents important techniques for one of the ultimate challenges in large scale systems, the determination of robust economic policies in the face of anticipated disturbances and in the presence of rival economic models, of which there are a plethora. It was in 1980 that Professor Lawrence R. Klein received the Nobel Prize in economics for his work in econometrics and optimal fiscal policy generation by the modern techniques of optimization. The advances represented by the differences between the *Econometrica* paper of 1949, referred to above, and Professor Klein's work, as well as the current work of other leading researchers in this important area, clearly manifest the substantive advances in techniques in this broad area of large scale systems. Therefore, Rustem's contribution in the area of robust optimal economic policies is a most welcome addition to this volume. The next contribution, "Methods in the Analysis of Multistage Commodity Markets," by Michèle Breton, Alain Haurie, and Georges Zaccour, presents significant advances in techniques in the large scale systems area of networks of interacting oligopolies. Methods for the characterization and computation of equilibrium solutions for such complex large scale systems problems are motivated by a significant applied example from which more general results and techniques are developed. Because of the pervasiveness of this category of large scale systems on the international scene, this contribution is also a most welcome addition to this volume.

Another major large scale systems area is production control and the determination of optimal production rates, which is a control variable in the stochastic environment which production systems represent. The next contribution, "A Control Problem for Processing Discrete Parts with Random Arrivals," by Chin-Ping Lee and N. Harris McClamroch presents rather general and powerful results for dealing with the rather complex problem. The formulation of the problem by the authors has a number of significant advantages, including generality, mathematical tractability, and a formulation which has the essential properties of a closed-loop production. A computationally effective algorithm (a nonlinear search routine) is then presented for the determination of the optimal production rate. This contribution concludes with a substantive example to illustrate the design methodology. Because of the great significance of optimal production control on the international scene, this chapter also constitutes a most welcome addition to this volume on large scale systems. The next contribution, "Model Reduction for Flexible Structures," by Wodek Gawronski and Jer-Nan Juang, presents many highly effective techniques for near optimal and robust model truncation techniques. Very important, also, is the fact that the techniques for model reduction, while developed within the context of flexible structures, are, in fact, applicable to any linear system. Even with the powerful advances in computer technology there is still a great need for model reduction of systems which would otherwise require a high order number of state variables for their "faithful" representation. What is really important is an adequately effective model representation of high order complex systems. Because of the comprehensive treatment of this important subject in this contribution, it represents an important element of this volume on large scale systems.

The increasing emergence of large structural systems and modern system performance requirements has resulted in an increasing utilization of active control techniques. To put it somewhat simplistically, in the "era" of simple rigid body control problem, the system design and control actuator problem was considerably simpler. However, the requirements dictated by more complex flexible structures or systems, both for reasons of system performance and structural alleviation, has brought with it a requirement for distributed control as well as the accompanying distributed sensors. This area of what could be described as enhanced control of systems is commonly referred to as active control systems. To put it another way, if one wants to do more in the way of control for distributed systems, for whatever reason, one has to systematically introduce additional sensors as well as related additional control actuators. The next contribution, "Distributed Transducers for Structural Measurement and Control," by Shawn E. Burke and James E. Hubbard, Jr., is a rather comprehensive treatment of this subject and also a most welcome contribution to this volume on large scale systems. As in the case of the previous contribution, it should also be pointed out that, although the techniques in this contribution are presented within the context of large scale structural systems, the techniques are more generally applicable to linear systems with similar fundamental characteristics.

The next contribution, "Robust Adaptive Identification and Control Design Techniques," by Gary A. McGraw is a comprehensive treatment of robust adaptive identification and control algorithms for disturbances and unmodeled system dynamics, a highly significant broad area in large scale systems. There are two basic approaches to coping with uncertainty in control system design. One approach is robust control, in which a single controller is designed to operate over all possible parameter variations and unmodeled system dynamics. The other approach is adaptive control, in which the structured parameters are estimated and the controller updated accordingly. An adaptive control should only be considered after a fixed robust design has been shown not to meet the design objectives. In this event, this contribution presents numerous powerful techniques which are verified in both computer simulations and in the laboratory in this contribution. In the final contribution, "Techniques in the Optimal Control of Distributed Parameter Systems," by Muibi Ade. Salami, the significant issue of practical means (algorithm development) of optimal control of distributed parameter systems is treated. Examples of the applied significance of the techniques presented in this chapter include such large scale systems areas as aerospace, defense, chemical, environmental, and infrastructural industries. As a result, this contribution is a most fitting chapter with which to conclude this volume.

The authors are all to be commended for their superb contributions, which will provide, in this volume, a significant reference source for practicing professionals and research workers in the broad area of large scale systems for many years to come.

MULTIOBJECTIVE DECISION-TREE ANALYSIS IN INDUSTRIAL SYSTEMS

YACOV Y. HAIMES and DUAN LI

Center for Risk Management of Engineering Systems
University of Virginia
Charlottesville, Virginia 22903

I. INTRODUCTION

The design and operation of industrial systems often lead to sequential decisionmaking problems. Since Howard Raiffa published his celebrated book on decision-tree analysis [1] in 1968, decision-tree analysis has been successfully applied in solving a general class of multistage decision problems with probabilistic nature. The analysis of a decision tree aims at evaluating quantitatively and iteratively how good each of its decision sequences achieves its objective. The optimal decision (in the single-objective sense) is commonly found by a method called averaging out and folding back -- an approach that is grounded on the concept of the expected value.

Today's industrial systems are becoming more and more complex. It is generally realized that a single criterion (such as cost minimization) is seldom an adequate measure of the performance of such systems. Noncommensurable and conflicting objectives characterize most real-world industrial design and operation problems. Multiobjective optimization (see Chankong and

Haimes [2]) thus needs to be performed in order to provide the decisionmaker(s) with a set of noninferior policies and with trade-offs between the different objectives.

The expected value of a system's measure -- the most commonly used scalar representation of risk -- commensurates the system's various possible outcomes and their associated probabilities. Often, the expected-value approach is inadequate to provide sufficient information for risk management. It is particularly deficient for addressing extreme events, since these events are concealed during the amalgamation of events of low probability and high consequence and events of high probability and low consequence. Multiple risk measures need to be introduced to quantify the risk at different levels. Risk management thus incorporates the selection of the most preferred option in a multiobjective framework.

Recently, a multiobjective decision-tree method has been developed by Haimes et al. [3] to extend the traditional decision-tree method to incorporate multiple objectives and various risk measures. In that paper, the multiobjective decision tree is explained through an example problem of a flood warning system. In this article, the usefulness and efficiency of multiobjective decision-tree analysis will be shown through some example problems in industrial systems.

II. DECISION TREE WITH MULTIPLE OBJECTIVES

Similar to a single-objective decision tree, a multiobjective decision tree is composed of decision nodes, chance nodes, and branches. At each decision node, designated by a square in the tree, the feasible alternatives are shown as branches emerging to the right side of the decision node; from among these branches the decisionmaker decides on the best alternative to perform. Each alternative branch may lead to another decision node, a chance node, or a terminal point. A chance node, designated by a circle in the tree, indicates that a chance event is expected at this point in the process; i.e., one of the states of nature may occur. The states of nature are shown on the tree as branches to the right of the chance nodes. A branch representing the state of nature may be followed by another chance node, a decision node, or a terminal point.

There exist, however, two major differences between the traditional single-objective decision tree and the multiobjective decision tree presented in Haimes et al. [3]. The new features of the multiobjective decision tree are (a) the vector-valued performance measure and (b) multiple-risk measures. These features enable us to make a significant extension of the averaging-out-and-folding-back strategies.

An alternative at a decision node in a multiobjective decision tree is evaluated by multiple criteria. Assume that we are performing vector minimization with k objective functions, f_1, f_2, $\cdots$, f_k. An alternative a^* is noninferior if there does not exist another feasible alternative $\bar{a}$ such that

$$f_i(\bar{a}) \le f_i(a^*) \qquad i = 1,2,\cdots,k \tag{1}$$

with strict inequality holding for at least one i. Clearly, a noninferior solution is not unique; the vector optimization at each decision node yields, in general, a set of noninferior solutions. Consider a decision node n in the tree. Assume that the set of k-dimensional vector-valued costs-to-go (the cumulative vector-valued objective function to the terminal points of the tree) is available at the end of each branch emerging to the right side of this decision node. Denote the set of vector-valued costs-to-go at the end of a branch associated with alternative a_j at decision node n by $\psi^{n+1}(a_j)$ with elements in the form of $\phi = [\phi_1^{n+1},\cdots,\phi_k^{n+1}]' \in \psi^{n+1}(a_j)$, where $'$ is the transpose operator; the "local vector-valued cost" associated with alternative a_j by $r(a_j) = [r_1(a_j),\cdots,r_k(a_j)]'$; and the k-dimensional binary operator which combines the vector-valued local cost with the vector-valued cost-to-go by $\overleftarrow{o}$. The operator $\overleftarrow{o}$ may represent different operations in its components and may vary from node to node. The folding-back process at decision node n thus becomes a multiobjective optimization problem,

$$\min\ [r(a_j)\overleftarrow{o}\psi^{n+1}(a_j)] \tag{2a}$$

where the element in the parentheses represents the operation between $r(a_j)$ and all elements in the set $\psi^{n+1}(a_j)$, i.e.,

$$[r_1(a_j)\overleftarrow{o}\phi_1^{n+1}, \cdots, r_k(a_j)\overleftarrow{o}\phi_k^{n+1}]' \tag{2b}$$

Solving Eq. (1) yields the noninferior set of vector-valued costs-to-go, ψ^n, at decision node n.

Consider a chance node m in the tree. Assume that the set of k-dimensional vector-valued costs-to-go (the cumulative vector-valued objective function to the terminal points of the tree) is available at the end of each branch emerging to the right side of this chance node. Denote the set of the vector-valued costs-to-go at the end of a branch associated with the jth state of nature at chance node m by $\psi^{m+1}(j)$ with elements in the form of $\phi = [\phi_1^{m+1}, \cdots, \phi_k^{m+1}]' \in \psi^{m+1}(j)$. Assume that the probabilities associated with the branches of the states of the nature are known. Those probabilities can be either prior probabilities or calculated posterior probabilities. The noninferior set of vector-valued costs-to-go at chance node m is identified by the following formula,

$$\min\ E^s\{\psi^{m+1}(j)\} \tag{3}$$

Solving Eq. (3) yields the noninferior set of vector-valued costs-to-go, ψ^m, at chance node m. The superscript s in E^s denotes the sth averaging-out strategy. If s is equal to 5, E^5 denotes the conventional expected-value operator over all possible states of nature. If s is equal to 4, E^4 denotes the operator of conditional expected value in the extreme region over all possible states of nature (which will be described in detail in section III). Unlike single-objective decision-tree analysis, where there is only an averaging-out process at the chance nodes, multiobjective optimization needs to be performed at the chance nodes in the multiobjective decision tree. If there are d_j elements in each $\psi^{m+1}(j)$, then there exist $\Pi_j\{d_j\}$ combinations of possible decision strategies following chance node m. In the process of averaging-out, the resulting inferior combinations need to be discarded from further consideration.

The general solution procedure for multiobjective decision trees can now be stated as follows:

Step 1. Assign a prior probability or calculate the posterior probability for each chance branch. Assign the vector-valued performance measure for each decision branch. (or map the vector-valued performance measure to each of the terminal points of the tree.)

Step 2. At each chance node m, use Eq. (3) to find the noninferior set of vector-valued costs-to-go.

Step 3. At each decision node n, use Eq. (2) to find the noninferior set of vector-valued costs-to-go.

Start from each terminal point of the tree and move backward on the tree. Repeat Steps 2 and 3 until the noninferior solutions at the starting point of the tree are obtained.

A. EXAMPLE 1: INFRASTRUCTURE PROBLEM

Consider the need to make a decision at the beginning of a planning period to replace, repair, or do nothing for a physical infrastructure that can fail and has been operating for a period. The objective is to determine the best maintenance policy (repair, replace, or do nothing) through the use of multiobjective decision-tree analysis.

Assume that the probability density function of the failure of a new system is of a known Weibull distribution,

$$P_{NEW}(t) = \lambda\alpha t^{\alpha-1}Exp[-\lambda t^{\alpha}] \qquad t \geq 0,\ \lambda > 0,\ \alpha > 0 \qquad (4)$$

and that a new system costs \$1000. At the beginning of this planning period, the system under investigation has been already operating for several years and the probability density function of its failure is of the form

$$p_{OLD}(t) = (\lambda+s)\alpha t^{\alpha-1}\text{Exp}[-(\lambda+s)t^{\alpha}] \qquad t \geq 0 \tag{5}$$

where the parameter s represents a declining factor of an aging system. The exact value of s is unknown. The value of s can be best described by an <u>a priori</u> distribution p(s), which is of a uniform distribution u[0.05,0.1]. A repair action that may be taken at the beginning of the planning period can recover the system's operational capability by updating its failure probability density function to

$$p_{REP}(t) = (\lambda + 0.05)\alpha t^{\alpha-1}\text{Exp}[-(\lambda+0.05)t^{\alpha}] \qquad t \geq 0 \tag{6}$$

The cost of the repair action is a function of the declining factor s,

$$C_{REP} = 200 + 4000(s - 0.05) \qquad 0.05 \leq s \leq 0.1 \tag{7}$$

To reduce the uncertainty of the value of s, assume that a test can be performed by an experiment which costs \$100. Three outcomes result from the experimentation: X_1, X_2, and X_3. The conditional probabilities of X_1, X_2, and X_3 are given as

$$P(X_1|s) = 10(0.1 - s) \qquad 0.05 \leq s \leq 0.1 \tag{8a}$$

$$P(X_2|s) = \begin{cases} 40(s - 0.05) & 0.05 \leq s \leq 0.075 \\ 40(0.1 - s) & 0.075 \leq s \leq 0.1 \end{cases} \tag{8b}$$

$$P(X_3|s) = 10(s - 0.05) \qquad 0.05 \leq s \leq 0.1 \tag{8c}$$

Thus, the posterior distribution of s can be obtained by the Bayesian formula,

$$p(s|X_i) = P(X_i|s)p(s)/P(X_i) \tag{9}$$

Specifically, we have

$$p(s|X_1) = 800(0.1 - s) \qquad 0.05 \leq s \leq 0.1 \tag{10a}$$

$$p(s|X_2) = \begin{cases} 1600(s - 0.05) & 0.05 \leq s \leq 0.075 \\ 1600(0.1 - s) & 0.075 \leq s \leq 0.1 \end{cases} \tag{10b}$$

$$p(s|X_3) = 800(s - 0.05) \qquad 0.05 \leq s \leq 0.1 \tag{10c}$$

Figure 1 presents the corresponding multiobjective decision tree for this example problem, with λ equal to 0.1 and α equal to 2.

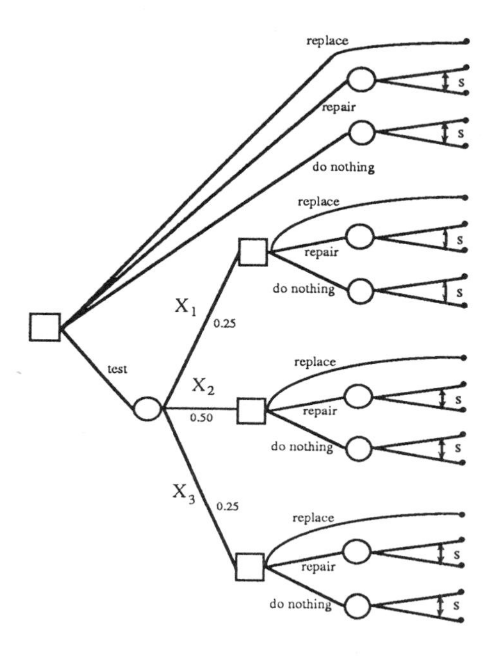

Fig. 1. Decision tree for Example 1.

If the alternative of preventive replacement is adopted, the system's mean time before failure is

$$E_{NEW}(T) = (1/0.1)^{0.5}\Gamma(1.5) = 2.802506. \tag{11}$$

where $\Gamma(t)$ is the Gamma function defined as $\Gamma(t) = \int_0^\infty u^{t-1}e^{-u}du$, $t > 0$.

If the repair alternative action is adopted, the system's mean time before failure is

$$E_{REP}(T) = (1/0.15)^{0.5}\Gamma(1.5) = 2.288236 \tag{12}$$

The expected cost for the repair action is calculated by

$$E(C_{REP}) = \int_{0.05}^{0.1} [200 + 4000(s-0.05)]p(s)ds = 300 \tag{13}$$

If the alternative of doing nothing is adopted, the system's mean time before failure is

$$E_{OLD}(T) = \int_{0.05}^{0.1} [1/(0.1+s)]^{0.5}\Gamma(1.5)p(s)ds = 2.123948 \quad (14)$$

For each outcome of the experiment, similar calculations can be performed for each alternative, except for replacing the prior distribution by the posterior distribution. Table 1 summarizes the results for the alternatives available after an experiment. At the chance node after the test, a total of 27 strategies can be adopted after the experiment. Using the triple notation $(\cdot,\cdot,\cdot)$, where the ith number ($i = 1,2,3$) in the triad represents the alternative should X_i occurs, number 1 denotes the doing-nothing action, number 2 denotes the repair action, and number 3 denotes the replacement action. By performing the multiobjective optimization -- minimization of expected cost and maximization of the mean time before failure -- we found that 18 out of 27 strategies are noninferior. Table 2 summarizes the after-experiment noninferior strategies.

Table 1. Scenarios of After-Experiment Alternatives for Example 1

Outcome	Alternative	Expected Cost	Mean Time before Failure
X_1	Do nothing	0	2.174778
	Repair	266.667	2.288236
	Replace	1000	2.802506
X_2	Do nothing	0	2.1211383
	Repair	299.95	2.288236
	Replace	1000	2.802506
X_3	Do nothing	0	2.07307
	Repair	333.33	2.288236
	Replace	1000	2.802506

Adding the cost of the experiment to each noninferior after-

Table 2. Noninferior After-Experiment Strategies for Example 1

Strategy	Expected Cost	Mean Time before Failure
(1,1,1)	0	2.1225312
(2,1,1)	66.66675	2.1508957
(1,1,2)	83.3325	2.1763227
(1,2,1)	149.975	2.20608
(2,2,1)	216.64175	2.2344445
(1,2,2)	233.3075	2.2598715
(1,1,3)	250	2.3048902
(2,1,3)	316.66675	2.3332547
(1,2,3)	399.975	2.388439
(2,2,3)	466.64175	2.4168035
(1,3,1)	500	2.463215
(2,3,1)	566.66675	2.4915795
(1,3,2)	583.3325	2.5170065
(2,3,2) (3,2,3)	649.99925	2.545371
(1,3,3)	750	2.645574
(2,3,3)	816.66675	2.6739385
(3,3,3)	1000	2.802506

experiment solution, rolling back to the initial decision node, and combining the alternatives without experiment, we finally obtain the noninferior solutions for the planning problem, which are given in Table 3. Figure 2 depicts the resulting noninferior solutions of Example 1 in the functional space. By evaluating the trade-offs between the cost and the mean time before failure (i.e., how much extra cost one is willing to afford to yield a unit improvement of the mean time before failure) the decisionmaker can find his/her most-preferred solution from among the set of noninferior solutions.

One interesting phenomenon can be observed in this example problem. In single-objective decision-tree analysis, whether or not an experiment should be performed depends on the value of the expected value of experimentation. In multiobjective decision-tree analysis, the value of experimentation is judged in a

Table 3. Noninferior Solutions for Example 1

Strategy	Expected Cost	Mean Time before Failure
Doing Nothing	0	2.123948
(2,1,1)	166.66675	2.1508957
(1,1,2)	183.3325	2.1763227
(1,2,1)	249.975	2.20608
Repair	300	2.288236
(1,1,3)	350	2.3048902
(2,1,3)	416.66675	2.3332547
(1,2,3)	499.975	2.388439
(2,2,3)	566.64175	2.4168035
(1,3,1)	600	2.463215
(2,3,1)	666.66675	2.4915795
(1,3,2)	683.3325	2.5170065
(2,3,2) (3,2,3)	749.99925	2.545371
(1,3,3)	850	2.645574
(2,3,3)	916.66675	2.6739385
Preventive Replacement	1000	2.802506

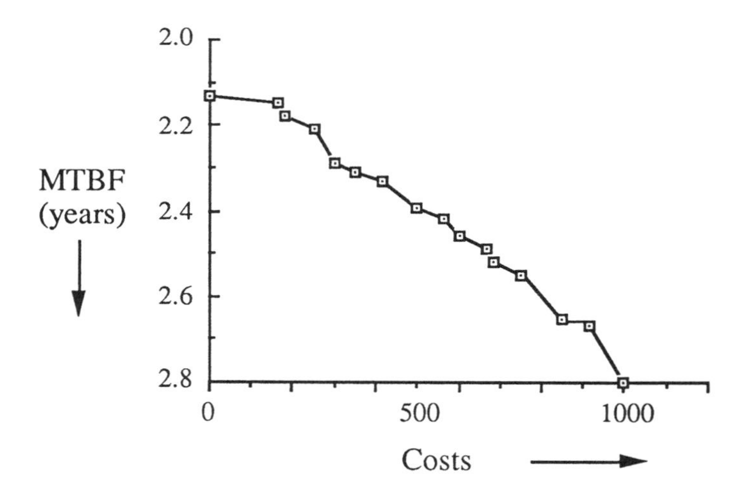

Fig. 2. Noninferior solutions for Example 1.

multiobjective way. In this example, the noninferior frontiers generated with and without experimentation do not dominate each other; they supplement each other to generate more solution candidates for the decisionmaker(s).

III. DECISION TREE WITH MUlTIPLE RISK MEASURES

In order to provide the decisionmaker(s) with more valuable information in risk management situations, Asbeck and Haimes developed the partitioned multiobjective risk method (PMRM) [4] to quantify risks at different levels. The PMRM generates a number of conditional expected-value functions (termed risk functions) that represent various risks given that the damage falls within specific ranges associated with certain exceedance probabilities. Assume the risk can be represented by a continuous random variable X with a known probability density function $p_X(x;s_j)$, where s_j $(j = 1,\cdots,q)$ is a control policy. The PMRM partitions the probability axis into three ranges. Denote the partitioned points on the probability axis by α_i $(i = 1,2)$. For each α_i and each policy s_j, it is assumed that there exists a unique damage β_{ij} such that

$$P_X(\beta_{ij};s_j) = \alpha_i \qquad (15)$$

where P_X is the cumulative distribution function of X. These β_{ij} (with β_{0j} and β_{3j} representing, respectively, the lower bound and upper bound of the damage) define the conditional expectation as follows:

$$f_i(s_j) = \frac{\int_{\beta_{i-2,j}}^{\beta_{i-1,j}} x p_X(x;s_j)dx}{\int_{\beta_{i-2,j}}^{\beta_{i-1,j}} p_X(x;s_j)dx} \qquad i = 2,3,4;\; j = 1,\cdots,q \qquad (16)$$

where f_2, f_3, and f_4 represent the risk with high probability of exceedance and low damage, the risk with medium probability of exceedance and medium damage, and the risk with low probability of exceedance and high damage, respectively. The unconditional (conventional) expected value of X is denoted by $f_5(s_j)$. The relationship between the conditional expected values (f_2, f_3, f_4) and the unconditional expected value (f_5) is given by

$$f_5(s_j) = \theta_2 f_2(s_j) + \theta_3 f_3(s_j) + \theta_4 f_4(s_j) \quad (17)$$

where θ_i (i = 2,3,4) is the denominator of Eq. (16). From the definition of β_{ij}, it can be seen that $\theta_i \geq 0$ is a constant (that is independent of the policy s_j), and $\theta_2 + \theta_3 + \theta_4 = 1$. It is easy to see that the conventional expected value is a weighting sum of the three conditional expected values.

Combining any one of the generated conditional expected risk functions or the unconditional expected risk function with the primary objective function (e.g., cost), f_1, creates a set of multiobjective optimization problems:

$$\min\ [f_1, f_i]', \quad i = 2,3,4,5. \quad (18)$$

By evaluating risk at various levels, this set of multiobjective formulations provides deeper insight into the probabilistic behavior of the problem than the single multiobjective formulation $\min\ [f_1, f_5]'$. The trade-offs between the cost function f_1 and any risk function f_i, $i \in \{2,3,4,5\}$ allow decisionmakers to consider the marginal cost of a small reduction in the risk objective, given a particular level of risk assurance for each of the partitioned risk regions and given the unconditional risk function, f_5. The relationship of the trade-offs between the cost function and the various risk functions is given by

$$1/\lambda_{15} = \theta_2/\lambda_{12} + \theta_3/\lambda_{13} + \theta_4/\lambda_{14} \quad (19)$$

where

$$\lambda_{1i} = -\partial f_1/\partial f_i, \quad \lambda_{1i} > 0, \quad i = 2,3,4,5 \quad (20)$$

is the trade-off value between the primary objective f_1 and one of the risk functions f_i in Eq. (18). This knowledge of the relationship among the marginal costs is important to help the decisionmaker(s) determine an acceptable level of risk.

A. EXAMPLE 2: RELIABILITY OF A DESIGN PROBLEM

Consider a design problem of a system's reliability. The failure rate λ of the system is an unknown constant and is assumed to be of a lognormal distribution $LN(\mu,\sigma)$. The probability that the system survives until time t, which is denoted by V(t), is related to λ by

$$V(t) = Exp[-\lambda t] \tag{21}$$

and the system's mean time before failure, which is denoted by m, can be calculated by

$$m = 1/\lambda \tag{22}$$

Thus, a high value of the failure rate λ results in poor system's reliability. There are three design alternatives that can be adopted to control the median value of the failure rate. For each alternative associated with one controlled median value, the decisionmaker may decide further whether or not to take a measure to control the error factor of the failure rate. The error factor is defined by 5-percentile = median/error factor and 95-

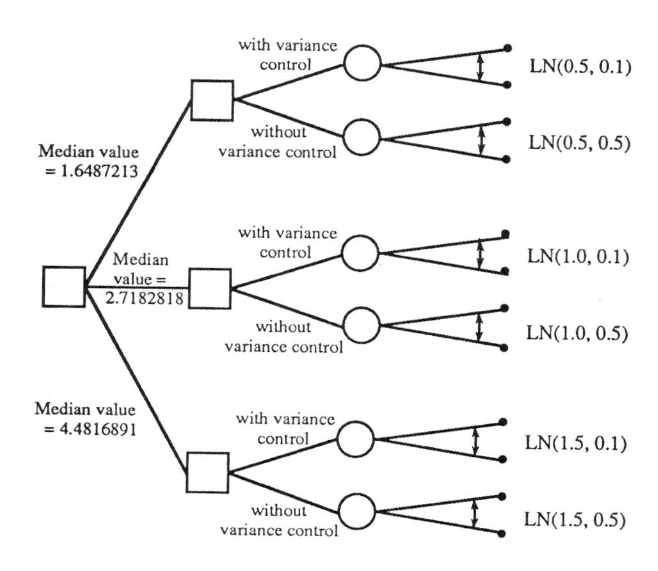

Fig. 3. Decision tree for Example 2.

Table 4. Design Alternatives and Their Costs for Example 2

Alternative #	Median Value	Error Factor	LN(μ,σ)	Cost
1	1.6487213	1.1788036	LN(0.5,0.1)	$600
2	1.6487213	2.2761832	LN(0.5,0.5)	$400
3	2.7182818	1.1788036	LN(1.0,0.1)	$500
4	2.7182818	2.2761832	LN(1.0,0.5)	$300
5	4.4816891	1.1788036	LN(1.5,0.1)	$300
6	4.4816891	2.2761832	LN(1.5,0.5)	$100

percentile = median × error factor. Figure 3 presents the decision tree for this problem. Table 4 summarizes data for the six available design alternatives. The relationships between the median value and the mean value and between the error factor and the variance can be obtained by

$$\mu = \text{Ln}[\text{median}] \tag{23a}$$

and

$$\sigma = \text{Ln}[\text{error factor}]/1.645 \tag{23b}$$

We can see from Table 4 that reducing the error factor from 0.5 to 0.1 costs an additional $200. Is this effort worthwhile? This question will be answered from the following two different perspectives of risk management through multiobjective decision-tree analysis.

(a) Minimization of cost and expected-value-of-failure rate:

Assume that we would like to minimize the cost and the expected-value-of-failure rate $E(\lambda)$ in this design problem. The expected-value-of-failure rate is obtained by the formula

$$E(\lambda) = \text{Exp}[\mu + \sigma^2/2]$$

(b) Minimization of cost and expected-value-of-extreme-values-of-failure rate:

Assume that we would like to minimize the cost and the

conditional expected-value-of-failure rate should extreme events occur. The conditional expected value f_4 is given by

$$f_4 = E[\lambda | LN^{-1}(\alpha) \leq \lambda \leq \infty] / Prob[LN^{-1}(\alpha) \leq \lambda \leq \infty]$$
$$= Exp[\mu + \sigma^2/2]/(1 - \alpha) \cdot \{1 - \phi[\phi^{-1}(\alpha) - \sigma]\} \quad (24)$$

where α is the partitioned point on the probability axis, which takes a value of 0.999 in this example.

Table 5 presents results of the cost, the expected-value-of-failure rate, and the expected-value-of-extreme-values-of-failure rate for each of six alternatives. We can see that alternatives 1, 2, 4, and 6 are noninferior for problem (a) while alternatives 1, 3, 5, and 6 are noninferior for problem (b). One important observation (see Fig. 4) is that although reducing the variance of the risk (the error factor of the failure rate) may not contribute much to reducing the expected value, it often reduces the conditional expected value of extreme values of risk.

REFERENCES

1. H. Raiffa, "Decision Analysis, Introductory Lectures on Choice under Uncertainty," Addision-Wesley, Reading, Massachusetts, 1968.
2. V. Chankong and Y. Y. Haimes, "Multiobjective Decision Making: Theory and Methodology," Elsevier-North Holland, New York, 1983.
3. Y. Y. Haimes, D. Li, and V. Tulsiani, "Multiobjective Decision Tree Method," to appear in Risk Analysis, 1989.
4. E. Asbeck and Y. Y. Haimes, "The Partitioned Multiobjective Risk Method," Large Scale Systems, **6**, 13-38 (1984).

ACKNOWLEDGMENTS

This work was supported in part by NSF Grant No. CES-8617984, under the title, "Hierarchical-multiobjective management of large scale infrastructure". The authors appreciate Vijay Tulsiani for his contribution to this paper and Virginia Benade for her editorial assistance.

Table 5. Noninferior Solutions for Example 2

Alternative #	Cost	E(λ)	f_4	Noninferior Solution
1	$600	1.656986	2.311495	For both (a) and (b)
2	$400	1.868246	8.965713	Only for (a)
3	$500	2.731907	3.811011	Only for (b)
4	$300	3.080217	14.78196	Only for (a)
5	$300	4.504154	6.283294	Only for (b)
6	$100	5.078419	24.37133	For both (a) and (b)

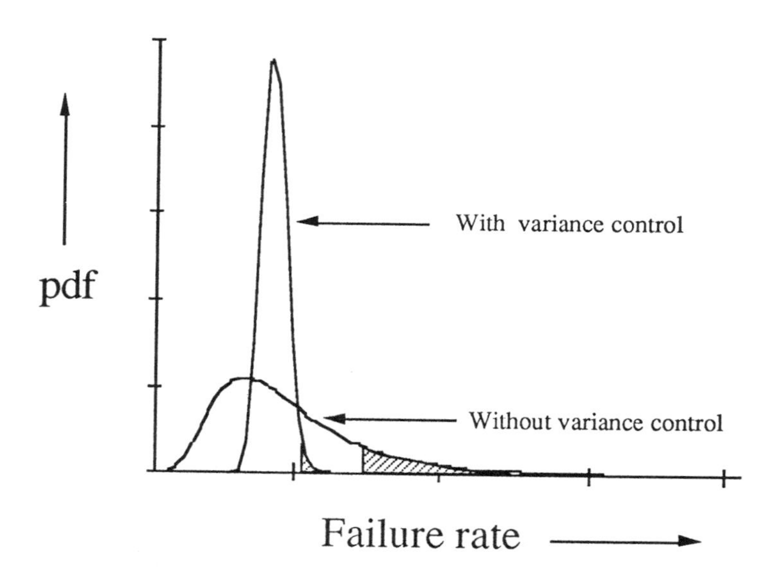

Fig. 4. Variance and region of extreme events.

METHODS FOR OPTIMAL ECONOMIC POLICY DESIGN

Berc Rustem

Department of Computing
Imperial College of Science, Technology & Medicine
180 Queen's Gate, London SW7 2BZ, U. K.

I. INTRODUCTION

Consider the policy optimization problem

$$\min \mathcal{E} \left\{ J\,(Y, U) \;\middle|\; F\,(Y, U, \epsilon) = 0 \right\} \tag{1}$$

where $\mathcal{E}$ denotes expectation, ϵ is a vector of random noise, Y, and U are respectively the endogenous or output variables and policy instruments or controls of the system. J is the policy objective function and F is the model of the economy which is affected by the random disturbance vector ϵ. In general, F is nonlinear with respect to Y, U and sometimes also with respect to ϵ. Problem (1) is essentially a static transcription of a dynamic optimization problem in discrete time where

$$U = \begin{bmatrix} u_1 \\ \vdots \\ u_t \\ \vdots \\ u_T \end{bmatrix}; \quad Y = \begin{bmatrix} y_1 \\ \vdots \\ y_t \\ \vdots \\ y_T \end{bmatrix}; \quad \epsilon = \begin{bmatrix} \epsilon_1 \\ \vdots \\ \epsilon_t \\ \vdots \\ \epsilon_T \end{bmatrix}$$

where

$$\bar{\ell}(x, c, \alpha) = J(x) + \frac{c}{2} \| F(x) \|_2^2 + \frac{1}{2c} \| (c\, h(x) + \alpha) \|_2^2. \tag{8}$$

$\bar{\ell}$ and ℓ differ only in their last terms corresponding to the inequality constraints. In contrast to (5), the Hessian of $\bar{\mathcal{L}}$ exists, provided J, F and h are suitably differentiable. The role of α in replacing the Hessian of (6) by that of (7) is crucial. Furthermore, this replacement resembles the *smearing* procedure adopted, for example, by Polak and Tits [10]. The main difference is in the choice of α and c. It is shown below that this choice allows the treatment of the active constraints *locally* as equality constraints.

The *successive quadratic programming* (SQP) quasi-Newton algorithm considered in this paper, involves the basic iterative procedure

$$x_{k+1} = x_k + \tau_k d_k \tag{9}$$

where $d_k \in \mathbb{R}^{nr}$ solves the *Quadratic Programming Subproblem* (QPS)

$$\min \left\{ q\,(d, c_k, \alpha_k(x_k)) \;\middle|\; \nabla F_k^T d + F_k = 0, \;\; \nabla h_k^T d + h_k \leq 0 \right\} \tag{10}$$

$\nabla F_k \in \mathbb{R}^{nr} \times \mathbb{R}^{T \times m}$, the multipliers of this QPS, corresponding to d_k, are denoted by λ_{k+1}, μ_{k+1}. It is assumed that the feasible set of (10) is nonempty and that the QPS does have a solution. The objective function is given by

$$q\,(d, c, \alpha(x_k)) = d^T \nabla \ell\,(x_k, c, \alpha(x_k)) + \tfrac{1}{2} d^T \hat{H}_k d \tag{11}$$

with $\nabla \ell = \nabla_x \ell$. $\hat{H}_k$ is a quasi-Newton approximation to the Hessian of $\bar{\mathcal{L}}$, with respect to x, at $x_k, \lambda_k, \mu_k, c_k, \alpha_k$. Thus, $H_k = \nabla_x^2 \bar{\mathcal{L}}(x_k, \lambda_k, \mu_k, c_k, \alpha_k)$. It should be noted that this $\hat{H}_k$ is different from the Hessian of the ordinary Lagrangian (4). We note that in Tapia [11] an observation is made concerning the positive effects of excluding from $\hat{H}$ the second derivative terms arising from the constraints in the penalty term. Since this term vanishes at a (feasible) solution of the problem, all subsequent results are equally applicable if this term is ignored in $\hat{H}$. The determination of $\alpha_k(x_k)$ and c_k are discussed in the next Section along with two alternative strategies for the stepsize τ_k.

In the quadratic objective function (11), $\nabla \ell$ is included instead of the more common ∇J only (see e.g. [12]; [13]). The reason for this is somewhat convoluted. As discussed Rustem ([14]; Theorem (4.1), the condition for $\{\tau_k\} - 1$ is dependent on the accuracy of $\hat{H}_k$[2]. In order to establish the convergence rate results with this $\hat{H}_k$, $\nabla \ell$ is required in (11). Thus, $\nabla \ell$ is incorporated in order to ensure the *consistency* of the algorithm when τ_k is approaching unity with the subsequent stage of convergence rate analysis.

Convergence to unit stepsizes can also be shown if $\nabla \ell$ is replaced by ∇J in the QPS. Once $\{\tau_k\} - 1$ is established, the superlinear convergence rates of both algorithms are shown, in Rustem [15], to depend on the accuracy of the approximate Hessian used in the QPS. The latter convergence rate result is equally applicable if $\hat{H}_k$ is chosen to approximate the Hessian of the ordinary (4) or the augmented (5) Lagrangian, with or without the above second order terms. However, as mentioned earlier, in order to achieve $\{\tau_k\} - 1$, the algorithm uses $\hat{H}_k$ as an approximation of the augmented Lagrangian Hessian and for this $\hat{H}_k$, $\nabla \ell$ is used in (11). It should be noted that the inclusion of the equality constraint penalty term in the quadratic objective function has also been considered, for different reasons, elsewhere in the literature (see e.g. Tapia, [16]; [11]).

Following the original suggestion by Wilson [18], several variants of the SQP algorithm have been proposed. These have proved to be effective in solving problem (3) (see, e.g. [17]). One common feature of all these algorithms is the formulation of a QPS whose objective function is based on a linear term depending on ∇J_k and a quadratic term depending an a quasi-Newton approximation of the Hessian of the Lagrangian (4). A second common feature among SQP algorithms is a strategy that determines the stepsize τ_k. However, it is essentially in the design of this strategy that most algorithms seem to differ from each other. Specifically, in Biggs [19], [20], [21], an equality constrained QPS is formulated and the stepsize strategy is based on a penalty function

---

[2]With or without the second order terms arising from the constraints in the penalty term.

which, in the case of inequality constraints, is differentiable only once. Han [12], Mayne and Polak [22] and Powell [13] consider a stepsize based on a nondifferentiable penalty function. Polak and Mayne [23] consider two stepsize strategies: one to ensure the decrease in the objective function J, and another, nondifferentiable strategy, to ensure feasibility. Gabay [24] considers, for the equality constrained problem, a reduced order algorithm which uses an equality constrained QPS based on the reduced Hessian of the Lagrangian (4) and two stepsize strategies: one based on a nondifferentiable penalty function and another based on a norm condition to ensure improved feasibility and subsequently Q-superlinear convergence rate. Schittkowski [25], [26] suggests the solution of an inequality constrained QPS and a stepsize based on an augmented Lagrangian function. Gill et al [27] discuss a stepsize strategy based on an augmented Lagrangian and demonstrate $\{ \tau_k \} \rightarrow 1$. In this approach, the penalty parameters c and α are determined using different criteria than those discussed below. In addition, the stepsize strategy used alters x, λ, μ, α instead of just x and $\alpha(x)$ as discussed below. Finally, Burke and Han [28] discuss an algorithm which allows the possibility of the QPS to be infeasible. A stepsize strategy and a penalty parameter updating scheme are utilized to ensure convergence. Both the stepsize and penalty parameter determination schemes are somewhat different from the approach adopted below. The latter extends the approach discussed in [29], [30] to ensure the attainment of unit stepsizes and superlinear convergence.

It has not been possible to establish convergence to unit stepsizes for some of the stepsize strategies above. Local convergence rate results have been obtained *assuming* that there exists a stage (i.e. $K \geq 0$) after which, due to attraction towards the solution, $\tau_k = 1$, $\forall\, k \geq K$ (see, e.g. Biggs [20]; Boggs et al [31]; Coleman and Conn [32]; Fontecilla et al [33]; Han [34]; Palomares and Mangasarian [35]; Powell [13]; Stoer and Tapia [36]). However, Chamberlain [37] and Maratos [38] have provided counter examples illustrating that strategies similar to those proposed by [12] and [13] may not converge to unity. In order to force convergence in such situations, Chamberlain et al [39] have proposed the "watchdog technique" which allows, whenever it becomes necessary, the relaxation of a given sufficient decrease criterion. For the remaining strategies, unit stepsizes can be established by invoking additional strategies or assumptions, which ensure that the multipliers, λ_k, μ_k, generated by the QPS (10), as well as $\{x_k\}$ and the Hessian approximations generated by the algorithms are sufficiently close to their respective values at the solution of (3) (see e.g. [25]; Lemma 6.1, Theorem 6.4)

In the above algorithms, all the stepsize strategies are well defined around the solution of (3). However, no strategy for inequality constrained problems is, in general, twice continuously differentiable. Indeed, as mentioned above, some strategies are not even once differentiable. Differentiability is not always important, but if it can be assured, convergence to unit stepsizes can be demonstrated with otherwise reasonably relaxed conditions. In the algorithms discussed below, twice continuous differentiability is maintained. For equality constraints, the chosen penalty function ensures this, whereas for inequality constraints the smearing technique proposed by Polak and Tits [10] is adapted and incorporated in the algorithms as a simple rule which maintains twice continuous differentiability. This property is utilized in establishing convergence to unit stepsizes and superlinear convergence rates.

The difference between the two algorithms discussed in Section III is only in the choice of the stepsize strategy. The basic algorithm in both cases is formulated, as in (9) - (11), in full, rather than reduced dimension. The equality constrained case could also be formulated in reduced dimension by adopting the approach of Gabay [24] for reducing the dimensionality of the QPS and approximating the Hessian in this reduced dimension. On the other hand, if slack variables are not to be used, the inequality constrained case is easier to treat in full dimension. As in [9], [13] and [21], the algorithm below uses a positive definite approximation to the Hessian. Departing from these, the Hessian used is an approximation to the Lagrangian augmented by a penalty function (7) such as, for example, in Tapia [16], with or without the second order terms arising from the constraints in the penalty function. The stepsize strategies are based on this penalty function.

The penalty function used by either stepsize strategy is directly incorporated in QPS (10). The descent property of the direction d_k generated by the QPS is ensured in a simple manner. It is also shown that the choice of the penalty parameters c and α in (5) ensure that these do not grow indefinitely. The convergence of the algorithms, (Rustem, [14]: Theorems (3.1) and (3.2)), is independent of the choice of the Hessian approximation. Convergence can be demonstrated for any uniformly bounded symmetric positive definite choice for this matrix. This choice is, however, important in the achievement of unit stepsizes and superlinear rates of convergence. The latter is discussed in [19]. The fact that $\hat{H}_k$ incorporates an approximation to the Hessian of the penalty function is important for establishing the condition for $\{ \tau_k \} \to 1$. It is at this stage that maintaining the existence of H_k is important, in particular for inequality constraints. Two different approaches may be used to establish conditions that ensure $\{ \tau_k \} \to 1$. The first approach relates the achievement of unit stepsizes to the accuracy of the approximate Hessian of the penalty function (8). For equality constraints only, this result is discussed in [30]. Its generalization to inequality constraints is straightforward and will not be explicitly stated in this study. The Hessian involved in this approach is independent of Lagrange multipliers. It is *part* of the Hessian of (7) used in the QPS. Unit stepsizes are thus shown to depend on a condition which is independent of the nearness of the Lagrange multipliers to their respective values at the solution of (3). The second approach, discussed in Rustem ([14]; Theorem (4.1)), relates $\{ \tau_k \} \to 1$ to the accuracy of the Hessian of (7) used in the QPS.

III. THE ALGORITHMS

Two stepsize strategies are considered for one basic algorithm involving the QPS (10) and rules for determining the penalty parameters α and c. The j th element of the vectors h and α are respectively denoted by h^j and α^j. The following is used in both algorithms

$$\alpha^j_{k+1}(x) = \begin{cases} -c_{k+1} h^j(x) & \text{if} \quad h^j(x) \leq 0 \\ \alpha^j_k(x_k) & \text{if} \quad h^j(x) > 0 \end{cases} \tag{12.a}$$

where α_0 is set in Step 0 with $c_0 \in (0, \infty]$ given as one of the inputs to the program choice and c_{k+1} is set in Step 3 below. We are mainly concerned with two specific values of α. These are

$$\alpha^j_{k+1}(x_k) = \begin{cases} -c_{k+1} h^j(x_k) & \text{if} \quad h^j(x_k) \leq 0 \\ \alpha^j_k(x_k) & \text{if} \quad h^j(x_k) > 0 \end{cases} \tag{12.b}$$

and

$$\alpha^j_{k+1}(x_{k+1}) = \begin{cases} -c_{k+1}\, h^j(x_{k+1}) & \text{if} \quad h^j(x_{k+1}) \leq 0 \\ \alpha^j_{k+1}(x_k) & \text{if} \quad h^j(x_{k+1}) > 0 \end{cases} \quad . \quad (12.c)$$

It should be noted that (12.c) differs from the definition (12.a). The reason for this is to ensure, in practice, that the first inequality in (13) below is satisfied for $c_{k+1} \neq c_k \neq c_*$. c_* is a constant value of the penalty parameter discussed in Rustem [14]. The convergence results are not affected since the above discrepancy is eliminated when $\alpha^j_{k+1}(x_k) = \alpha^j_k(x_k)$ when $c_{k+1} = c_k = c_*$.

Algorithm 1

Given $x_0, c_0 \in \mathbb{R}_+, \delta \in (0, \infty), \rho_1 \in (0, 1 - \frac{1}{2\epsilon_1}), \epsilon_1 \in (\frac{1}{2}, 1], \gamma \in (0, 1), \hat{H}_0$.

<u>**Step 0:**</u> Set $k = 0$, $\alpha^j_0 = \max\{-c_0\, h^j(x_0), 0\}$, $j = 1, \ldots, i$.

<u>**Step 1:**</u> Compute ∇J_k, ∇F_k, ∇h_k. Solve the QPS (10) to obtain d_k and the associated multiplier vectors λ_{k+1}, μ_{k+1}.

<u>**Step 2:**</u> Test for optimality: if optimality is achieved, <u>stop</u>. Else go to step 3.

<u>**Step 3:**</u> If

$$\nabla J_k^T d_k + \epsilon_1 d_k^T \hat{H}_k d_k - (h_k)_+^T \alpha_k(x_k) - c_k [\| F_k \|_2^2 + \|(h_k)_+\|_2^2] \leq 0 \quad (13)$$

then $c_{k+1} = c_k$. Else set

$$c_{k+1} = \max\left\{ \frac{\nabla J_k^T d_k + \epsilon_1 d_k^T \hat{H}_k d_k - (h_k)_+^T \alpha_k(x_k)}{\| F_k \|_2^2 + \|(h_k)_+\|_2^2},\; c_k + \delta \right\}. \quad (14)$$

Calculate $\alpha_{k+1}(x_k)$ using (12.b).

<u>**Step 4:**</u> Find the smallest nonnegative integer j_k such that $\tau_k = \gamma^{j_k}$ with $x_{k+1} = x_k + \tau_k d_k$ satisfying the inequality

$$\ell(x_{k+1}, c_{k+1}, \alpha_{k+1}(x_{k+1})) - \ell(x_k, c_{k+1}, \alpha_{k+1}(x_k)) \leq \tau_k \rho_1 \nabla \ell(x_k, c_{k+1}, \alpha_{k+1}(x_k))^T d_k$$

(15)

$\alpha_{k+1}(x_{k+1})$ is computed using (12,c).

Step 5: Update $\hat{H}_k$ to compute $\hat{H}_{k+1}$.

Step 6: Set $k = k + 1$ and go to Step 1.

Algorithm 2

Given x_0, $c_0 \in \mathbb{R}_+$, $\delta \in (0, \infty)$, $\rho_2 \in (0, 1)$, $\epsilon_2 \in (0, \frac{1}{2}]$, $\gamma \in (0, 1)$, $\hat{H}_0$.

Steps 0-2: As in Algorithm 1.

Step 3: If

$$\nabla J_k^T d_k + (\tfrac{1}{2}+\epsilon_2)\, d_k^T \hat{H}_k d_k - (h_k)_+^T \alpha_k(x_k) - c_k\, [\| F_k \|_2^2 + \|(h_k)_+\|_2^2] \leq 0$$

then $c_{k+1} = c_k$. Else set

$$c_{k+1} = \max \left\{ \frac{\nabla J_k^T d_k + (\frac{1}{2} + \epsilon_2)\, d_k^T \hat{H}_k d_k - (h_k)_+^T \alpha_k(x_k)}{\| F_k \|_2^2 + \|(h_k)_+\|_2^2},\ c_k + \delta \right\}$$

Calculate $\alpha_{k+1}(x_k)$ using (12,b).

Step 4: Line search: choose the stepsize $\tau_k = \gamma^{j_k}$ such that the inequality

$$\ell(x_{k+1}, c_{k+1}, \alpha_{k+1}(x_{k+1})) - \ell(x_k, c_{k+1}, \alpha_{k+1}(x_k)) \leq \tau_k \rho_2\, q(d_k, c_{k+1}, \alpha_{k+1}(x_k))$$

is satisfied with $x_{k+1} = x_k + \tau_k d_k$. $\alpha_{k+1}(x_{k+1})$ is computed using (12).

Steps 5-6: As in Algorithm 1.

An alternative to δ in Step 3 of both algorithms is $\delta = \beta c_k$ for some $\beta \in (0, \infty)$.

The next few results are concerned with the properties of $\alpha_{k+1}(x)$ given by (12) and used in Step 4, c_{k+1} computed in step 3 and the solution of the QPS, d_k.

Lemma (1)

For $\alpha_k(x)$ computed using (12) and for $c_k \geq 0$ we have

$$h^j(x) \leq 0 \quad \Rightarrow \quad (c_k h^j(x) + \alpha_k^j(x))_+ = (c_k h^j(x) + \alpha_k^j(x)) = 0 \quad (16.a)$$

$$h^j(x) > 0 \quad \Rightarrow \quad (c_k h^j(x) + \alpha_k^j(x))_+ = (c_k h^j(x) + \alpha_k^j(x)) \geq 0 \quad (16.b)$$

$$(c_k h^j(x) + \alpha_k^j(x))_+ = (c_k h^j(x) + \alpha_k^j(x)) \geq 0\,. \quad (16.c)$$

Proof

Using (12), it can be seen that when $h^j(x) \leq 0$, $\alpha_k(x) = -c_k h^j(x) \geq 0$ and when $h^j(x) > 0$, $\alpha_k(x) = \alpha_k(x_{k-1}) \geq 0$. (16,c) follows directly from (16.a,b). □

Lemma (2)

For $\alpha_{k+1}(x)$ computed and fixed using (12) such that $\alpha_k = \alpha_k(x_k)$ and $\alpha_{k-1} = \alpha_{k-1}(x_{k-1})$ and for $c_{k+1} \geq c_k \geq 0$ we have

$$(c_k h_k^j + \alpha_k^j)_+ \leq (c_{k+1} h_k^j + \alpha_{k+1}^j(x_k))_+ \quad (17.a)$$

$$(c_k h_k^j + \alpha_k^j)_+ \leq (c_k h_k^j + \alpha_k^j(x_{k-1}))_+ \quad (17.b)$$

and

$$(h_k)_+^T \alpha_{k+1}(x_k) = (h_k)_+^T \alpha_k. \quad (17.c)$$

Proof

The relation (17.a) follows directly from (12) and with $\alpha_k = \alpha_k(x_k)$

$$h_k^j \leq 0 \quad \Rightarrow (c_k h_k^j + \alpha_k^j)_+ = (c_k h_k^j + \alpha_k^j)$$

$$= (\, c_{k+1} h_k^j + \alpha_{k+1}^j(x_k)\,)_+$$

$$= (c_{k+1}h_k^j + \alpha_{k+1}^j(x_k))$$

$$= 0;$$

$$h_k^j > 0 \quad \Rightarrow (c_k h_k^j + \alpha_k^j)_+ = (c_k h_k^j + \alpha_k^j)$$

$$\leq (c_{k+1}h_k^j + \alpha_{k+1}^j(x_k))_+$$

$$= (c_{k+1}h_k^j + \alpha_{k+1}^j(x_k))$$

$$= (c_{k+1}h_k^j + \alpha_k^j(x_k)).$$

Inequality (17,b) follows from (12) since with $\alpha_k = \alpha_k(x_k)$

$$h_k^j \leq 0 \quad \Rightarrow 0 = (c_k h_k^j + \alpha_k^j)_+ \leq (c_k h_k^j + \alpha_k^j(x_{k-1}))_+$$

$$h_k^j > 0 \quad \Rightarrow 0 \leq (c_k h_k^j + \alpha_k^j)_+ = (c_k h_k^j + \alpha_k^j(x_{k-1}))_+.$$

Equality (17,c) also follows from (12) since $\alpha_{k+1}^j(x_k) \neq \alpha_k^j(x_k)$ if $h_k^j \leq 0$, in which case $(h_k^j)_+ = 0$. □

Lemma (3) The descent property of d_k.

Let $F_k \neq 0$ and $(h_k)_+ \neq 0$ (the case when $F_k = 0$ and $(h_k)_+ = 0$ is discussed in Lemma 4). The direction d_k satisfying the equality constraints of the QPS and the penalty parameter c_{k+1} chosen as in Step 3 of Algorithm 1 ensure the inequality

$$\nabla \mathcal{L} (x_k, c_{k+1}, \alpha_{k+1}(x_k))^T d_k \leq -\epsilon_1 d_k^T \hat{H}_k d_k; \qquad \epsilon_1 \in (\tfrac{1}{2}, 1] \tag{18}$$

and d_k with c_{k+1} chosen as in Step 3 of Algorithm 2 ensure

$$q(d_k, c_{k+1}, \alpha_{k+1}(x_k)) \leq -\epsilon_2 d_k^T \hat{H}_k d_k; \qquad \epsilon_2 \in (0, \tfrac{1}{2}]. \tag{19}$$

Remark

The first order optimality conditions of QPS (10) are given by

$$\hat{H}_k d_k + \nabla J_k + \nabla F_k [c_k F_k + \lambda_{k+1}] + \nabla h_k [(c_k h_k + \alpha_k(x_k))_+ + \mu_{k+1}] = 0 \tag{20,a}$$

$$\mu_{k+1}^T [\nabla h_k^T d_k + h_k] = 0; \qquad \mu_{k+1} \geq 0 \tag{20,b}$$

$$\nabla h_k^T d_k + h_k \leq 0; \qquad \nabla F_k^T d_k + F_k = 0. \tag{20,c}$$

Proof [of Lemma (3)]

In order to establish the required result, d_k is only required to satisfy the linearized constraints (20,c). The other optimality conditions are required for subsequent results. Using (20,c), (16,c) in Lemma (1) and (17,c) in Lemma (2) we can write

$$\begin{aligned} d_k^T [\nabla J_k &+ \epsilon_1 \hat{H}_k d_k + c_{k+1} \nabla F_k F_k + \nabla h_k (c_{k+1} h_k + \alpha_{k+1}(x_k))_+] \\ &\leq d_k^T \nabla J_k + \epsilon_1 d_k^T \hat{H}_k d_k - c_{k+1} \| F_k \|_2^2 \\ &\qquad - h_k^T (c_{k+1} h_k + \alpha_{k+1}(x_k))_+ \\ &= d_k^T \nabla J_k + \epsilon_1 d_k^T \hat{H}_k d_k - c_{k+1} (\| F_k \|_2^2 + \|(h_k)_+\|_2^2) \\ &\qquad - (h_k)_+^T \alpha_k(x_k). \end{aligned} \tag{21}$$

In order to ensure (18), (21) must be nonpositive and c_{k+1} chosen in step 3 of Algorithm 1 ensures the latter. The corresponding result for (19) can be obtained by replacing ϵ_1 in (21) by $(\epsilon_2 + \frac{1}{2})$. □

We discuss briefly the effect of the penalty parameter on the direction and Lagrange multipliers generated by the QPS. It is well known that, *for a given value of* $\hat{H}_k$, the direction generated by the algorithm, for equality constraints only, is independent of c_k (see [16], [40]). For certain choices of the Hessian, the independence from c_k can be further relaxed. Consider the Newton version of the algorithm for solving the *equality constrained problem only.* The necessary condition of the subproblem becomes

$$\left[\nabla^2 J(x_k) + c_k \nabla F_k \nabla F_k^T + \sum_i \nabla^2 F^i(x_k) \left[c_k F_k^i + \lambda_k^i \right] \right] d_k$$

$$+ c_k \nabla F_k F_k + \nabla F_k \lambda_{k+1} = 0$$

and $\nabla F_k^T d_k + F_k = 0$. If the second derivative term, $\nabla^2 F^i(x_k) [c_k F_k^i]$, due to the constraint curvatures, is ignored, then it can be verified that the penalty parameter does not affect the multiplier. However, this result is only clear when the Hessian or Hessian approximation used in the QPS does not include the term $\nabla^2 F^i(x_k) [c_k F_k^i]$. For other choices of the Hessian used in the QPS and also for inequality constrainted problems, this conclusion may not apply and hence c *may affect the direction and the Lagrange multipliers* of the QPS.

It is shown in the next Lemma that (21) is nonpositive when $F_k = 0$ and $(h_k)_+ = 0$.

Lemma (4) The finiteness of the penalty parameter

Let J, F and h be once differentiable and let d_k, λ_{k+1}, μ_{k+1} generated by the QPS be bounded. Then

(i) If for some or all k the constraints of (3) are satisfied, i.e. $F(x_k) = 0$ and $h(x_k) \leq 0$, then the descent properties of (18) and (19) of Lemma (3) are satisfied for any choice of $c_k \in [0, \infty)$.

(ii) Let c_k be bounded and $F(x_k) = 0$ and $h(x_k) \leq 0$, for some k. Then, Step 3 of both algorithms choose $c_{k+1} = c_k$. For this choice, the direction d_k ensures the descent property of Lemma (3).

(iii) $\forall$ k, $\exists\, c_{k+1} \in [0, \infty)$ satisfying Step 3 of both algorithms.

(iv) Let the sequence $\{x_k\}$ generated by either algorithm be bounded. Then c_k is increased finitely often and $\exists$ an integer $k_* \geq 0$ and $\exists$ $c_* \in [0, \infty)$ such that the algorithm chooses $c_k = c_*$, $\forall$ $k \geq k_*$.

Proof (Rustem,[14]).

In [14] it is shown that the above algorithms converge to a local optimum and that they also generate stepsizes τ_k that converge to unity. The local convergence rates near the solution is Q- or two-step Q superlinear, depending on the accuracy of the approximate Hessian used by the algorithm (see [15]).

IV. SENSITIVITY ANALYSIS AND ROBUST POLICY DESIGN

Consider the feedback policy optimization problem

$$\min_{Y, U, \theta} \left\{ \hat{J}(Y, U, \beta) \,\middle|\, F(Y, U, \epsilon) = 0;\; U = f(Y, \theta) \right\} \tag{22}$$

and let

$$\hat{\epsilon} = \begin{bmatrix} \beta \\ \epsilon \end{bmatrix}$$

be the vector of uncertain parameters, forecasts or residual variables. The role of $\beta \in \mathbb{R}^1$ will be discussed in Section V. For $\beta = 0$, we have

$$\hat{J}(Y, U, 0) = J(Y, U).$$

The relationship $U = f(Y, \theta)$ represents the parameterized feedback laws written in stacked form and $\theta \in \mathbb{R}^\theta$ is the vector of parameters which are determined optimally. Such feedback laws are utilized, for example, in Karakitsos and Rustem [41] in connection with a large macroeconometric model of the UK economy.

Given the values of ϵ and θ, the model solution program simultaneously solves the model equations F (Y, U, ϵ) = 0 and determines the value of the feedback U = f (Y, θ) to yield the corresponding values of [Y, U]. Thus, we can write

$$\begin{bmatrix} Y \\ U \end{bmatrix} = \begin{bmatrix} g_1(\theta, \epsilon) \\ g_2(\theta, \epsilon) \end{bmatrix}. \tag{23}$$

Using (23) to eliminate [Y, U] from (22) yields an unconstrained optimization problem in θ

$$\min_{\theta} \left\{ \hat{G}(\theta, \hat{\epsilon}) \right\} \tag{24}$$

where

$$\hat{G}(\theta, \hat{\epsilon}) = \hat{J}(g_1(\theta, \epsilon), g_2(\theta, \epsilon), \beta)$$

which, when $\beta = 0$ becomes

$$G(\theta, \epsilon) = \hat{G}\left(\theta, \begin{bmatrix} 0 \\ \epsilon \end{bmatrix}\right) = \hat{J}(g_1(\theta, \epsilon), g_2(\theta, \epsilon), 0).$$

We are interested in the behaviour of a local solution $\theta(\hat{\epsilon}_0)$ of (24) when the base value of $\hat{\epsilon}$, denoted by $\hat{\epsilon}_0$, is subject to perturbation. For simplicity of notation, without loss of generality, we shall assume that $\hat{\epsilon}_0 = 0$. Clearly, when $\hat{\epsilon}_0 = 0$, the solution of (24) yields the optimum deterministic solution to the feedback optimization problem $\min_{\theta} \left\{ G(\theta, 0) \right\}$. We now apply some of the results of Fiacco [42] to establish the basic sensitivity results around the solution of (24) with $\hat{\epsilon}_0 = 0$.

Proposition 1

Let θ_* denote a local isolated solution of $\min_{\theta} \left\{ G(\theta, 0) \right\}$ satisfying the second order sufficient conditions for a local optimum. If $\hat{G}$ defining the problem $\min_{\theta} \left\{ \hat{G}(\theta, \hat{\epsilon}) \right\}$ is twice continuously differentiable in $[\theta, \hat{\epsilon}]$ in a neighbourhood of $[\theta_*, 0]$, then for $\hat{\epsilon}$ in a neighbourhood of 0, there exists a unique once continuously differentiable vector function $\theta(\hat{\epsilon})$ satisfying the second order sufficient conditions for a local minimum of $\min_{\theta} \left\{ \hat{G}(\theta, \hat{\epsilon}) \right\}$ such that $\theta(0) = \theta_*$ and, hence, $\theta(\hat{\epsilon})$ is a locally unique minimum of $\min_{\theta} \left\{ \hat{G}(\theta, \hat{\epsilon}) \right\}$. □

The proof of the above result follows from Fiacco ([42]: Theorem 2.1). The following corollary of the above result relates the perturbations to a first order approximation of the perturbed solution in the neighbourhood of $\hat{\epsilon}_0 = 0$ ([42]: Corollaries 2.1, 3.1).

Corollary

Under the assumptions of the above proposition, a first order approximation of $\theta(\hat{\epsilon})$ in a neighbourhood of $\hat{\epsilon} = \hat{\epsilon}_0 = 0$ is given by

$$\theta(\hat{\epsilon}) = \theta_* - \left[\nabla_\theta^2 \hat{G}(\theta_*, 0) \right]^{-1} \frac{\partial \left[\nabla_\theta \hat{G}(\theta_*, 0) \right]}{\partial \hat{\epsilon}} \hat{\epsilon} + O(\| \hat{\epsilon} \|) \tag{25.a}$$

and the derivative of θ with respect to $\hat{\epsilon}$ can be approximated by

$$\frac{d\,\theta(\hat{\epsilon})}{d\,\hat{\epsilon}} = -\left[\nabla_\theta^2 \hat{G}(\theta_*, \hat{\epsilon}) \right]^{-1} \frac{\partial \left[\nabla_\theta \hat{G}(\theta_*, \hat{\epsilon}) \right]}{\partial \hat{\epsilon}} \tag{25.b}$$

and

$$\frac{d\,\theta(0)}{d\,\hat{\epsilon}} = -\left[\nabla_\theta^2 \hat{G}(\theta_*, 0) \right]^{-1} \frac{\partial \left[\nabla_\theta \hat{G}(\theta_*, 0) \right]}{\partial \hat{\epsilon}}. \tag{25.c}$$

□

In the context of linear continuous time dynamical systems, the derivative of the state trajectory with respect to perturbed parameters of the system is discussed in Becker et al [43] and Wuu et al [44].

We seek to formulate the policy optimization problem such that the sensitivity of the solution to perturbations will be minimized simultaneously with the original policy objective function. The sensitivities are characterized as

$$\frac{\partial J(Y, U)}{\partial \epsilon} \quad \text{and} \quad \frac{\partial Y}{\partial \epsilon}.$$

The former is discussed in Section V and the latter in Rustem [4]. The relations of these sensitivities to the variances of the associated stochastic problems are also discussed. The sensitivities of interest can clearly be evaluated in terms of (25) using the chain law.

It is important to note, from the point of view of computational implementation, that the Hessian required in (25, c) for the evaluation of

$$\frac{d\theta(0)}{d\hat{\epsilon}},$$

will already be available if a Newton-type algorithm is used to compute the point θ_*. An estimate of this matrix will be available if a quasi-Newton algorithm is used. Thus, much of the information required to calculate the sensitivity information will already be available. The second matrix

$$\frac{\partial \nabla_\theta \hat{G}(\theta, \hat{\epsilon})}{\partial \hat{\epsilon}}$$

is known once the problem is specified and need only be evaluated at $[\theta_*, 0]$.

The disadvantage of the robust approach discussed below is that the perturbation results on which it is based restrict it to a neighbourhood of the robust optimal solution. Nevertheless, this is the robust solution of the nonlinear problem and, as shown in the examples in Rustem [4], this neighbourhood can be sufficiently large. The advantages are that the approach is implementable for nonlinear systems, it is a deterministic approach that also has a stochastic, mean-variance, interpretation in the linear case and that it is relatively easy to compute and does not involve Monte Carlo simulations.

V. ROBUSTNESS WITH RESPECT TO THE POLICY OBJECTIVE FUNCTION

In order to ensure that the optimal policy takes into account the possibility of perturbations in the parameters, we reformulate the deterministic, unperturbed problem

$$\min_{Y, U, \theta} \left\{ J(Y, U) \mid F(Y, U, 0) = 0; \; U = f(Y, \theta) \right\}.$$

Consider the ***policy objective sensitivity*** term

$$J_\epsilon = \left. \frac{\partial J(Y, U)}{\partial \epsilon} \right|_{\epsilon = 0} = \left. \frac{\partial Y}{\partial \epsilon} \frac{\partial J(Y, U)}{\partial Y} + \frac{\partial U}{\partial \epsilon} \frac{\partial J(Y, U)}{\partial U} \right|_{\epsilon = 0} \tag{26. a}$$

where

$$\frac{\partial Y}{\partial \epsilon} = \frac{\partial g_1(\theta, \epsilon)}{\partial \epsilon} \quad \text{and} \quad \frac{\partial U}{\partial \epsilon} = \frac{\partial g_2(\theta, \epsilon)}{\partial \epsilon} \tag{26. b}$$

or equivalently, using the model $F(Y, U, \epsilon)$ and the feedback law $U = f(Y, \theta)$ directly

$$\frac{\partial F}{\partial \epsilon} = \frac{\partial F}{\partial Y} \frac{\partial Y}{\partial \epsilon} + \frac{\partial F}{\partial U} \frac{\partial U}{\partial \epsilon}, \quad \frac{\partial U}{\partial \epsilon} = \frac{\partial U}{\partial Y} \frac{\partial Y}{\partial \epsilon} \tag{26. c}$$

$$\frac{\partial Y}{\partial \epsilon} = \left[\frac{\partial F}{\partial Y} + \frac{\partial F}{\partial U} \frac{\partial U}{\partial Y} \right]^{-1} \frac{\partial F}{\partial \epsilon}. \tag{26. d}$$

In order to generate a ***robust*** policy, an additional penalty term can be imposed on the objective function in order to minimize the policy objective sensitivity as measured by the column vector (26). Consider, therefore, the deterministic problem

$$\min_{Y,\, U,\, \theta} \left\{ \hat{J} (Y, U, \beta) \,\middle|\, F (Y, U, 0) = 0; \;\; U = f (Y, \theta) \right\} \qquad (27, a)$$

where $0 \leq \beta \leq 1$ and

$$\hat{J} (Y, U, \beta) = (1 - \beta) J (Y, U) + \beta < J_{\epsilon} , \Sigma \; J_{\epsilon} > \qquad (27, b)$$

which is a convex combination of the original objective function and the cost sensitivity term $<J_{\epsilon} , \Sigma \; J_{\epsilon}>$ for a specified value of β. The matrix Σ is a positive semi definite weighting matrix. It is shown below that the choice of Σ has a natural choice when the problem is reexamined from the risk aversion point of view. For $\beta > 0$, problem (27) has a solution at which the objective function $J (Y, U)$ is less sensitive to departures from zero perturbation, $\epsilon = 0$, than in the case $\beta = 0$. Using (23), $[Y, U]$ can be eliminated from the policy optimization problem to yield $G(\theta, 0) = J (g_1 (\theta, 0), g_2 (\theta, 0)$

$$\hat{G} \left(\theta, \begin{bmatrix} \beta \\ 0 \end{bmatrix} \right) = (1 - \beta) G(\theta, 0) + \beta < J_{\epsilon} , \Sigma \; J_{\epsilon} >$$

and the minimization problem becomes

$$\min_{\theta} \left\{ \hat{G} \left(\theta, \begin{bmatrix} \beta \\ 0 \end{bmatrix} \right) \right\}$$

As an example, consider the quadratic policy objective function

$$J (Y, U) = \tfrac{1}{2} < Y - Y^d, H_y (Y - Y^d) > + \tfrac{1}{2} < U - U^d, H_u (U - U^d) >$$

where H_y is a symmetric positive semi-definite matrix and H_u is a symmetric positive definite matrix. The superscript d denotes the desired, or bliss, values of the variables. In this case, the sensitivity term (26) is given by

$$J_\epsilon = \left[\frac{\partial g_1(\theta, \epsilon)}{\partial \epsilon} \right]^T H_y (Y - Y^d) + \left[\frac{\partial g_2(\theta, \epsilon)}{\partial \epsilon} \right]^T H_u (U - U^d) \Bigg|_{\epsilon = 0} . \tag{28}$$

In general, the matrices $\frac{\partial g_1(\theta, \epsilon)}{\partial \epsilon}$ and $\frac{\partial g_2(\theta, \epsilon)}{\partial \epsilon}$ are not constant and, therefore, the corresponding $\hat{J}(Y, U, \beta)$ is not a quadratic function. Apart from a purely linear model, one special case that subsumes the linear model framework and admits constant matrices is when the model (23) takes the special form

$$\begin{bmatrix} Y \\ U \end{bmatrix} = \begin{bmatrix} g_1(\theta) + \hat{N}_y \epsilon \\ g_2(\theta) + \hat{N}_u \epsilon \end{bmatrix} \tag{29}$$

where

$$\hat{N}_y = \frac{\partial g_1(\theta, \epsilon)}{\partial \epsilon} \in \mathbb{R}^{m \times T} \times \mathbb{R}^{\epsilon \times T} \text{ and } \hat{N}_u = \frac{\partial g_2(\theta, \epsilon)}{\partial \epsilon} \in \mathbb{R}^{n \times T} \times \mathbb{R}^{\epsilon \times T}$$

are constant matrices. The algorithm summarized below circumvents this problem by invoking Proposition 1. By solving the robust policy problem (27) for $\beta = 0$ initially and then solving the problem for gradually increasing values of β, it is possible to regard these matrices as constant during each optimization. The solution of each optimization can be made to lie arbitrarily close to the previous solution, and by Proposition 1, this closeness is ensured for sufficiently small changes in β from one optimization to the next. Thus, the above matrices can be taken as constants.

Algorithm: Computing robust policies while keeping $\partial g_1/\partial\epsilon$ and $\partial g_2/\partial\epsilon$ constant during each optimization

Step 0: Given a small number $\Delta\beta \in (0, 1)$; set $k=0$, $\beta = \beta_0 = 0$, and solve the optimization problem (27, a).

Step 1: If $\beta_k = 1$, stop ; else set $\beta_{k+1} = \beta_k + \Delta\beta$ and $k = k + 1$.

Step 2: Compute $\partial g_1/\partial\epsilon \,\big|_{\epsilon=0}$ and $\partial g_2/\partial\epsilon \,\big|_{\epsilon=0}$ along the current optimal trajectory.

Step 3: Given $\beta = \beta_k$ solve (27,a) to compute the current optimal trajectory with constant $\partial g_1/\partial\epsilon$ and $\partial g_2/\partial\epsilon$ in J_ϵ given by (28).

Step 4: If two successive optimal trajectories satisfy prescribed convergence criteria, go to Step 1. Otherwise, go to Step 2.

Proposition 1 and its Corollary ensure that the optimal solution is differentiable with respect to β. Thus, by choosing sufficiently small increases to β, we can in turn ensure that successive solutions of (27,a) are close to each other such that in this interval $\partial g_1/\partial\epsilon$ and $\partial g_2/\partial\epsilon$ can be regarded as constants.

The risk aversion or mean-variance interpretation of the above robust strategy can be discussed using the model (29). For linear models, [6] and [7] have also taken a similar risk aversion approach. Consider ϵ as a Gaussian random vector $\epsilon \sim \mathcal{N}(0, \Omega)$. Using the model (29), the mean value of the quadratic cost function is given by

$$\mathcal{E}\left[J(Y, U) \right] = J(g_1(\theta), g_2(\theta)) + \tfrac{1}{2}\,\mathrm{tr}(\Omega\, \hat{N}_y^T H_y \hat{N}_y) + \tfrac{1}{2}\,\mathrm{tr}(\Omega\, \hat{N}_u^T H_u \hat{N}_u) \tag{30}$$

Similarly, the variance of $J(Y, U)$ is given by

$$\begin{aligned}\mathrm{var}\left[J(Y, U) \right] &= \mathcal{E}\left[J(Y, U) - \mathcal{E}\left[J(Y, U) \right] \right]^2 \\ &= \langle (g_1(\theta) - Y^d),\, H_y \hat{N}_y \Omega \hat{N}_y^T H_y (g_1(\theta) - Y^d) \rangle\end{aligned}$$

$$+ < (g_2(\theta) - U^d), H_u \hat{N}_u \Omega \hat{N}_u^T H_u (g_2(\theta) - U^d) >$$

$$+ < (g_1(\theta) - Y^d), H_y \hat{N}_y \Omega \hat{N}_u^T H_u (g_2(\theta) - U^d) >$$

$$+ \tfrac{1}{2} tr(\Omega \hat{N}_y^T H_y \hat{N}_y)(\Omega \hat{N}_u^T H_u \hat{N}_u)$$

$$+ \tfrac{1}{2} tr(\Omega \hat{N}_y^T H_y \hat{N}_y) + \tfrac{1}{2} tr(\Omega \hat{N}_u^T H_u \hat{N}_u). \tag{31}$$

The mean variance objective function defined by

$$(1 - \beta) \mathcal{E} [J (Y, U)] + \beta \text{ var} [J (Y, U)] \tag{32}$$

is the same as the robust policy objective function (27.b) for quadratic objectives with J_ϵ defined by (28) when the last two terms in (30) and the last three terms in (31) can be treated as constants. This is clearly true for constant $\hat{N}_y$, $\hat{N}_u$. Obviously, for linear models, this constancy is trivially satisfied. The equivalence of the mean-variance (32) and robust (27, b) objectives follows if Σ in (27.b) is identified with the covariance matrix Ω.

When the model is linear and the objective function is quadratic, problem (1) can be solved using the stochastic and adaptive control techniques described in Kendrick [45]. In the nonlinear stochastic framework discussed above, we can refine the computation of $\mathcal{E} [J (Y, U)]$ in (32) by using the approach of Hall et al [46]. This is aimed to take account of any bias due to the nonlinearity of the model in computing the expectation. To do this, let

$$\mathcal{E} [Y] = \hat{Y} + \mathcal{B}$$

where the expected value of Y equals to the deterministic solution, $\hat{Y}$, plus the expected deviation of this deterministic value from $\mathcal{E} [Y]$, $\mathcal{B}$. Equivalently, we can write

$$\mathcal{E} [Y - Y^d] = \hat{Y} - Y^d + \mathcal{B} = \hat{Y} - \hat{Y}^d$$

where $\hat{Y}^d = \mathcal{B} - Y^d$. Using this, we can derive the relationship

$$\mathcal{E}[< Y - Y^d, H_y (Y - Y^d) >] = < \mathcal{E}[Y - Y^d], H_y \mathcal{E}[Y - Y^d] >$$

$$+ \operatorname{tr}\left(H_y \mathcal{E}([Y - Y^d][Y - Y^d]^T)\right)$$

$$= < \hat{Y} - \hat{Y}^d, H_y (\hat{Y} - \hat{Y}^d) >$$

$$+ \operatorname{tr}\left(H_y \mathcal{E}([Y - Y^d][Y - Y^d]^T)\right).$$

We assume that the last term is constant. With this assumption, the trace term can be eliminated from the optimization computation. The expected value of the original decision making problem (1) thus becomes

$$\min_{Y, U} \mathcal{E} \left\{ J (Y, U) \,|\, F (Y, U, \epsilon) = 0 \right\}$$

$$= \min_{Y, U} \mathcal{E} \left\{ \tfrac{1}{2} < Y - Y^d, H_y (Y - Y^d) > + \tfrac{1}{2} < U - U^d, H_u (U - U^d) > \right|$$

$$F (Y, U, \epsilon) = 0 \Big\}$$

$$= \min_{\hat{Y}, U} \left\{ \tfrac{1}{2} < (\hat{Y} - \hat{Y}^d), H_y (\hat{Y} - \hat{Y}^d) > + \tfrac{1}{2} < U - U^d, H_u (U - U^d) > \right.$$

$$\left. + \operatorname{tr}\left(H_y \mathcal{E}([Y - Y^d][Y - Y^d]^T)\right) \,\middle|\, F (\hat{Y}, U, 0) = 0 \right\}.$$

The procedure for computing the optimal control strategy can be described as follows:

Algorithm: Computing optimal controls with nonlinear bias compensation

Step 1: Compute the initial $\hat{Y}$, and U by computing the optimal deterministic solution of (1), with $\epsilon \equiv 0$.

Step 2: Perform a set of Monte Carlo simulations of the model around the base defined by the deterministic optimal solution U and compute $\mathcal{E}[Y]$. Thus compute $\mathcal{B} = \mathcal{E}[Y] - \hat{Y}$.

Step 3: Use $\mathfrak{B}$ to formulate and solve the problem

$$\min_{Y,U} \left\{ \tfrac{1}{2} < (Y - \hat{Y}^d), H_y (Y - \hat{Y}^d) > + \tfrac{1}{2} < U - U^d, H_u (U - U^d) > \;\middle|\; F(Y, U, 0) = 0 \right\}$$

and return to Step 1. The convergence criterion is that if two successive solutions computed in this step are sufficiently close to each other, the algorithm stops.

This method has a number of weaknesses. The first is that the trace term above and $\mathfrak{B}$ are assumed to be constant whereas they may be functions of the control strategy. This limits the algorithm to a first order estimate of the nonlinear effects of the stochastic disturbances. The second is that the algorithm does not have a stepsize strategy to ensure convergence and that it is not certain whether the algorithm would be computationally viable if it had such a strategy. Nevertheless, it is a small but effective step in the right direction for solving large scale nonlinear problems. The extension of the algorithm to the feedback case can be easily designed by considering (23) as the model, instead of $F(Y, U, \epsilon) = 0$. In this case, in addition to $\mathfrak{B}$, it would also be necessary to estimate the bias between the deterministic U and its expected value.

VI. THE SPECIFICATION OF THE QUADRATIC OBJECTIVE FUNCTION

An important problem in policy optimization is the determination of an objective function that generates optimal policies which are acceptable to the policy maker. In this and the next two sections, we discuss the specification of quadratic objective functions. We provide an iterative algorithm, that interacts with the decision maker, to specify diagonal objective functions. We also provide an algorithm for the specification of the equivalent non-diagonal objective. By a diagonal quadratic function, we mean one with a diagonal Hessian. The specification of diagonal and non-diagonal Hessians is also utilized in [47] to discuss the arbitrariness of the shadow prices, or Lagrange multipliers. The criterion used for the specification of the objective function is the acceptability of the optimal solution to the decision maker. The possibility of formulating optimization problems conveniently by an appropriate Lagrangian formulation is not new (see

Geoffrion [48]; Ponstein [49]). However, in [47] the converse problem is considered in which, given the formulation, it is shown that the Lagrange multiplier, or shadow price, values can be altered as desired by the decision maker without altering the optimal solution.

Using the notation introduced in (2), consider the linear-quadratic optimization problem

$$\min \left\{ < x - x^d, D (x - x^d) > \Big| \nabla F^T x = F \right\}. \tag{33}$$

where $x \in \mathbb{R}^{n_x}$, $F \in \mathbb{R}^{T \times m}$ is a fixed vector, D is a diagonal matrix with nonnegative elements, x^d is the desired or bliss value of x or the unconstrained optimum of the quadratic objective function. The columns of the constant matrix $\nabla F \in \mathbb{R}^{n_x} \times \mathbb{R}^{T \times m}$ are assumed, without loss of generality, to be linearly independent. The diagonal weighting matrix in (33) occurs in economic decision making and in general in linear-quadratic optimal control problems. Economic decision problems that can be formulated as (33) include dynamic policy design problems of macroeconomics and management decision making under conflicting objectives.

The desirability of diagonal weighting matrices is for computational reasons and for the interpretation of the problem and the optimal solution. Evidently, a general, symmetric matrix, which we denote by H and assume that it is positive definite, can be factorized such that $H = L D L^T$. The matrix L is a lower diagonal matrix. The transformation $y = L^T x$, $y^d = L^T x^d$ yields the diagonal quadratic objective function

$$< y - y^d, D (y - y^d) >.$$

However, the transformed constraints $\nabla F^T L^{-T} y = F$ may pose computational problems. This is in particular true when the constraints are nonlinear. We shall not dwell on this aspect as we have introduced the problem with only linear constraints. The same difficulty also occurs in any orthogonal decomposition of H. A more important aspect is the difficulty of assigning an interpretation to the off-diagonal elements of a general symmetric matrix. These elements are usually understood to represent trade-offs between achieving alternative objectives. However, they are difficult to assign an interpretation as distinct from the diagonal elements, which represent the relative importance of each objective. In other words, the diagonal elements are perceived to serve almost the same purpose. Another reason for preferring (33) is that in dynamic problems solved via dynamic programming, maintaining the block diagonality of D, in terms of the time structure, is desirable at least for the physical interpretation of the resultant optimal linear feedback laws.

The precise value of the weighting matrix is not known in most economic decision making problems under conflicting objectives. Thus, even a judicious choice of the weighting matrix need not necessarily yield a solution of problem (33) that is acceptable to the decision maker. In the discussion below, we shall discuss an iterative algorithm, involving interactions with the decision maker, to specify a diagonal quadratic objective function that generates a solution of (33) which is ***acceptable*** to the decision maker. The algorithm involves the modification of the desired, or bliss, values of the objective function. For dynamic policy optimization problems, this objective function permits the generation of linear optimal feedback laws inherent in the solution, in terms of the original, rather than transformed variables. This contrasts with the case when non-diagonal weighting matrices have to be factorized and the original variables have to be transformed in order to obtain feedback laws which will usually apply only to the transformed variables.

The diagonal matrix specification method is shown to be equivalent to the respecification of the general non-diagonal symmetric matrix discussed in Rustem, Velupillai, Westcott [50] and Rustem and Velupillai [51] where the desired values remain fixed but the weighting matrix that is generated does not maintain a diagonal structure. This correspondence is helpful both computationally, as diagonal matrices are simpler to compute with, and also provides the non-diagonal weighting matrix that would generate the same optimal solution with the original desired values as the solution obtained with the modified desired values and the diagonal weighting matrix. In addition, the correspondence can be used to discuss the complexity of the algorithm discussed below, using the results in Rustem and Velupillai [52], to establish the termination property of the algorithm in polynomial time.

VII. SPECIFYING DIAGONAL QUADRATIC OBJECTIVE FUNCTIONS

Let (33) be solved for a given weighting matrix D and a given initial vector of desired values x_0^d. The solution is denoted by

$$x_0 = \arg\min \left\{ < x - x_0^d, D (x - x_0^d) > \middle| \nabla F^T x = F \right\}.$$

The solution is presented to the decision maker who is required to respond by either declaring that $x_0 \in \Omega$, where Ω is the set of acceptable policies, or if $x_0 \notin \Omega$, the decision maker is required to specify the modified form of x_0 that is in Ω. The set Ω is the decision maker's set of acceptable policies. It is supposed to exist only in the mind of the policy maker. We assume that there does not exist any analytical characterization of this set. Clearly, if such a characterization was possible, it could be used to augment the constraints of the decision problem and the resultant problem can be easily solved. However, such a characterization is both inherently difficult, or impossible, and sometimes also undesirable from the decision maker's point of view.

The decision problem can now be formulated as the computation of the policy, optimally determined via (33), but which also is acceptable to the policy maker and is hence in the set Ω. We assume that

$$\left\{ x \middle| \nabla F^T x = F \right\} \cap \Omega \neq \emptyset.$$

The decision maker's preferred alternative to x_0 is denoted by x_p. By definition, we have $x_p \in \Omega$ *but not necessarily*

$$x_p \in \{ x \mid \nabla F^T x = F \} \cap \Omega.$$

If x_p satisfies all these feasibility restrictions, this preferred alternative would conceptually solve the decision problem. Let

$$\delta_0 = x_p - x_0$$

where δ_0 is the *correction* vector that needs to be added to x_0 in order to ensure that

$$x_0 + \delta_0 \in \Omega.$$

Using δ_0, we can revise the desired, or bliss, value as

$$x_1^d = x_0^d + \alpha_0 \delta_0$$

where $\alpha_0 \geq 0$ is a scalar. Using this new desired value, problem (33) is solved once again to yield a new optimal solution, x_1. This solution is shown below to have desirable characteristics. However, as there is no guarantee that $x_1 \in \Omega$, the above procedure may need to be repeated. The resulting algorithm is summarized below. The complexity and termination properties of the algorithm are discussed in Rustem [47].

Algorithm: Updating Desired Values with a Fixed Diagonal Weighting Matrix

__Step 0:__ Given D, x_0^d, the sequence $\{\alpha_k\}$ and the constraints, set $k = 0$.

__Step 1:__ Compute the solution of the optimization problem

$$x_k = \arg\min \left\{ < x - x_k^d, D (x - x_k^d) > \middle| \nabla F^T x = F \right\}.$$

__Step 2:__ Interact with the decision maker. If $x_k \in \Omega$, __stop__. Otherwise, the decision maker is required to specify the preferred value x_p, and hence,

$$\delta_k = x_p - x_k.$$

__Step 3:__ Update the desired values

$$x_{k+1}^d = x_k^d + \alpha_k \delta_k. \tag{34}$$

set $k = k + 1$ and go to Step 1.

The effect of changing the desired values as described above alters the optimal solution systematically. The relationship of x_{k+1}, x_k and α_k, δ_k is summarized in the following results. It must be noted that the possibility of altering the desired values in order to change the solution is known in the practice of optimal control (see [53]). It is also the case that the choice of the objective function may be based on criteria other than $x_k \in \Omega$. For example, in the linear stochastic dynamical systems, the weighting matrices might be chosen to yield a stable minimum variance controller (see [9]). However, in the present study we consider the equivalence properties of the approach adopted in this section to respecification, discussed in Section VIII, of nondiagonal weighting matrices while keeping x^d fixed. This equivalence also provides the key to the complexity and convergence of the policy design process. In addition, as it is possible to decompose a symmetric matrix into a sequence of rank one updates (see Fiacco and McCormick [54]), the method in this section and the equivalence result allow the possibility of expressing non diagonal weighting matrices in terms of diagonalized objective functions.

Proposition 2

Assume that D is nonsingular, then the two successive solutions of the algorithm are related by

$$x_{k+1} = x_k + \alpha_k P \delta_k \tag{35}$$

where

$$P = I - D^{-1} \nabla F (\nabla F^T D^{-1} \nabla F)^{-1} \nabla F^T$$

and the corresponding change in the Lagrange multipliers or shadow prices is given by

$$\lambda_{k+1} = \lambda_k - \alpha_k (\nabla F^T D^{-1} \nabla F)^{-1} \nabla F^T \delta_k . \tag{36}$$

When D is positive semi-definite, the optimal solution and Lagrange multipliers can be written as

$$x_{k+1} = x_k + \alpha_k Z (Z^T D Z)^{-1} Z^T D \delta_k \tag{37}$$

where $Z \in \mathbb{R}^{n_x \times (n-m)}$ is an orthogonal matrix such that $Z^T \nabla F = 0$, and

$$\lambda_{k+1} = \lambda_k - \alpha_k (\nabla F^T \nabla F)^{-1} \nabla F^T D \Big[I - Z (Z^T D Z)^{-1} Z^T D \Big] \delta_k . \quad (38)$$

Proof (Rustem, [47]; Proposition 1) □

Remark

The choice of the orthogonal matrix Z is discussed further in Rustem and Velupillai [51]. The numerically stable way of generating Z is by considering the QR decomposition of ∇ F. The matrix Z is given by the last n-m columns of the matrix H of this decomposition (see [55]).

The matrix P is a projection operator that projects vectors onto the subspace

$$\{ x \mid \nabla F^T x = 0 \}.$$

Thus, P δ_k is the ***nearest feasible alternative*** of δ_k. When D is nonsingular, the operators $Z(Z^T DZ)^{-1}Z^T D$ and P are, in effect, identical (see [51]; Lemma 2). This discussion can be extended to demonstrate the equivalence of (35), (37) and (36), (38).

VIII. THE EQUIVALENCE OF DIAGONAL AND NON-DIAGONAL QUADRATIC OBJECTIVE FUNCTIONS

We consider the diagonal equivalent of the algorithm discussed in [51]. The complexity of the algorithm in Section VII is established in [47], [52] by exploiting this equivalence. The following algorithm uses the same δ_k as in the algorithm in Section VII. It keeps the desired values fixed but updates the weighting matrix of the quadratic optimization problem .

Algorithm : Fixed Desired Values and Non-diagonal Weighting Matrix

__Step 0:__ Given a positive semi-definite weighting matrix H_0, the sequence μ_k, the desired values x^d and the constraints, set k = 0.

__Step 1:__ Compute the solution of the optimization problem

$$x_k = \arg\min \Big\{ < x - x^d, H_k (x - x^d) > \Big| \nabla F^T x = F \Big\}. \quad (39)$$

Step 2: Interact with the decision maker. If $x_k \in \Omega$, stop. Otherwise, the decision maker is required to specify the preferred value x_p, and hence,

$$\delta_k = x_p - x_k.$$

Step 3: Update the weighting matrix

$$H_{k+1} = H_k + \mu_k \frac{H_k \delta_k \delta_k^T H_k}{< \delta_k, H_k \delta_k >}. \tag{40}$$

Set k = k + 1 and go to Step 1.

The matrix H_k computed in the algorithm above is in general non-diagonal. It is shown below that starting with an initial diagonal matrix, the above algorithm and the algorithm in Section VII are equivalent. At each stage, the non-diagonal version above has a constant diagonal equivalent in the algorithm of Section VII. The equivalent results to Proposition 2 related to the above algorithm are discussed in [51]. We summarize these results. When H_k is positive definite, each subsequent iterate of the above algorithm is given by

$$x_{k+1} = x_k + \hat{\alpha}_k P_k \delta_k \tag{41.a}$$

$$\lambda_{k+1} = \lambda_k - \hat{\alpha}_k (\nabla F^T H_k^{-1} \nabla F)^{-1} \nabla F^T \delta_k \tag{41.b}$$

$$P_k = I - H_k^{-1} \nabla F (\nabla F^T H_k^{-1} \nabla F) \nabla F^T \tag{42.a}$$

$$\hat{\alpha}_k = - \frac{\mu_k < \delta_k, H_k (x_k - x^d) >}{< \delta_k, H_k \delta_k > + \mu_k < H_k \delta_k, P_k H_k^{-1} (H_k \delta_k) >} \tag{42.b}$$

and when H_k is positive semi-definite, (41) becomes

$$x_{k+1} = x_k + \hat{\alpha}_k Z (Z^T H_k Z)^{-1} Z^T H_k \delta_k \tag{42.c}$$

$$\lambda_{k+1} = \lambda_k - \hat{\alpha}_k (\nabla F^T \nabla F)^{-1} \nabla F^T H_k (I - Z (Z^T H_k Z)^{-1} Z^T H_k) \delta_k \tag{42.d}$$

$$\hat{\alpha}_k = - \frac{\mu_k < \delta_k, H_k (x_k - x^d) >}{< \delta_k, H_k \delta_k > + \mu_k < H_k \delta_k, Z (Z^T H_k Z)^{-1} Z^T H_k \delta_k >} \tag{42.e}$$

(see, Rustem and Velupillai [51]: Theorems 1, 2, Lemma 2). The value of $\hat{\alpha}_k$ in (42.e), can be shown to be identical to that given by (42.b) when H_k is positive definite. Similarly, (41.a.b) and (42.a.b) are identical when H_k is positive definite.

We can now show the equivalence of the solution of a diagonal quadratic optimization problem with only the desired or bliss values modified, and the quadratic optimization in which only the diagonal matrix has been modified to a non-diagonal form.

Proposition 3

Let H_k be nonsingular. Then, there exist μ_k, α_k and $\hat{\alpha}_k$ such that for

$$x^d_{k+1} = x^d_k + \alpha_k \delta_k; \tag{43.a}$$

$$H_{k+1} = H_k + \mu_k \frac{H_k \delta_k \delta_k^T H_k}{< \delta_k, H_k \delta_k >} \tag{43.b}$$

we have

$$\arg\min \left\{ < x - x^d_{k+1}, H_k(x - x^d_{k+1}) > \,\middle|\, \nabla F^T x = F \right\}$$

$$= \arg\min \left\{ < x - x^d_k, H_{k+1}(x - x^d_k) > \,\middle|\, \nabla F^T x = F \right\}.$$

Moreover, we have $\hat{\alpha}_k = \alpha_k$.

If α_k , μ_k are restricted such that $\alpha_k, \mu_k \geq 0$, then the choice of δ_k is restricted by the inequality

$$< \delta_k, H_k (x_k - x_k^d) > \leq 0.$$

Proof ([47]: Proposition 3) □

Remark

The above proposition clearly holds for $H_k = D$ and H_{k+1}, as given above, is D with a rank one update and hence it is no longer, in general, diagonal.

When H_k is *singular*, it can be shown that (43) can be written as

$$H_k x_{k+1}^d = H_k (x_k^d + \alpha_k \delta_k)$$

from which the relevant parts of x_{k+1}^d can be recovered. For example, when $H_k = D$, a diagonal matrix, clearly only those elements of x_{k+1}^d corresponding to nonzero diagonal elements of D can be recovered. The correspondence of $\hat{\alpha}_k$ given by (42.e) and α_k can be established in the same way as in the following proof except that (42, c) is used for x_{k+1}.

The extension of the above result to nonlinear constraints is straightforward. The method can thus be easily extended to the nonlinear constrained case when the diagonal equivalent of a non-diagonal quadratic function is desired. The useful analytical equivalence of α and $\hat{\alpha}$ cannot be established exactly in the nonlinear case. Clearly, if the departure of x_{k+1} from x_k is sufficiently small, then this equivalence can also be established by invoking a mean value theorem and thereby using a local linear representation of the constraints (see e.g. [51]: Theorem 5). The following corollary summarizes the straightforward extension which can be refined, if required, as described above.

Corollary [The Extension to Nonlinear Constraints]

Let x_{k+1}^d and H_{k+1} be defined by (43) and let the constraints be given by

$$\mathbb{F} = \left\{ x \in \mathbb{R}^{nx} \;\middle|\; F(x) = 0 \right\}$$

where F is twice differentiable and $F : \mathbb{R}^{nx} \rightarrow \mathbb{R}^{m \times T}$. Then the equivalence

$$\arg\min \left\{ < x - x^d_{k+1}, H_k(x - x^d_{k+1}) > \;\middle|\; x \in \mathbb{F} \right\}$$

$$= \arg\min \left\{ < x - x^d_k, H_{k+1}(x - x^d_k) > \;\middle|\; x \in \mathbb{F} \right\}$$

holds for α_k given in Proposition 3.

Proof

The proof follows from the equivalence of the first order optimality conditions of both problems. □

The extension to nonlinear constraints is thus easily implementable as the basic ingredients that enter α_k are δ_k and x_{k+1}. Both of these vectors are known when any one of the two quadratic problems have already been solved.

The above result relates the effect of a single update of H_k that yields H_{k+1} or a single update of x^d_k that yields x^d_{k+1}. We now consider the sequence $\{H_k\}$ generated by the algorithm in this section and the corresponding sequence $\{x^d_k\}$ generated by the algorithm in Section VII. The diagonalizibility result stated below is essentially a corollary of the above discussion.

Theorem 1 [The Diagonalizibility of Quadratic Forms]

Let the sequence $\{H_k\}$ be generated by (40) and $\{x^d_k\}$ be generated by (34), let $H_0 = D$ then the equivalence between the diagonal and non-diagonal quadratic optimizations

$$\arg\min \left\{ < x - x^d_k, D (x - x^d_k) > \;\middle|\; \nabla F^T x = F \right\}$$

$$= \arg\min \left\{ < x - x^d_0, H_k (x - x^d_0) > \;\middle|\; \nabla F^T x = F \right\} \tag{44.a}$$

holds if

$$D\, x_k^d = D\, x_0^d - \sum_{i=0}^{k-1} \mu_i \frac{< \delta_i, H_i\, (x_k - x_0^d\,) >}{< \delta_i, H_i\, \delta_i >} H_i\, \delta_i . \tag{44.b}$$

Remark

If H_0 is nonsingular, then clearly the entire vector x_k^d can be recovered from (44,b) . Otherwise, only the part of x_k^d which is relevant to computing the optimal solution can be recovered.

Using the matrix update formula (40), expression (44,b) can be expressed in terms of H_0 only. However, in computational terms, this does not appear to be of significant advantage since from each iteration only the vector $H_i\, \delta_i$ needs to be stored. It can be verified that this vector, appropriately normalized by the scalar terms, is sufficient to construct x_k^d, given the current solution x_k and and the initial desired value x_0^d.

Proof

The equivalence of the optimality conditions of both problems yields

$$D\,(\, x_k - x_k^d\,) \;=\; H_k\,(\, x_k - x_0^d\,)$$

$$= \left[\, H_{k-1} + \mu_{k-1} \frac{H_{k-1}\delta_{k-1}\delta_{k-1}^T H_{k-1}}{< \delta_{k-1}, H_{k-1}\,\delta_{k-1}>} \,\right] (x_k - x_0^d\,)$$

$$= \left[\, D + \sum_{i=0}^{k-1} \mu_i \frac{H_i\,\delta_i\,\delta_i^T\, H_i}{< \delta_i, H_i\,\delta_i>} \,\right] (x_k - x_0^d\,)$$

from which (44,b) follows. □

The comlexity of the algorithms for diagonal and nondiagonal matrices can be discussed by invoking their equivalence to Khachian's [56], [57] algorithm for linear programming. This is done in Rustem and Velupillai [52] and Rustem [47].

IX. A MIN-MAX APPROACH FOR ROBUST POLICIES IN THE PRESENCE OF RIVAL MODELS OF THE SAME SYSTEM

The formulation of the policy optimization problem (1) is, in practice, an oversimplification. Originating from rival economic theories, there exist rival models purporting to represent the same system. The problem of forecasting under similar circumstances has been approached by forecast pooling by Granger and Newbold [58] and, more recently, by Makridakis and Winkler [59] and Lawrence et al., [60]. In the presence of rival models, the *policy maker may also wish to take account of all existing rival models in the design of optimal policy.* One strategy in such a situation is to adopt the worst case design problem

$$\min_{Y^1,\ldots,Y^{m_{mod}},\,U} \; \max_{i} \left\{ J^i(Y^i, U) \;\middle|\; F^i(Y^i, U) = 0;\; i = 1,\ldots,m_{mod} \right\} \tag{45}$$

where there are $i=1,\ldots,m_{mod}$ rival models, with Y^i, F^i respectively denoting the dependent (or endogenous) variable vector and the equations of the i th model. This strategy is the extension of a suboptimal approach originally discussed by Chow [61]. Problem (45) seeks the optimal strategy corresponding to the most adverse circumstance due to choice of model. All rival models are assumed to be known. The solution of (45) clearly does not provide an insurance against the eventuality that an unknown $(m_{mod}+1)$st model happens to represent the economy: it is just a robust strategy against known competing "scenarios". A similar, less extreme, formulation is also discussed below, utilizing the dual approach to (45).

Using the terminology introduced in (2), the min-max problem can be reformulated as follows

$$\min_{x} \; \max_{i} \left\{ J^i(x) \;\middle|\; F(x) = 0,\; i = 1,\ldots,m_{mod} \right\} \tag{46}$$

where F subsumes all the models. The formulation above is slightly more general than the original min-max problem above. Other equivalent formulations are discussed in Rustem [62], [63], [64].

Algorithms for solving (46) have been considered by a number of authors, including Charalambous and Conn [65], Coleman [66], Conn [67], Demyanov and Malomezov [68], Demyanov and Pevnyi [69], Dutta and Vidyagasar [70], Han [71], [72], Murray and Overton [73]. In the constrained case, discussed in some of these studies, convergence results to unit steplengths, global convergence and local convergence rates have not been established (e.g. [66], [70]). In this and the next section, the dual approach to (46), adopted originally by Medanic and Andjelic [74], [75] and Cohen [76] is utilized initially. Subsequently, both the dual and primal approaches are used to formulate the algorithm.

As in the previous sections, let $x \in \mathbb{R}^{nx}$ and let $F : \mathbb{R}^{nx} — \mathbb{R}^{m \times T}$ be twice continuously differentiable functions with $\hat{J} = [J^1, J^2, ..., J^{m_{mod}}]^T$. Furthermore, let $\mathit{1}$ be the m_{mod}-dimensional vector whose elements are all unity and

$$\mathbb{R}_+^{m_{mod}} = \left\{ \alpha \in \mathbb{R}^{m_{mod}} \;\middle|\; <\alpha, \mathit{1}> = 1, \alpha \geq 0 \right\}. \tag{47}$$

It should also be noted that (46) can be solved by the nonlinear programming problem

$$\min_{x, v} \left\{ v \;\middle|\; \hat{J}(x) \leq \mathit{1}\, v; \quad F(x) = 0 \right\}$$

where $v \in \mathbb{R}^1$. The following two results are used to introduce the dual approach to this problem.

Lemma (5)

Problem (46) is equivalent to

$$\min_{x} \max_{\alpha} \left\{ <\alpha, \hat{J}(x)> \;\middle|\; F(x) = 0, \alpha \in \mathbb{R}_+^{m_{mod}} \right\} \tag{48}$$

Proof

This result, initially proved by Medanic and Andjelic [74], [75] and also Cohen [76], follows from the fact that the maximum of m_{mod} numbers is equal to the maximum of their convex combination. □

In Medanic and Andjelic [74], [75], the model is assumed to be linear and the solution of (48) without the constraints $F(x) = 0$, is obtained using an iterative algorithm that projects α onto $\mathbb{R}_+^{m_{mod}}$. In Cohen [76], the iterative nature of the projection is avoided by dispensing with the equality constraint of (47) but including a normalization in a transformed objective function. Although the resulting objective function is not necessarily concave in the maximization variables, the algorithm proposed ensures convergence to the saddle point. The algorithm proposed in Cohen [76] is for nonlinear systems, utilizes a simple projection procedure but is essentially first order.

The vector α_* is also the shadow price in (46). An important feature of (48) which makes it preferable to (46) is that α^i can also be interpreted as the importance attached by the policy maker to the model $F^i(x) = 0$. *There may be cases in which the min-max solution α_* may be too extreme to implement. The policy maker may then wish to assign a value to α, in the neighbourhood of α_*, and determine a more acceptable policy by minimizing $<\alpha, \hat{J}(x)>$, with respect to x, for the given α.* Another interpretation of (48) is in terms of the robust character of min-max policies. This is discussed in the following Lemma.

Lemma (6)

Let there exist a min-max solution to (48), denoted by (x_*, α_*) and let f and h be once differentiable at (x_*, α_*). Further, let strict complementarity hold for $\alpha \geq 0$ at this solution. Then, for $i, j, \ell \in \{1, 2, \ldots, m_{mod}\}$

(i) $\quad J^i(x_*) = J^j(x_*), \qquad \forall\, i, j\ (i \neq j)$ iff $\alpha_*^i, \alpha_*^j \in (0, 1)$;

(ii) $\quad J^i(x_*) = J^j(x_*) > J^\ell(x_*), \qquad \forall\, i, j, \ell\ (\ell \neq i, j)$
iff $\alpha_*^\ell = 0$ and $\alpha_*^\ell = \alpha_*^j \in (0, 1)$;

(iii) $J^i(x_*) > J^j(x_*)$, $\forall\, j, (j \neq i)$ iff $\alpha_*^i = 1$;

(iv) $J^i(x_*) < J^j(x_*)$, $\forall\, j, (j \neq i)$ iff $\alpha_*^i = 0$.

Proof

The necessary conditions of optimality for (48) are

$$\nabla_x \hat{J}\,(x_*)\,\alpha_* + \nabla F_x(x_*)\,\lambda_* = 0;\;\; F(x_*) = 0;\;\; \hat{J}\,(x_*) + \mu_* + \mathit{1}\,\eta_* = 0; \tag{49}$$

$$<\mathit{1}, \alpha_* > = 1; \qquad \alpha_* \geq 0;$$
$$<\mu_{*}, \alpha_* > = 0; \qquad \mu_* \geq 0. \tag{50}$$

Where λ_*, μ_*, η_* are the multipliers of $F(x) = 0$, $\alpha \geq 0$ and $<\mathit{1}, \alpha> = 1$ respectively.

Case (i) can be shown by considering (50) which, for $\alpha_*^i, \alpha_*^j \in (0, 1)$ yields $\alpha_*^i \mu_*^i = \alpha_*^j \mu_*^j = 0$, and then $\mu_*^i = \mu_*^j = 0$. Using (49) we have $J^i(x_*) = J^j(x_*)$. The only if part is established using $J^i(x_*) = J^j(x_*)$ and noting that

$$\eta_* = - < \hat{J}\,(x_*), \alpha_*>.$$

Premultiplying the last equality in (50) by $\mathit{1}$ and using this equality yields

$$0 = <\mathit{1}, \hat{J}\,(x_*)> + <\mathit{1}, \mu_*> + <\mathit{1}, \mathit{1}>\,\eta_* = <\mathit{1}, \mu_*>.$$

By (50), $\mu_* = 0$. Furthermore, strict complementarity implies that $\alpha_*^i \in (0, 1)$, $\forall i$.

Case (ii) can be shown by considering (50) for $\alpha_*^i, \alpha_*^j \in (0, 1)$, $\alpha_*^\ell = 0$. We have $\alpha_*^i \mu_*^i = \alpha_*^j \mu_*^j = \alpha_*^\ell \mu_*^\ell = 0$, thence $\mu_*^i = \mu_*^j = 0$ and, by strict complementarity, $\mu_*^\ell > 0$. From (49) we have

$$0 = J^m(x_*) + \eta_* + \mu_*^m: \quad m = i, j \tag{51a}$$

$$0 = J^\ell(x_*) + \eta_* + \mu_*^\ell \tag{51b}$$

and combining these yields

$$J^{\ell}(x_*) - J^{m}(x_*) = -\mu_*^{\ell} < 0; \quad m = i, j.$$

To show the only if part, let $J^i(x_*) = J^j(x_*) > J^{\ell}(x_*)$. Combining (51) and using (50) we have

$$\alpha_*^{\ell}\left(J^{\ell}(x_*) - J^{m}(x_*)\right) = \alpha_*^{\ell}(\mu_*^{m} - \mu_*^{\ell}) = \alpha_*^{\ell}\mu_*^{m} \geq 0.$$

Since $J^{\ell}(x_*) - J^{m}(x_*) < 0$, we have $\alpha_*^{\ell} = 0$. Given that $\alpha_*^{\ell} = 0$, $\forall \ell$, $J^{\ell}(x_*) < J^{m}(x_*)$, we can use (45) for those i, j for which $J^i(x_*) = J^j(x_*)$ to establish $\mu_*^i = \mu_*^j = 0$. By strict complementarity this implies that $\alpha_*^i, \alpha_*^j \in (0, 1)$.

Case (iii) can be established noting that for $\alpha_*^i = 1$, we have $\mu_*^i = 0$, $\alpha_*^j = 0$, $\forall j \neq i$ and, by strict complementarity, $\mu_*^j > 0$. From (49) we thus obtain

$$J^j(x_*) - J^i(x_*) \leq \mu_*^i - \mu_*^j = -\mu_*^j < 0.$$

Conversely, $J^i(x_*) > J^i(x_*)$ implies

$$\alpha_*^j\left(J^j(x_*) - J^i(x_*)\right) = \alpha_*^j\mu_*^i \geq 0$$

and thus $\alpha_*^j = 0$, $\forall j \neq i$. Case (iv) can be established as the converse of (iii). □

The above result illustrates the way in which α_* is related to $\hat{J}(x_*)$. When some of the elements of α_* are such that $\alpha_*^i \in (0, 1)$ for some $i \in M \subset \{1, 2, \ldots, m_{mod}\}$, it is shown that $J^i(x_*)$ have the same value. ***In this case, the optimal policy x_* yields the same objective function value whichever model happens to represent the economy.*** Thus, x_* is a ***robust policy.*** In other circumstances, the policy maker is ensured that implementing x_* will yield an objective function value which is at least as good as the min-max optimum. This noninferiority of x_* may, on the other hand, amount to a cautious approach with high political costs. The policy maker can, in such circumstances, use α_* as a guide and seek in its neighbourhood, a slightly less cautious scheme which is politically more acceptable. As also mentioned above, this can be done by minimizing $\langle \alpha, \hat{J}(x) \rangle$ for a given value of α.

In a numerical example of the min-max approach (48) two models of the UK economy have been considered. One of these models is the HM Treasury model (α^1) and the other is the NIESR model (α^2). The min-max solution is found to be $\alpha_k^1 = .6$ and $\alpha_k^2 = .4$ (see Becker et al. [77]).

In the algorithm discussed below, a stepsize strategy is described that directly aims at measuring progress towards the min-max solution. The algorithm defines the direction of progress as a quasi-Newton step obtained from a quadratic subproblem. An augmented Lagrangian function is defined and a procedure is formulated for determining the penalty parameter. The growth in the penalty parameter is required only to ensure a descent property. It is shown in Rustem [64] that this penalty parameter does not grow indefinitely.

In the unconstrained case, the algorithm below is similar to that of Han ([71], [72]). One slight difference is the stepsize strategy (58, a) used below. Although this is different from the strategy used by Han, it can be made equivalent to Han's strategy with a simple modification (see [64]). Both in the unconstrained and the constrained cases, the conditions for the attainment of unit stepsizes are established below. These are related to the accuracy of the projection of the Hessian approximation used by the quadratic subproblem.

The characterization of the solution of a constrained min-max problem such as (45) as a saddle point of the Lagrangian function is known to be heavily dependent on the convexity properties of the underlying problem (see Demyanov and Malomezov [68]; Arrow, Gould and Howe, [78]; Rockafeller, [79]). Motivated by the discussions in Arrow and Hurwicz [80] and Arrow and Solow [81], it has been shown that, in the case of the nonlinear programming problem (i.e. just the min case of the min-max problem below), these convexity assumptions can be relaxed via a modified Lagrangian approach (see [78]; [79]). This modified Lagrangian approach is essentially a convexification procedure for more general problems. We invoke these results to characterize the solution of the above min-max problem as a saddle-point. The saddle-point formulation and the convexification are subsequently used by the algorithm discussed in the next Section.

Let the Lagrangian function associated with (45) be given by

$$L(x, \alpha, \lambda, \mu, \eta) = <\hat{J}(x), \alpha> + <F(x), \lambda> + <\alpha, \mu> + (<1, \alpha> - 1)\,\eta \tag{52}$$

where $\lambda \in \mathbb{R}^{m \times T}$, $\mu \in \mathbb{R}_+^{m_{mod}} = \{ \mu \in \mathbb{R}^{m_{mod}} \mid \mu \geq 0 \}$ and $\eta \in \mathbb{R}^1$ are the multipliers associated with $F(x) = 0$, $\alpha \geq 0$ and $<1, \alpha> = 1$ respectively. The characterization of the min-max solution of (45) as a saddle point requires the relaxation of convexity assumptions (see [68], [76]). In order to achieve this characterization, we modify (52) by augmenting it with a penalty function. Hence, we define the augmented Lagrangian by

$$L^a(x, \alpha, \lambda, \mu, \eta, c) = L(x, \alpha, \lambda, \mu, \eta) + \frac{c}{2} <F(x), F(x)> \tag{53}$$

where the scalar $c \geq 0$ is the penalty parameter.

In nonlinear programming algorithms, the penalty parameter c is either taken as a constant is increased by a prefixed rate or is adapted as the algorithm progresses. Of the latter type of strategy, [10], [19], [23] are specific examples. We also adopt an adaptive strategy below. The departure of the strategy below from other works is mainly in the relationship of c and the descent property of the direction of search, discussed in Rustem ([64]; Lemmas (3.2) and (3.4)). In particular, c is only adjusted to ensure that the direction of search is a descent direction for the penalty function that regulates the stepsize strategy (58) below. This approach is an extension of a strategy for nonlinear programming discussed in Rustem [14], [30].

We let $H(\cdot)$ and $\mathrm{H}(\cdot)$ denote the Hessian of L and L^a, with respect to x, evaluated at $(\cdot)$. We also denote by $\nabla F(.)$ the matrix

$$\nabla F(x) = [\, \nabla_x F^1(x), \ldots, \nabla_x F^e(x) \,].$$

Sometimes, $\nabla F(x)$ evaluated at x_k will be denoted by ∇F_k and $F(x_k)$ will be denoted by F_k. Thus, a local linearization of $F(x)$ at x_k can be written as

$$F(x) = F_k + \nabla F_k^T \,[\, x - x_k \,].$$

Assumption (1)

The columns of ∇F_k are assumed to be linearly independent. □

This assumption is used to simplify the quadratic subproblem used in the algorithm in Section X for solving (45) and ensure that the system $F_k + \nabla F_k^T [x - x_k]$ has a solution, $\forall\ x_k$. This assumption can be relaxed by increasing the complexity of the quadratic subproblem.

Theorem (2)

Suppose $(x_*, \alpha_*, \lambda_*, \mu_*, \eta_*)$ satisfy the second order sufficiency conditions for x_*, α_* to be an isolated local solution of the min-max problem (45). That is, we assume

$$\nabla_{x,\alpha} L(x_*, \alpha_*, \lambda_*, \mu_*, \eta_*) = 0;$$
$$F(x) = 0;$$
$$<v, H(x_*, \alpha_*, \lambda_*, \mu_*, \eta_*)\, v> \; > 0;\ \forall\ v \neq 0 \in \mathbb{R}^{nr}: \nabla F\,(x_*)^T\, v = 0;$$
$$\alpha_* \geq 0;\quad \mu_* \geq 0;\quad <\mu_*, \alpha_*> = 0;$$
$$< 1, \alpha_* > = 1.$$

Then, if c is sufficiently large, (i) the function $L^a(x_*, \alpha_*, \lambda_*, \mu_*, \eta_*, c)$ has an unconstrained local minimum with respect to x and maximum with respect to α; (ii) $\forall$ $\lambda \in \mathbb{R}^{m \times T}$ sufficiently close to λ_*, $L^a(x, \alpha, \lambda, \mu, \eta, c)$ is strictly convex in x close to x_* and concave in λ, μ, η and α.

Remark

The equivalent result for the min-max formulation (46) follows directly from [78] and [79].

Proof

The proof follows directly from Arrow, Gould and Howe ([78], Theorem 2.2), Rockafeller ([79], Theorem 3.1) for x, λ and Demyanov and Malomezov ([68]; Theorem 5.1) for x, α, μ, η. □

We now state the saddle point property of $L^a(\cdot)$ in the neighbourhood of the min-max solution.

Theorem (3)

If c in (53) is sufficiently large, then under the conditions of Theorem 2

$$L^a(x_*, \alpha, \lambda, \mu, \eta, c) \leq L^a(x_{**}, \alpha_{**}, \lambda_{**}, \mu_{**}, \eta_{**}, c) \leq L^a(x, \alpha_{**}, \lambda_{**}, \mu_{**}, \eta_{**}, c)$$

for every x in some neighbourhood of x_* and for every α, λ, μ, η.

Remark

The equivalent result for the min-max formulation (46) follows directly from [78] and [79].

Proof

The proof follows directly from Arrow, Gould and Howe ([78], Theorems 2.2, 3.1), Rockafeller ([79], Theorem 3.1) for the nonlinear programming part involving x, λ and Demyanov and Malomezov ([68]; Theorem 5.2) for x, α, μ, η. □

X. THE MIN-MAX ALGORITHM FOR RIVAL MODELS

Consider the objective function

$$\mathcal{F}(x, \alpha) = < \alpha, \hat{J}(x) > \tag{54}$$

and its linear approximation, with respect to x, at a point x_k,

$$\mathcal{F}_k(x, \alpha) = < \alpha, \hat{J}(x_k) + \nabla \hat{J}(x_k)^T (x - x_k) >. \tag{55.a}$$

where

$$\nabla \hat{J}(x) = \left[\nabla J^1(x), \ldots, \nabla J^m(x)\right].$$

We shall sometimes denote $\hat{J}(x)$ and $\nabla \hat{J}(x)$, evaluated at x_k, by $\hat{J}_k$ and $\nabla \hat{J}_k$ respectively. Thus, for $d = x - x_k$ (55,a) can be written as

$$\mathcal{F}_k(x_k + d, \alpha) = \langle \alpha, \hat{J}_k + \nabla \hat{J}_k^T d \rangle. \tag{55,b}$$

The quadratic objective function used to compute the direction of progress is given by

$$q_k(x, \alpha, c) = \mathcal{F}_k(x, \alpha) + c \langle \nabla F_k F_k, (x - x_k) \rangle + \tfrac{1}{2} \langle x - x_k, \hat{H}_k (x - x_k) \rangle$$

or alternatively,

$$q_k(d, \alpha, c) = \mathcal{F}_k(x_k + d, \alpha) + c \langle \nabla F_k F_k, d \rangle + \tfrac{1}{2} \langle d, \hat{H}_k d \rangle.$$

The matrix $\hat{H}_k$ is a positive semi-definite approximation to the Hessian

$$H_k = \sum_{i=1}^{m} \alpha_k^i \nabla^2 J^i(x_k) + \sum_{j=1}^{e} \lambda_k^j \nabla^2 F^j(x_k) + c \nabla F_k \nabla F_k^T \tag{56}$$

It should be noted that the second derivatives due to the penalty term in the augmented Lagrangian (i.e. $c \sum_{j=1}^{e} \nabla^2 F^j(x_k) F^j(x_k)$) are not included in (56). The reason for this is discussed in [11]. Furthermore, as $F^j(x_*) = 0$ at the solution x_*, ignoring this term does not affect the asymptotic properties of the algorithm. The values α_k and λ_k are given by the solution of the quadratic subproblem in the previous iteration. The direction of progress at each iteration of the algorithm is determined by the quadratic subproblem

$$\min_d \max_\alpha \left\{ q_k(d, \alpha, c_k) \;\middle|\; \nabla F_k^T d + F_k = 0 , \alpha \in \mathbb{R}_+^{m_{mod}} \right\}. \tag{57,a}$$

Since the min-max subproblem is more complex, we also consider the quadratic programming subproblem

$$\min_{d,\, v} \left\{ v + c_k <\nabla F_k F_k, d> + \tfrac{1}{2} <d, \hat{H}_k d> \;\middle|\; \nabla F_k^T d + F_k = 0, \; \nabla \hat{J}_k^T d + \hat{J}_k \leq 1\, v \right\}.$$

(57,b)

The two subproblems are equivalent. Also, (57,b) involves fewer variables. It is shown below that the multipliers associated with the inequalities are the values α and that the solution of either subproblem satisfies common convergence properties.

Let the value of d, α, v solving (57) be denoted by $d_k, \alpha_{k+1}, v_{k+1}$. The stepsize along d_k is defined using the equivalent min-max formulation (46). Thus, consider the function

$$\psi(x) = \max_{i \in \{1, 2, \ldots, m_{mod}\}} \{ J^i(x) \}$$

and its linear approximation

$$\psi_k(x) = \max_{i \in \{1, 2, \ldots, m_{mod}\}} \{ J^i(x_k) + <\nabla J^i(x_k), x - x_k> \}.$$

Let $\psi_k(x_k + d_k)$ be given by

$$\psi_k(x_k + d_k) = \max_{i \in \{1, 2, \ldots, m_{mod}\}} \{ J^i(x_k) + <\nabla J^i(x_k), d_k> \}.$$

The stepsize strategy determines τ_k as the largest value of $\tau = (\gamma)^j, j = 0, 1, 2, \ldots$ such that x_{k+1} given by

$$x_{k+1} = x_k + \tau_k d_k$$

satisfies the inequality

$$\psi(x_{k+1}) + \frac{c_{k+1}}{2} < F_{k+1}, F_{k+1} > - \psi(x_k) - \frac{c_{k+1}}{2} < F_k, F_k >$$

$$\leq \rho \, \tau_k \, \Phi(d_k, c_{k+1}) \tag{58.a}$$

where $\rho \in (0, 1)$ is a given scalar and $\Phi(d_k, c_{k+1})$

$$\Phi(d_k, c_{k+1}) =$$

$$\psi_k(x_k + d_k) - \psi(x_k) + c_{k+1} < F_k, \nabla F_k^T d_k > + \tfrac{1}{2} < d_k, \hat{H}_k d_k > \tag{58.b}$$

The stepsize τ_k determined by (58) basically ensures that x_{k+1} simultaneously reduces the main objective and maintains or improves the feasibility with respect to the constraints. The penalty term used to measure this feasibility is a quadratic and consistent with the augmented Lagrangian (53). It is shown in Rustem ([64]: Theorem (4.1)) that (58) can always be fulfilled by the algorithm.

The determination of the penalty parameter c is an important aspect of the algorithm. This is discussed in the following description.

Algorithm: Min-max robust control with rival models

<u>Step 0:</u> Given x_0, $c_0 \in [0, \infty)$, and small positive numbers δ, ρ, ϵ, γ such that $\delta \in (0, \infty)$, $\rho \in (0, 1)$, $\epsilon \in (0, \frac{1}{2}]$, $\gamma \in (0, 1)$, $\hat{H}_0$, set k=0.

<u>Step 1:</u> Compute $\nabla \hat{J}_k$ and ∇F_k. Solve the quadratic subproblem (57) (choosing (57.a) or (57.b) defines a particular algorithm) to obtain d_k, α_{k+1}, and the associated multiplier vector λ_{k+1}. In (57.a), we also compute μ_{k+1}, η_{k+1} and in (57.b) we also compute v_{k+1}.

<u>Step 2:</u> Test for optimality: If optimality is achieved, <u>stop</u>. Else go to **Step 3.**

<u>Step 3:</u> If

$$\psi_k(x_k + d_k) - \psi(x_k) - c_k < F_k, F_k > + (\epsilon + \tfrac{1}{2}) < d_k, \hat{H}_k d_k > \leq 0.$$

then $c_{k+1} = c_k$. Else set

$$c_{k+1} = \max\left\{ \frac{\psi_k(x_k + d_k) - \psi(x_k) + (\epsilon + \frac{1}{2}) < d_k, \hat{H}_k d_k >}{< F_k, F_k >}, c_k + \delta \right\} \quad (59)$$

Step 4: Find the smallest nonnegative integer j_k such that $\tau_k = \gamma^{j_k}$ with $x_{k+1} = x_k + \tau_k d_k$ such that the inequality (58) is satisfied.

Step 5: Update $\hat{H}_k$ to compute $\hat{H}_{k+1}$, set $k = k + 1$ and go to Step 1.

In Step 3, the penalty parameter c_{k+1} is adjusted to ensure that progress towards feasibility is maintained. In particular, c_{k+1} is chosen to make the direction d_k computed by the quadratic subproblem is a descent direction for the penalty function $\psi(x_k) - \frac{c_{k+1}}{2} < F_k, F_k >$. We now summarise the optimality conditions of the quadratic subproblem (57). Subsequently, we establish the descent property of d_k and that c_k determined by (59) is not increased indefinitely.

In Rustem [64], it is shown that the above algorithm converges to a local solution of the min-max problem and that it also generates stepsizes τ_k that converge to unity and the local convergence rate near the solution is Q- or two-step Q superlinear, depending on the accuracy of the approximate Hessian used by the algorithm.

11. CONCLUDING REMARKS

Methods for optimal economic policy design are constrained with the size and nonlinearity of the econometric models. This is the basis of the essentially deterministic approach adopted above. The basic tool used is a powerful deterministic optimization algorithm. Such a computational technique is outlined in the discussion of the sequential quadratic programming algorithms in Sections II and III.

The robust strategy discussed in Sections IV and V ensure that the objective function value at the optimum is relatively insensitive to perturbations. The robust strategy is motivated by the sensitivity analysis of nonlinear problems. On the other hand, the mean-variance interpretation of this approach is based on a linear model. The question, therefore, arises whether the mean-variance interpretation applies for nonlinear models. This may be resolved by considering the linearized model around the nonlinear robust optimum. Clearly, all the robust properties associated with the strategies are valid for the linearized model. Thus robust control is at the very least the correct strategy for first order effects. We also know that if the linearization was done at an arbitrary point (e.g. the initial, or base, trajectory used to initialize the algorithm for the nonlinear problem) and not the robust optimum of the nonlinear problem, the robust objective function could be further reduced by moving to the robust optimal solution of the nonlinear model. Thus, the robust optimal solution of the nonlinear problem is central for the first order test mentioned above.

The problem of specifying the policy objective function can be resolved by utilizing the approaches discussed in Sections VI - VIII for tailoring the weighting matrix of the quadratic objective function or, for constant diagonal matrices, tailoring the desired values. Both approaches are equivalent. The possibility of constructing quadratic objective functions with diagonal weighting matrices is desirable both in terms of computational convenience and interpretability. Furthermore, it provides a possibility for defining the Pareto weights in decision making under conflicting objectives. Under conflicting objectives, the attainment of each element of the desired value vector is presumed to conflict with the attainment of other desired value elements. The extension of these results to nonlinear constraints permits the wider applicability of the results.

The existence of rival models of the same economic system can be resolved by adopting a robust policy design approach that avoids major errors due basing the policy on the wrong model. The min-max approach formulated in Section IX and the min-max algorithm discussed in Section X provide a robust model pooling method in the optimal value of α_{**}. The value of α_{*} can also be used to search for less robust alternative policies in the neighbourhood of α_{**}.

ACKNOWLEDGEMENTS

The financial support of ESRC is gratefully acknowledged.

REFERENCES

1. H.J. Kushner and D.S. Clark, "Stochastic Approximation Methods for Constrained and Unconstrained Systems", Springer-Verlag, Berlin, 1978.

2. B.T. Polyak, "Nonlinear Programming Methods in the Presence of Noise", Math. Prog., 14, 87-98 (1978).

3. W. Syski, "A Method of Stochastic Subgradients with Complete Feedback Stepsize Rule for Convex Stochastic Approximation Problems", JOTA, 59, 487-504 (1988).

4. Rustem, B., "Optimal, Time Consistent Robust Feedback Rules under Parameter, Forecast and Behavioural Uncertainty", in "Dynamic Modelling and Control of National Economies", (N. Christodoulakis, Ed.), Pergamon Press, Oxford, 1989.

5. Karakitsos, E., B. Rustem and M. Zarrop (1980). "Robust Economic Policy Formulation Under Uncertainty", *PROPE DP. 38*, Imperial College.

6. Stöppler, S. (1979). "Risk Minimization by Linear Feedback", *Kybernetes*, 8, 171-184.

7. Mitchell, D.W. (1979). "Risk Aversion and Macro Policy", *The Economic Journal*, 89, 913-918.

8. Whittle, P. (1982). *Optimization Over Time, Vol. 1*, J. Wiley & Sons, New York.

9. Engwerda, J.C. and P.W. Otter (1989). "On the Choice of Weighting Matrices in the Minimum Variance Controller", *Automatica*, 25, 279-285.

10. E. Polak, nd A.L. Tits (1980). "A Globally Convergent, Implementable Multiplier Method with Automatic Penalty Limitation", *Appl. Math. and Optimization*, 6, pp. 335-360.

11. R.A. Tapia (1986). "On Secant Updates for Use in General Constrained Optimization", *Technical Report 84-3*, Rice University.

12. S-P Han, (1977). "A Globally Convergent Method for Nonlinear Programming", *JOTA*, 22, 297-309.

13. M.J.D. Powell (1978). "A fast Algorithm for Nonlinearly Constrained Optimization Calculations", in Numerical Analysis Proccedings, Dundee 1977, G.A. Watson, ed., Springer-Verlag, Berlin.

14. B. Rustem, "Equality and Inequality Constrained Optimization Algorithms with Convergent Stepsizes", Department of Computing, Imperial College, PROPE Discussion Paper 70 (1989).

15. B. Rustem, "On the Q- and Two-Step Q-Superlinear Convergence of Successive Quadratic Programming Algorithms", Department of Computing, Imperial College, PROPE Discussion Paper 89 (1989).

16. R.A. Tapia, "Diagonalized Multiplier Methods and quasi-Newton Methods for Constrained Optimization", JOTA, 22, 135-194 (1977).

17. Y. Fan, S. Sarkar, L. Lasdon, "Experiments with Successive Quadratic Programming Algorithms", JOTA, 56, 359-382, (1988).

18. R. B. Wilson, "A Simplicial Algorithm for Concave Programming", Ph. D. Dissertation, Grad. School of Business Admin., Harvard University (1963).

19. M.C.B. Biggs, "The Development of a Class of Constrained Minimization Algorithms and their Application to the Problem of Power Scheduling", Ph.D. Thesis, University of London (1974)

20. M.C.B. Biggs, "On the Convergence of Some Constrained Minimization Algorithms Based on Recursive Quadratic Programming", J. Inst. Maths. Applics., 21, 67-81 (1978)

21 M.C.B. Biggs, "Line Search Procedures for Nonlinear Programming Algorithms with Quadratic Programming Subproblems", Hatfield Polytechnic, Numerical Optimization Centre, Report 116 (1981).

22. D.Q. Mayne AND E. Polak, "A Superlinearly Convergent Algorithm for Constrained Minimization Problems", Math. Prog. Study, 16, 45-61 (1982).

23. E. Polak and D.Q. Mayne, "A Robust Secant Method for Optimization Problems with Inequality Constraints", JOTA, 33, 463-477 (1981).

24. D. Gabay, "Reduced quasi-Newton Methods with Feasibility Improvement for Nonlinearly Constrained Optimization", Math. Prog. Study, 16, 18-44 (1982).

25. K. Schittkowski, "The Nonlinear Programming Method of Wilson, Han and Powell with an Augmented Lagrangian Type Line Search Function, Part 1: Convergence Analysis", Numer. Math., 38, 84-114 (1981).

26. K. Schittkowski, "The Nonlinear Programming Method of Wilson, Han and Powell with an Augmented Lagrangian Type Line Search Function, Part 2: An Efficient Implementation with Linear Least Squares Subproblems", Numer. Math., 38, 115-127 (1981).

27. P.E. Gill, W. Murray, M.A. Saunders, M.H. Wright, "Some Theoretical Properties of an Augmented Lagrangian Merit Function", Stanford University, Technical Report SOL 86-6R (1986).

28. J.V. Burke and S.-P. Han, "A Robust Sequential Quadratic Programming Method", Math. Prog., 43, 277-303 (1989).

29, B. Rustem, "A Class of Superlinearly Convergent Projection Algorithms with Relaxed Stepsizes", Appl. Math. and Optimization, 12, 29-43 (1984).

30. B. Rustem, "Convergent Stepsizes for Constrained Optimization Algorithms",

JOTA, 49. 135-160 (1986).

31. P.T. Boggs, J.W. Tolle and P. Wang, "On the Local Convergence of Quasi-Newton Methods for Constrained Optimization", SIAM J. Control and Optimization, 20, 161-171 (1982).

32. T.F. Coleman and A.R. Conn, "On the Local Convergence of a quasi-Newton Method for the Nonlinear Programming Problem", SIAM J. Numerical Analysis, 21, 755-769 (1984).

33. R. Fontecilla, T. Steihaug and R.A. Tapia, "A Convergence Theory for a Class of quasi-Newton Methods for Constrained Optimization", SIAM J. Num. Anal., 24, 1133-1151 (1987).

34. S-P. Han, "Superlinearly Convergent Variable Metric Algorithms for General Nonlinear Programming Problems", Math. Prog., 11, 263-282 (1976).

35. U.M.G. Palomares and O.L. Mangasarian, "Superlinearly Convergent quasi-Newton Algorithms for Nonlinearly Constrained Optimization Problems", Math. Prog., 11, 1-13 (1976).

36. J. Stoer and R.A. Tapia, "On the Characterization of Q-Superlinear Convergence of quasi-Newton Methods for Constrained Optimization", Rice University, Technical Report 84-2 (1986).

37. R.M. Chamberlain, "Some Examples of Cycling in Variable Metric Methods for Constrained Minimization", Math. Prog., 16, 378-383 (1979).

38. N. Maratos, "Exact Penalty Function Algorithms for Finite Dimensional and Control Optimization Problems", Ph. D. Thesis, University of London (1978).

39. R.M. Chamberlain, M.J.D. Powell, C. Lemarechal, H.C. Pedersen, "The Watchdog Technique for Forcing Convergence in Algorithms for Constrained Optimization", Math. Prog. Study, 16, 1-17 (1982).

40. M.J.D. Powell, "Gradient Conditions and Lagrange Multipliers in Nonlinear Programming", in "Nonlinear Optimization", (L.C.W. Dixon, E. Spedicato and G.P. Szego, eds.), Birkhauser, Boston, (1980).

41. E. Karakitsos and B. Rustem, "Optimal Fixed Rules and Simple Feedback Laws for Nonlinear Econometric Models", Automatica, 21, 169-180 (1985).

42. A.V. Fiacco, "Sensitivity Analysis for Nonlinear Programming Using Penalty Methods", Math. Prog., 10, 267-311 (1976).

43. R.G. Becker, A.J. Heunis and D.Q. Mayne, "Computer Aided Design of Control Systems via Optimization", Proc. IEE, 126, 573-578 (1979).

44. T.L. Wuu, R.G. Becker and E. Polak, "On the Computation of Sensitivity Functions of Linear Time-Invariant Systems Responses via Diagonalization", IEEE Trans., AC-31, 1141-1143 (1986).

45. D.A. Kendrick, "Stochastic Control for Econometric Models", McGraw Hill, New York, 1981.

46. S.G. Hall, I.R. Harnett and M.J. Stephenson, "Optimal Control of Stochastic Non-linear Models" in "Dynamic Modelling and Control of National Economies", (N. Christodoulakis, Ed.), Pergamon Press, Oxford, 1989.

47. B. Rustem, "On the Diagonalizability of Quadratic Forms and the Arbitrariness of Shadow Prices", in "Dynamic Modelling and Control of National Economies", (N. Christodoulakis, Ed.), Pergamon Press, Oxford, 1989.

48. A.M. Geoffrion, "Duality in Nonlinear Programming: A Simplified Applications Oriented Development", SIAM Review, 13, 1-37 (1971).

49. J. Ponstein, "Applying Some Modern Developments to Choosing Your Own Lagrange Multipliers", SIAM Review, 25, 183-199 (1983).

50. B. Rustem, K. Velupillai, J. Westcott, "Respecifying the Weighting Matrix of a Quadratic Objective Function", Automatica, 14, 567-582 (1978).

51. B. Rustem and K. Velupillai, "Constructing Objective Functions for Macroeconomic Decision Models: A Formalization of Ragnar Frisch's Approach", Imperial College PROPE Discussion Paper No. 69, (1988).

52. B. Rustem and K. Velupillai, "Constructing Objective Functions for Macroeconomic Decision Models: On the Complexity of the Policy Design Process", Imperial College, PROPE Discussion Paper No. 84, (1988).

53. A.J. Hughes-Hallett and H.J.B. Rees, "Quantitative Policies and Interactive Planning", Cambridge University Press, Cambridge, 1983.

54. A.V. Fiacco and G. P. McCormick, "Nonlinear Programming: Sequential Unconstrained Minimization Techniques" J. Wiley, New York 1968.

55. P.E. Gill, W. Murray, M.H. Wright "Practical Optimization", Academic Press, London and New York, 1981.

56. L.G. Khachian, "A polynomial Algorithm in Linear Programming", Soviet Mathematics Doklady, 20, 191-194 (1979).

57. L.G. Khachian, "Polynomial Algorithm in Linear Programming", USSR Computational Mathematics and Mathematical Physics, 20, 53-72 (1980).

58. C. Granger, and P. Newbold, "Forecasting Economic Time Series", Academic Press New York , 1977.

59. S. Makridakis and R. Winkler, "Averages of Forecasts: Some Empirical Results", Management Science, 29, 987-996 (1983).

60. M.J. Lawrence, R.H. Edmunson and M.J. O'Connor, "The Accuracy of Combining Judgemental and Statistical Forecasts", Management Science, 32, 1521-1532 (1986).

61. G.C. Chow, "Effective Use of Econometric Models in Macroeconomic Policy Formulation" in "Optimal Control for Econometric Models" (S. Holly, B. Rustem, M. Zarrop, eds.), Macmillan, London, 1979.

62. B. Rustem, "Methods for the Simultaneous Use of Multiple Models in Optimal Policy Design", in "Developments in Control Theory for Economic Analysis", (C. Carraro and D. Sartore, eds.), Maritnus Nijhoff Kluwer Publishers, Dordrecht, 1987.

63. B. Rustem, "A Superlinearly Convergent Constrained Min-Max Algorithm for Rival Models of the Same System", Comp. Math. Applic. 17, 1305-1316, (1988).

64. B. Rustem, "A Constrained Min-Max Algorithm for Rival Models of the Same Economic System", Department of Computing, Imperial College, PROPE Discussion Paper 98 (1989).

65. C. Charalambous and A.R. Conn, "An Efficient Algorithm to Solve the Min-Max Problem Directly", SIAM J. Num. Anal., 15, 162-187 (1978).

66. T.F. Coleman, "A Note on 'New Algorithms for Constrained Minimax Optimization' ", Math. Prog., 15, 239-242, (1978).

67. A.R. Conn, "An efficient second order method to solve the constrained minmax problem", Department of Combinatorics and Optimization, University of Waterloo, Report January (1979).

68. V.F. Demyanov and V.N. Malomezov, "Introduction to Minmax", J. Wiley, New York , 1974.

69. V.F. Demyanov and A.B. Pevnyi, "Some estimates in Minmax Problems", Kibernetika, 1, 107-112 (1972).

70. S.R.K. Dutta and M. Vidyasagar, "New Algorithms for Constrained Minmax Optimization", Math. Prog., 13, 140-155 (1977).

71. S-P. Han, "Superlinear Convergence of a Minimax Method", Dept. of Computer Science, Cornell University, Technical Report 78-336, (1978).

72. S-P. Han, "Variable Metric Methods for Minimizing a Class of Nondifferentiable Functions", Math. Prog., 20, 1-13, (1981).

73. W. Murray and M.L. Overton, "A Projected Lagrangian Algorithm for Nonlinear Minmax Optimization", SIAM J. Sci. Stat. Comput., 1, 345-370 (1980).

74. J. Medanic and M. Andjelic, "On a Class of Differential Games without Saddle-point Solutions", JOTA , 8, 413-430 (1971).

75. J. Medanic and Andjelic, "Minmax Solution of the Multiple Target Problem", IEEE Trans. , AC-17, 597-604 (1972).

76. G. Cohen, "An Algorithm for Convex Constrained Minmax Optimization Based on

Duality", Appl. Math. Optim. 7, 347-372 (1981).

77. R.G. Becker, B. Dwolatzky, E. Karakitsos and B. Rustem, "The Simultaneous Use of Rival Models in Policy Optimization", The Economic Journal, 96, 425-448 (1986).

78. K.J. Arrow, F.J. Gould and S.M. Howe, "General Saddle Point Result for Constrained Optimization", Math. Prog., 5, 225-234 (1973).

79. R.T. Rockafeller, "A Dual Approach to Solving Nonlinear Programming Problems by Unconstrained Minimization", Math. Prog., 5, 354-373 (1973).

80. K.J. Arrow and L. Hurwicz, "Reduction of Constrained Maxima to Saddle-Point Problems, in "3rd Berkeley Symposium of Mathematical Statistics and Probability", (Ed. J. Neyman) 1956.

81. K.J. Arrow and R.M. Solow, "Gradient Methods for Constrained Maxima, with Weakened Assumptions", in "Studies in Linear and Nonlinear Programming", (Arrow, L. Hurwicz and H. Uzawa, eds.), Stanford University Press, Stanford, 1958.

Methods in the Analysis of Multistage Commodity Markets

Michèle Breton[1]
GERAD and HEC, Montréal, Canada

Alain Haurie[2]
University of Geneva, Switzerland

Georges Zaccour[3]
GERAD and HEC, Montréal, Canada

I. Introduction

We are concerned with the characterization and computation of equilibrium solutions in networks of interacting oligopolies. These systems arise typically in the analysis of resource commodity markets.

As a motivating example we may consider the marketing of natural gas. The resource is extracted from the ground, then processed and distributed by a finite number of firms to different regional markets through a common transportation system (pipeline) which can be subject to congestion. This distribution takes place over a long time horizon, permitting the agents to adjust their production capacity through investment, but also exposing them to some unescapable randomness due to economic and political uncertainties. This economic process has thus the double characteristic of being played on a transportation network and of occurring in a sequence of stages corresponding to the different nodes of a decision tree.

[1]Supported by grants from NSERC-Canada and FCAR-Quebec
[2]Supported by grants from NSERC-Canada, FCAR-Quebec, (obtained through GERAD and HEC), and FNRS-Switzerland
[3]Supported by grants from NSERC-Canada and PCIAC-Canada

In other circumstances, a natural resource is extracted from the ground and then processed through different stages of transformation until it reaches a useful form (e.g. oil, refined products, fuels, petrochemicals, etc; or phosphate rock, phosphorous acid, fertilizers, etc). Various firms, sometimes located in different countries, are involved in the processing of the resource at the successive transformation stages. Again this economic system has the combined structure of a transportation network and a sequential decision process. However in this case this sequential decision process is present even in a static version of the model. It is implied by the hierarchy among firms and economic agents induced by the different levels (stages) of transformation of the resource.

These markets often involve only a very limited number of competitors, thus putting the characterization of equilibria in the realm of game theory ([1, 2]).

The modeling frameworks proposed here are based on the combination of a transportation network with a multi-agent decision tree (or *dynamic game*) structure. The transportation network models the resource transformation stages as well as the spatial distribution of these activities. The dynamic game models the interaction, through time or through the different hierarchy levels of the transformation system, among different agents involved in the processing and marketing of the resource.

Depending on the market structure and the associated information and decision structures, different types of equilibria will tend to occur. In the present work we shall focus on two particular cases. In the first one, the players compete by installing capacities and sending flows from production sites toward market regions. In the second case, the market has a sequential multistage structure where different players control the different links of the transportation network that correspond to various transformation stages of the resource.

In the first case, we introduce a concept of market equilibrium, called *S-adapted equilibrium*. This equilibrium retains some features of the concept of *feedback equilibrium* used in the theory of dynamic games [3] since it permits the players to adapt their actions to the sample values of the stochastic process perturbing the system. This equilibrium can be related to some recent extensions of the *Nash-Cournot equilibrium* concept [1] to spatial oligopoly theory (e.g. see [4, 5, 6]). It is also close to the concept of *open loop equilibrium* since the players are not allowed to adapt their actions to previous decisions taken by other players. An S-adapted equilibrium is a set of contingency plans such that no player has an incentive to unilaterally modify his plan. It has been introduced in [7].

In the second case we propose an equilibrium concept, called *multilevel market equilibrium.* As in the concept of *feedback equilibrium*, used in dynamic game theory [3], we base its definition on a sequential dynamic programming argument. This equilibrium concept can also be related to the economic theory of *derived demands* [8, 9, 10]. It permits the consideration of a hierarchy among players, due to their respective location on the network. For example, in the case of the phosphate rock market, a limited number of firms are involved in the ore extraction, they sell it to a limited number of firms which produce phosphorous acid; these firms, in turn, sell it to fertilizer producing firms which, finally, satisfy the final demand arising from farming activities. In a sequential market equilibrium, each firm takes into account the *rational reaction function* of the firms located at the next level in the resource transformation hierarchy. Such a function describes the complete course of actions of a player in reaction to the decisions announced by the players located at the upstream level. At equilibrium, all reaction functions are rational, i.e. no player has an incentive to modify unilaterally his reaction function. This type of equilibrium has been studied in [11, 12] and more recently in [13].

The paper is organized as follows: In section II we propose two modeling frameworks combining a transportation network and a sequential game structure. In section III we introduce the concept of *S-adapted equilibria.* These equilibria are characterized by optimality conditions in the form of variational inequalities [14, 15]. We present a numerical example illustrating this concept. Section IV deals with the concept of *multilevel market equilibria* where different players intervene at different stages of the transformation process. These equilibria are characterized by a set of dynamic programming equations defined over the stages of the transformation process. We also present a numerical example illustrating this concept. In section V we conclude by indicating possible future developments concerning the modeling of these types of economic structures.

II. Two Modeling Frameworks

In this section we set the stage by defining two structures involving a network of oligopolies.

A. A Dynamic Network Game

In this modeling framework we retain the following features which are typical of many resource commodity markets

- Competitors and markets are spatially separated. The demand for transportation is derived from the demand and supply characteristics of these markets.
- Through capacity expansion, the competitors can modify the characteristics of the transportation network.
- Random perturbations can affect the demand laws at successive time periods.

1. Market and Information Structures

Let T be a discrete set of time periods. We consider a dynamical system described at time $t \in T$ as a network $(N_t, \mathcal{A}_t)$ where N_t is a set of nodes and $\mathcal{A}_t$ a set of arcs. A commodity flow $(q_a(t))_{a \in \mathcal{A}_t}$ circulating on the admissible arcs corresponds to a traded good passing, at time t, through successive transformation or distribution stages.

Let $J = \{1, \ldots, m\}$ be a set of m economic agents, also called *players*. With each arc $a \in \mathcal{A}_t$ and player $j \in J$ is associated a nonnegative capacity $\gamma_a^j(t)$. The set of *active players* on a, noted $J_a(t)$, is composed of those players who have a positive capacity on that arc. An active player j sends a flow $q_a^j(t) \in [0, \gamma_a^j(t)]$ on arc a.

Let E be a finite set and ε_t, $t \in T$, a stochastic process (e.g. a Markov chain) with values in E. Let $M_t \in N_t$ be the *market nodes*. With each node $n \in M_t$ is associated an *inverse demand law*, defined by $p_n = f_n(q_{\cdot n}(t), \varepsilon_t)$, where $q_{\cdot n}(t)$ is the total flow arriving at node n in period t and p_n is the clearing price at this market location.

A player j incurs a variety of costs on arc a in period t. These are

- an *investment cost* $IC_a^j(I_a^j(t))$ associated with the capacity expansion $I_a^j(t) = \gamma_a^j(t+1) - \gamma_a^j(t)$;
- a *transformation cost* $TC_a^j(q_a^j(t))$ associated with the commodity flow sent by player j on arc a;
- an arc *utilization cost* $UC_a^j(q_a(t), q_a^j(t))$ depending on both the flow $q_a^j(t)$ sent by player j and the total flow $q_a(t) = \sum_{j \in J_a(t)} q_a^j(t)$ circulating on arc a; this takes into account possible congestion effects.

The profit of player j, in each time period t, is the difference between the revenue generated by the flows sent by him to the market nodes and the sum of costs associated with every arc of the network

$$\begin{aligned}\Pi^j(\sigma, \varepsilon_t) \quad = \quad & \sum_{n \in M_t} q_{\cdot n}^j(t) f_n(q_{\cdot n}(t), \varepsilon_t) - \\ & \sum_{a \in \mathcal{A}_t} \{IC_a^j(I_a^j(t)) + TC_a^j(q_a^j(t)) + UC_a^j(q_a(t) q_a^j(t))\} \quad (1)\end{aligned}$$

where $q_{\cdot n}^j(t)$ represents the flow sent by player j on all arcs arriving at node $n \in M^t$, i.e. the total quantity player j sends to market n, and $q_{\cdot n}(t) = \sum_{j \in J} q_{\cdot n}^j(t)$ is the total flow arriving at node n.

Players interact in time and space. They can be located on different arcs of the network and, through their investment decisions, they can modify the network configuration.

To represent the interaction of players over time, we place the network in an *event tree*. The evolution of the game is then represented as in Figure 1. At any time period the network has a given topology. Players choose the flows circulating on the arcs and decide of their investments. This defines their profits for that period and determines the network topology for the next period. Then an event takes place, affecting the demand laws. This process is repeated for all $t \in T$.

The information structure studied in the context of this dynamical network game is symmetrical. Players act simultaneously, at any time period t, with only a knowledge of the current sample value s_t of the random process

ε_t, without having access to the decisions actually taken by their competitors at t or at previous periods. This information structure will be described in more detail when we introduce the concept of *S-adapted equilibrium.* In the next subsection we give an example of an economic structure modelled as a dynamical network game.

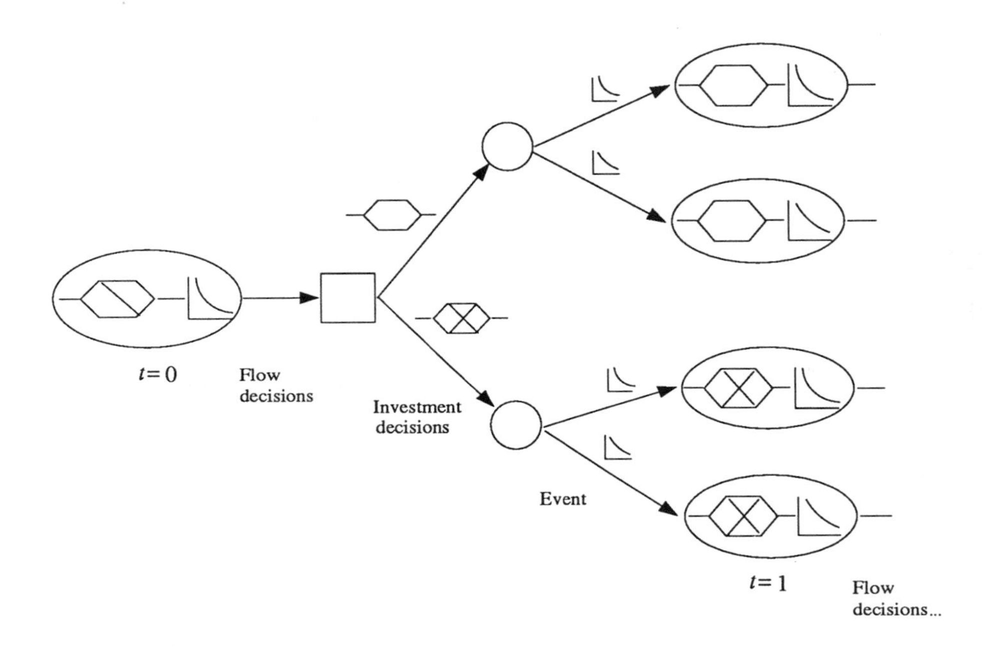

Figure 1: Evolution of the Game

2. Example: A Spatial Oligopoly Model

The following example is inspired from [4, 5, 6] and [16]. It describes a situation where a finite set of firms, (e.g. gas producers), supply a given set of markets, (e.g. regions), through a distribution network, (e.g. pipelines), subject to congestion.

The network topology at period t is described in Figure 2. We give below more detail on each component of the network:

- The set of nodes is $N = \{n_0, \ldots, n_7\}$ where n_1 and n_2 correspond to two different production sites and n_5, n_6, n_7 correspond to three different demand markets, charecterized by their inverse demand laws. The nodes n_3 and n_4 correspond to the origins of the distribution systems used by the producers in order to supply markets.

- The set of producers is J. The values γ_1^j, $j \in J$, on arc a_1 linking nodes n_1 and n_3, correspond to the production capacities installed by the different firms on location 1, similarly for location 2 and arc a_2.
- A transformation cost, corresponding to the production cost at each facility, is also defined on arcs a_1 and a_2.
- A utilization cost is defined on arcs a_3 to a_8 corresponding to the distribution system links (we assume infinite capacity for each player, since the congestion effect limits the possible use of the distribution system).

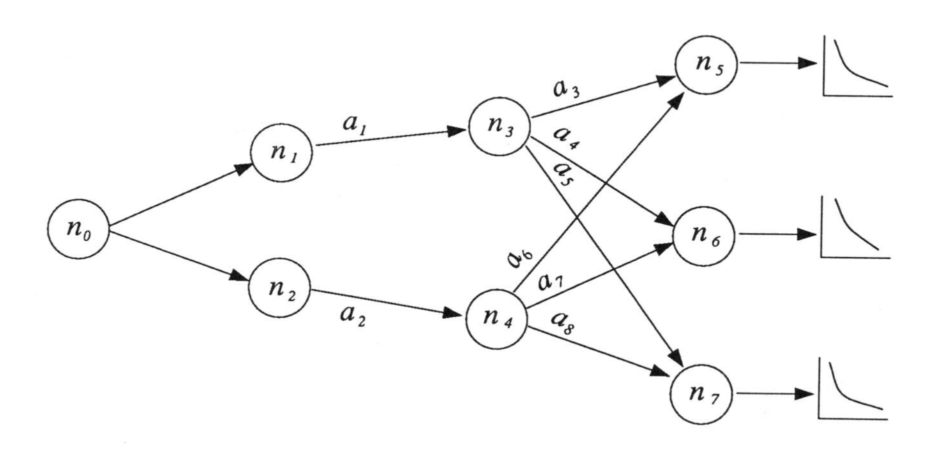

Figure 2: Spatial Oligopoly

We then consider a random process described by the event tree of Figure 3. Combining the network shown in Figure 2 with this event tree we obtain a dynamical network game as illustrated on Figure 1. This type of model has been used in [16] to model the competition on the European gas market.

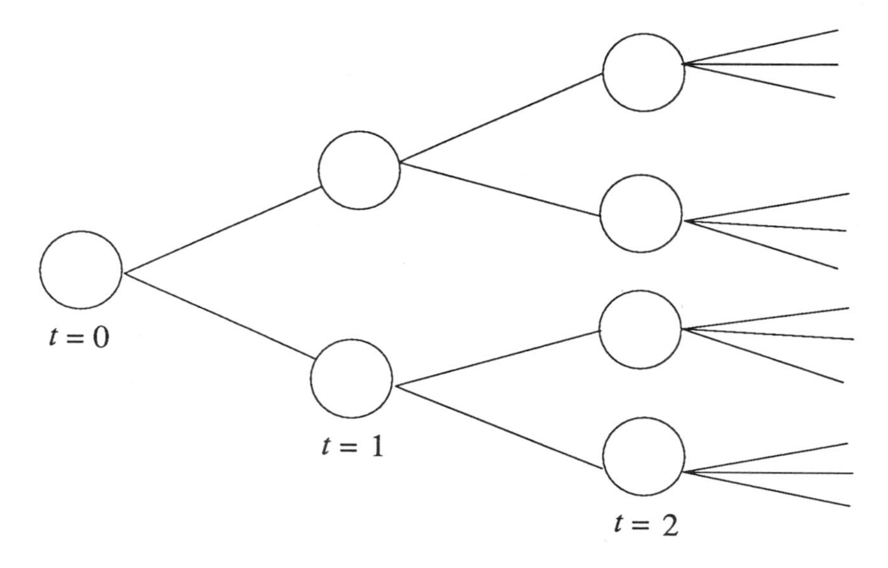

Figure 3: Event tree

B. A Multilevel Network Game

In this modeling framework we retain the following features appearing in many resource commodity markets

- Competitors are spatially separated and are associated with the successive transformation stages (semi-finished products) of the resource. The demand for a semi-finished product supplied by a firm is derived from the demand and supply characteristics of competitors located downstream on the network.
- There is a competitive demand price process occuring on the final product market.

Dynamic effects and random perturbations are not considered for the moment.

1. Market and Information Structures

We consider a network $(N, \mathcal{A})$ where the set of nodes N is composed of three types of nodes

- a *source* node n_0;
- a set $N_j = \{n_j : j \in J\}$ of nodes in correspondance with the set of players;
- a set M of *market* nodes representing spatially separated markets for the finished product.

Let L be the number of transformation stages between the extraction of the resource and its final form as a finished product. Let $\{J^\ell, \ell \in \{1, \ldots, L\}\}$ be a partition of the set of players indexed over the successive stages defined above and define $J^0 \equiv n_0$. A player $j \in J^\ell$ is an active player in stage ℓ. The set of arcs $\mathcal{A}$ is composed of two classes of arcs

- the set of transportation arcs $a = (n_i, n_j)$, where necessarily $n_i \in J^{\ell-1}$ and $n_j \in J^\ell$, $\ell = 1, \ldots, L$;
- the set of market arcs $a = (n_i, n_j)$ where necessarily $n_i \in J^L$ and $n_j \in M$.

A flow q_a circulating on an arc $a \in \mathcal{A}$ corresponds to a traded commodity, passing from one stage of transformation to the next, and finally put on a market as a final product.

With each market node $n \in M$ is associated an inverse demand law $p_n = f_n(q_{\cdot n})$, where $q_{\cdot n}$ is the total flow arriving at node n and p_n is the clearing price at this market location.

With each transportation arc $a = (n_i, n_j)$ is associated a unit price p_{ij} for the commodity circulating on that arc.

A player $j \in J^\ell$, $\ell = 1, \ldots, L-1$, controls

- the inflows $q_{\cdot j} = (q_{ij})_{i \in J^{\ell-1}}$,
- the prices of the outflows $p_{j\cdot} = (p_{jk})_{k \in J^{\ell+1}}$.

A player $j \in J^L$ controls

- the inflows $q_{\cdot j} = (q_{ij})_{i \in J^{L-1}}$,

- the outflows $(q_{jn})_{n\in M}$ to the different market nodes $n \in M$.

Associated with each player $j \in J^\ell$, $\ell = 1, \ldots, L$ is a transformation cost $TC_j(q_{\cdot j}, q_{j\cdot})$ which depends on both the inflow and the outflow vectors. The profit of player $j \in J^\ell$ is the difference between the revenue generated by the outflows and the costs due to the purchase of inflows and to the transformation process. The profit function is given by

$$\Pi^j(p_{\cdot j}, q_{\cdot j}, p_{j\cdot}, q_{j\cdot}) = \sum_{k\in J^{\ell+1}} q_{jk}p_{jk} - \sum_{i\in J^{\ell-1}} q_{ij}p_{ij} - TC^j(q_{\cdot j}, q_{j\cdot}) \quad (2)$$

when $j \in J^\ell$, $\ell = 1, \ldots, L-1$ and

$$\Pi^j(p_{\cdot j}, q_{\cdot j}, q_{j\cdot}) = \sum_{n\in M} q_{jn}f_n(q_{\cdot n}) - \sum_{i\in J^{L-1}} q_{ij}p_{ij} - TC^j(q_{\cdot j}, q_{j\cdot}) \quad (3)$$

when $j \in J^L$.

On Eqs.(2, 3) it is apparent that the profit of player j depends on the decisions of players located in the preceding and in the succeeding stage.

These concepts are detailed later on in the paper when we discuss the concept of *multilevel market equilibrium.*

2. Example: A Static Hierarchical Oligopoly Model

This example is inspired from [11, 12]. The network represents the transformation process of a commodity from its extraction to its final form through the successive stages of the production system. At each level, a limited number of firms behave in non-cooperative fashion, purchasing their input, transforming it and selling their output either to firms located at the next level or on the final demand market. We assume that firms can discriminate among the various buyers and sellers of the commodity.

A network representation of this model is depicted in Figure 4.

Players are located on nodes n_1 to n_4. At stage 1 of the transformation process, firms 1 and 2 are in competition, extracting the resource from a common source n_0, transforming it into a semi-product and selling it to firms 3 and 4 at the second stage of the transformation process. Arcs a_1 and a_2 model the transportation of the resource to the locations of firms 1 and 2 respectively; on these arcs are indicated prices p_{01} and p_{02}, representing the unit price (or royalty) paid by firms 1 and 2 at the extraction site.

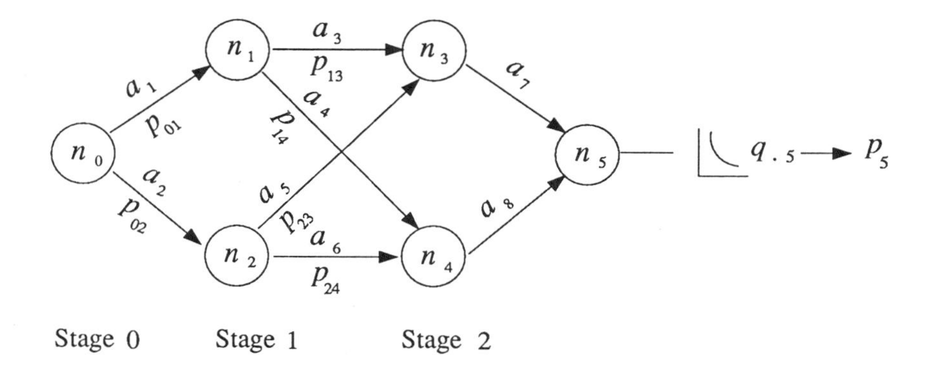

Figure 4: Multilevel Oligopoly Network

At the second stage of the transformation process, firms 3 and 4 are located on nodes n_3 and n_4. Arcs a_3 to a_6 represent the transportation of the commodity from firms 1 and 2 to firms 3 and 4. On these arcs are indicated the unit prices charged by the firms in level 1 to each firm in level 2.

Finally, node n_5 is a market node for the finished product, characterized by an inverse demand law $q_{.5} \mapsto p_5$. The market arcs a_7 and a_8 thus model the selling of this finished product on the market by firms 3 and 4 respectively.

III. S-adapted Equilibria

In this section we define an equilibrium concept corresponding to a situation where all the players have symmetric information and can adapt their decisions to the random events occurring at successive time periods. We first proceed with a formal definition of the concept of *S-adapted equilibrium*. We then give a set of optimality conditions permitting the computation of such equilibria, and we conclude the section with a numerical example.

A. S-adapted Strategies and Equilibria

The mathematical description of the information structure is borrowed from [7]. A player $j \in J$ defines, for each time period t and each possible sample value of ε_t, a set of capacity expansions (investments) $\tilde{I}_a^j(t, \varepsilon_t)$ and a set of flows $\tilde{q}_a^j(t, \varepsilon_t)$. We say that the players use *S-adapted* strategies $\tilde{\sigma}^j = (\tilde{I}_a^j, \tilde{q}_a^j)$, since they can adapt their actions to the *sample path* of the random process ε_t. We denote $\tilde{\sigma} = (\tilde{\sigma}^j)_{j\in J}$ the strategy vector of all players. This strategy class is, in a sense, "half way" between open-loop and feedback controls, two types of information structure commonly used in dynamic system theory. The players adopt, at time $t = 0$, a program of actions for each time period and each possible event. As the event sequence is not affected by players' decisions, it is possible to control the system in this "semi" open-loop way.

For a given sample value ε_t, the profit of player j in period t, denoted $\Pi^j(\tilde{\sigma}(t, \varepsilon_t), \varepsilon_t)$ is a function of the vector of strategies used by all players and of the sample value of the stochastic process perturbating the system.

The reward $\phi^j(\tilde{\sigma})$ of player j, when all players have selected an S-adapted strategy, is given by the discounted expected profit at time $t = 0$. To express this expectation let us introduce the following notations: SP is the set of all possible sample paths s in the event tree of the ε_t process; s_t is the sample value of ε_t along the sample path s. Let $P(s)$ be the elementary probability of the sample path s in the event tree. Then the rewards of player j can be written

$$\phi^j(\tilde{\sigma}) = \sum_{s\in SP} P(s) \sum_{t\in T} \beta_j^t \Pi^j(\tilde{\sigma}(t, s_t),\, s_t) \tag{4}$$

where $0 \leq \beta_j \leq 1$ is a given discount factor for player j and the profit in each node of the event tree is defined by

$$\begin{aligned} \Pi^j(\tilde{\sigma}(t, s_t)\, s_t) \quad = \quad & \sum_{n\in M_t} \tilde{q}_{.n}^j(t, s_t) f_n(\tilde{q}_{.n}(t, s_t), s_t) - \\ & \sum_{a\in \mathcal{A}_t} \{IC_a^j(\tilde{I}_a^j(t)) + \\ & TC_a^j(\tilde{q}_a^j(t, s_t)) + UC_a^j(\tilde{q}_a(t, s_t), \tilde{q}_a^j(t, s_t))\}. \end{aligned} \tag{5}$$

We say that a strategy is *admissible* if all capacities and flows are non-negative, on every arc the flow does not exceed the installed capacity and, at every node, flow conservation constraints are satisfied.

Definition 1 *The strategy* $\tilde{\sigma}_* = (\tilde{\sigma}_*^j)_{j \in J}$ *is an S-adapted equilibrium if each* $\tilde{\sigma}_*^j$ *is admissible and the following holds for any* $j \in J$ *and any admissible* $\tilde{\sigma}^j$

$$\phi^j(\tilde{\sigma}_*) \geq \phi^j(\tilde{\sigma}_{*(j)}) \tag{6}$$

where

$$\tilde{\sigma}_{*(j)} = (\tilde{\sigma}_*^1, \ldots, \tilde{\sigma}_*^{j-1}, \tilde{\sigma}^j, \tilde{\sigma}_*^{j+1}, \ldots, \tilde{\sigma}_*^m). \tag{7}$$

An S-adapted equilibrium corresponds to a situation where the players adopt, at time $t = 0$, a contingency plan, for the whole time horizon T and all possible sample values. At equilibrium no player has any incentive to modify unilaterally his plan of action. The following proposition gives conditions for the existence of S-adapted equilibria.

Proposition 1 *Denote by* Ω^j *the set of admissible strategies for player* j *and* Ω *the set of admissible strategy vectors* $\tilde{\sigma}$. *Under the following conditions, there exists at least one S-adapted equilibirum.*

i) J *is finite.*

ii) Ω^j *is compact and convex.*

iii) $\phi^j(\tilde{\sigma})$ *is scalar valued, continuous and bounded on* $\Omega = \Omega^1 \times \cdots \times \Omega^m$.

iv) $\phi^j(\tilde{\sigma})$ *is quasi-concave w.r.t.* $\tilde{\sigma}^j$, $\forall j \in J$.

Proof: See Theorem 7.7 in [1].

B. Optimality Conditions

From the definition given above, player $j \in J$ considers the following optimization problem:

$$\max_{\tilde{\sigma}^j \in \Omega^j} \phi^j(\tilde{\sigma}_*^1, \ldots, \tilde{\sigma}_*^{j-1}, \tilde{\sigma}^j, \tilde{\sigma}_*^{j-1}, \ldots, \tilde{\sigma}_*^m) \tag{8}$$

Assuming that $\phi^j(\tilde{\sigma})$ is concave w.r.t. $\tilde{\sigma}^j$ and Ω^j is convex and compact, then the Kuhn-Tucker conditions associated with problem (8) are necessary and sufficient:

$$\begin{aligned} \nabla_{\tilde{\sigma}_*^j} \phi^j(\tilde{\sigma}_*) &\leq 0 \\ [\tilde{\sigma}_*^j]^T \, [\nabla_{\tilde{\sigma}_*^j} \phi^j(\tilde{\sigma}_*)] &= 0 \\ \tilde{\sigma}_*^j \in \Omega^j, & \qquad \forall\, j \in J, \end{aligned} \tag{9}$$

where ∇ denotes the gradient operator.

Any vector of strategies $\tilde{\sigma}_*$ such that, $\forall\, j \in J$, $\tilde{\sigma}_*^j$ satisfies the above conditions is an S-adapted equilibrium. Further, this non-linear complementarity problem is equivalent to the following variational inequality ([4, 6, 17, 18, 19]):

Find $\tilde{\sigma}_* \in \Omega = \prod_{j \in J} \Omega^j$ such that:

$$F(\tilde{\sigma}^*)^T(\tilde{\sigma} - \tilde{\sigma}^*) \leq 0 \qquad \forall\, \tilde{\sigma} \in \Omega \tag{10}$$

where $F^j(\tilde{\sigma}^j) = \nabla_{\tilde{\sigma}^j} \phi^j(\tilde{\sigma})$ and $F(\tilde{\sigma}) = \{F^j(\tilde{\sigma}^j)\}_{j \in J}$.

There exist numerous algorithms to solve the above variational inequality. For a general discussion of these algorithms, we refer the interested reader to [14] and [15], to [4, 18] and [6, 19] for applications of these algorithms to the computation of Nash equilibria in the cases of static one-market oligopoly and static network oligopoly problems and to [20] for the general dynamic and stochastic network oligopoly problem, where demand laws are affine.

We illustrate how problem (8) can be solved by a variational inequality approach by presenting the very simple case of a static oligopoly problem.

Denote by J the set of players, by q^j the quantity put on the market by player j and let $q = (q^1, \ldots, q^m)$. Let $P(Q)$ be the inverse demand law, where $Q = \sum_{j \in J} q^j$. We assume that $C^j(q^j)$, the production cost function of j, is convex and continuously differentiable and that $P(Q)$ is affine i.e. $P(Q) = a - bQ$, $a, b > 0$.

Player j faces the following optimization problem:

$$\max_{q^j \geq 0} \quad \phi^j(q_*^1, \ldots, q_*^{j-1}, q^j, q_*^{j+1}, \ldots, q_*^m) = \qquad (11)$$
$$q^j P(q^j + \sum_{\substack{i \in J \\ i \neq j}} q_*^i) - C^j(q^j).$$

The variational inequality associated to problem (11) can be written as follows:

Find $q_* \geq 0$ such that

$$\sum_{j \in J} [P(Q_*) + q_*^j P'(Q_*) - C^{j'}(q_*^j)]\,[q^j - q_*^j] \leq 0 \qquad \forall\, q^j \geq 0. \qquad (12)$$

Substituting $a - bQ_*$ for $P(Q_*)$ in (12) leads to:

$$\sum_{j \in J} [\,(a - bQ_*) - bq_*^j - C^{j'}(q_*^j)\,]\,[q^j - q_*^j] \leq 0 \qquad (13)$$

which is equivalent to:

$$(a - bQ_*)\,(Q - Q_*) - \sum_{j \in J} [bq_*^j + C^{j'}(q_*^j)\,]\,[q^j - q_*^j] \leq 0. \qquad (14)$$

A solution to the above inequality is given by:

$$\max_{Q,q} \quad \left[\int_0^Q (a - bz)dz - \sum_{j \in J} \int_0^{q^j} (bz^j + C^{j'}(z^j))dz^j\right] \qquad (15)$$
$$= \max_{Q,q} \quad aQ - \frac{1}{2}bQ^2 - \sum_{j \in J} \left(\frac{b}{2}(q^j)^2 + C^j(q^j)\right)$$
$$\text{Subject to} \quad Q = \sum_{j \in J} q^j$$
$$q^j \geq 0.$$

Notice that in this particular case, the determination of an equilibrium vector is equivalent to finding a maximum of a non-linear program.

C. Numerical Illustration

We consider a two-player, three-period game, i.e. $J = \{1, 2\}$ and $T = \{0, 1, 2\}$. The players are producers of an homogeneous commodity sold on a competitive market.

The network corresponding to this structure is given in Figure 5. This is the most simple case with a single arc a linking node n_0, a dummy origin, and node n_1 which corresponds to the market location. Two firms are active on a. Notice that no congestion effects and no arc utilization costs are considered in this example.

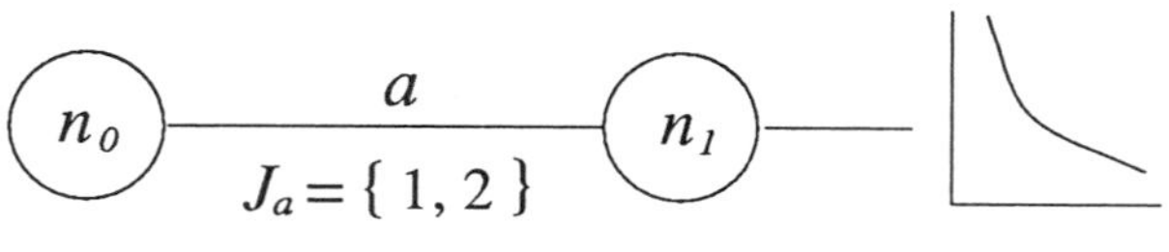

n_0 — a, $J_a = \{1, 2\}$ — n_1

Figure 5: The simple duopoly structure

The inverse demand law at node n_1 is

$$p(t, s^t) = f(q^1(t, s^t) + q^2(t, s^t), s^t) \tag{16}$$

where s^t is the sample value at period t of a random perturbation of the market at period t, $q^1(t, s^t) + q^2(t, s^t)$ is the total quantity put on the market, and $p(t, s^t)$ is the clearing market price at period t, for the realization s^t of the random perturbation. The function $f(\cdot, \cdot)$ is assumed to be affine, with negative slope w.r.t. its first argument, and the random perturbations are described by an event tree shown in Figure 6.

Players' actions correspond to the quantities they put on the market at each of the three periods, together with their investment decisions to change (here, to increase) their production capacities. Player j is described by the following data:

- The initial production capacity $\gamma^j(0, s^0)$.
- A transformation cost function $TC^j(\tilde{q}^j(t, s^t))$, where $\tilde{q}^j(t, s^t)$ denotes the quantity put on the market at period t, and for sample value s^t.
- An investment cost function $IC^j(\tilde{I}^j(t, s_t))$.

Let S^t denote the set of possible sample values of the random perturbation at period t, let $\alpha(s^t) \in S^{t-1}$ denote the unique predecessor of $s^t \in S^t$, $\quad t = 1, 2$, and $B(s^t) \subset S^{t+1}$, $\quad t = 0, 1$, denote the set of successors of s^t on the event tree. Let $P(s^t|\alpha(s^t)) \geq 0$ be the conditional probability associated with the arc $(\alpha(s^t), s^t)$ in the event tree, with

$$\sum_{s^{t+1} \in B(s^t)} P(s^{t+1}|s^t) = 1. \tag{17}$$

The set S^0 reduces to the singleton s^0 called the *root* of the event tree.

Player j's reward is given by the following expression which is an equivalent formulation of the expectation defined in Eq. (4). In this expression a sample path is decomposed into the successive sample values at periods 0,1,2 and the elementary probability of the sample path is expressed by composing the conditional transition probabilities.

$$\begin{aligned}
\phi^j(\tilde{\sigma}) \;=\; & \tilde{q}^j(0,s^0) f(\tilde{q}^1(0,s^0) + \tilde{q}^2(0,s^0), s^0) \\
& -TC^j(\tilde{q}^j(0,s^0)) - IC^j(\tilde{I}^j(0,s^0)) \\
& +\beta^j \sum_{s_k^1 \in S^1} P(s_k^1|s^0) \{ (\tilde{q}^j(1,s_k^1) f(\tilde{q}^1(1,s_k^1) \\
& +q^2(1,s_k^1), s_k^1) - TC^j(\tilde{q}^j(1,s_k^1)) - IC^j(I^j(1,s_k^1)) \\
& +\beta^j \sum_{s_l^2 \in B(s_k^1)} P(s_l^2|s_k^1) [(\tilde{q}^j(2,s_l^2) f(\tilde{q}^1(2,s_l^2) \\
& +q^2(2,s_l^2), s_l^2) - TC^j(\tilde{q}^j(2,s_l^2))] \}.
\end{aligned} \tag{18}$$

The optimization of the reward given by Eq. (18) is performed subject to the following constraints:

expansion of production capacity

$$\gamma^j(t,s^t) = \gamma^j(t-1,\alpha(s^t)) + \tilde{I}^j(t,\alpha(s^t)) \quad s^t \in S^t, \quad t = 1, 2 \tag{19}$$

capacity constraint

$$\tilde{q}^j(t,s^t) \leq \gamma^j(t,s^t) \quad s^t \in S^t, \quad t = 0, 1, 2 \tag{20}$$

nonnegativity constraints

$$\tilde{q}^j(t,s^t) \;\geq 0, \qquad s^t \in S^t, t = 0,1,2 \tag{21}$$

$$\tilde{I}^j(t,s^t) \;\geq 0, \qquad s^t \in S^t, t = 0,1. \tag{22}$$

Using an adaptation of Theorems 7.1 and 7.7 in [1] one can easily show that, if all the cost functions are strictly convex, then the reward functions are strictly concave and there exists a unique S-adapted equilibrium for the market game defined by Eqs. (16, 22).

As a numerical illustration, consider the case where the cost functions are defined as follows:

$$\begin{aligned} TC^1(q^1) &= 3\,(q^1)^2 \\ TC^2(q^2) &= 2\,(q^2)^2 \\ IC^1(I^1) &= 8\,(I^1)^2 \\ IC^2(I^2) &= 7\,(I^2)^2 \end{aligned} \tag{23}$$

Figure 6 gives a representation of the demand laws at each node of the event tree. The computational technique outlined in Section III.B has been applied to this problem, i.e. a non-linear program, equivalent to the solution of the equilibrium problem, has been defined and solved using a general NLP code. Tables 1 and 2 give the results of the computation of S-adapted stochastic equilibria for two different values of β^j, $j = 1.2$. Notice, in the results shown if Tables 1 and 2, the modification of production and investment decisions in this equilibrium solution, when the sample values vary.

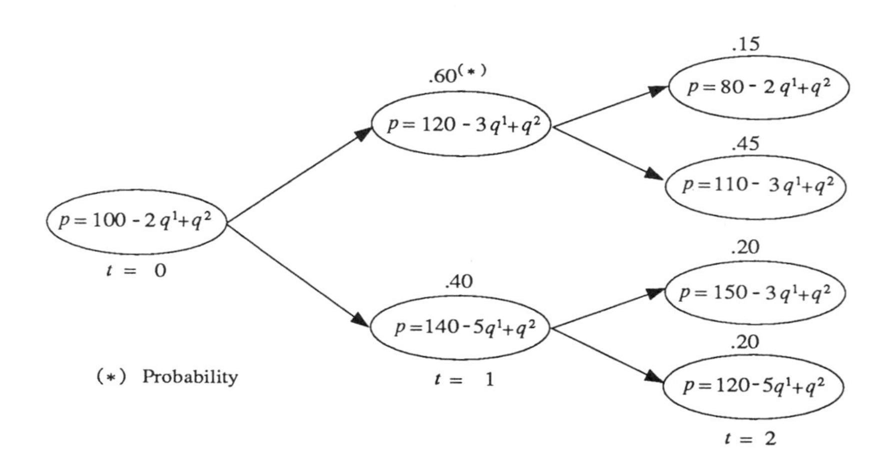

Figure 6: Demand laws at nodes of the event tree

TABLE 1. Numerical results for $\beta^1 = \beta^2 = 1/1.1$.

Node	$(0, s^0)$	$(1, s^1)$	$(1, s^2)$	$(2, s^1)$	$(2, s^2)$	$(2, s^3)$	$(2, s^4)$
Price	82.000	84.741	85.085	56.192	72.902	108.633	71.720
q^1	5.000	5.808	5.038	5.381	5.843	6.595	3.914
q^2	4.000	5.945	5.945	6.523	6.523	7.194	5.742
IC^1	5.000	5.808	5.808	5.843	5.843	6.595	6.595
IC^2	4.000	5.945	5.945	6.523	6.523	7.194	7.194
I^1	.808	.035	.787				
I^2	1.945	.578	1.249				
$\pi^{1(*)}$	329.777	355.425	315.963	178.102	267.392	484.258	194.011
π^2	269.519	391.601	385.658	232.596	322.678	560.060	285.847

(*) Profits are in present values.

TABLE 2. Numerical results for $\beta^1 = \beta^2 = 1/1.6$.

Node	$(0, s^0)$	$(1, s^1)$	$(1, s^2)$	$(2, s^1)$	$(2, s^2)$	$(2, s^3)$	$(2, s^4)$
Price	82.000	85.568	85.995	56.882	74.459	111.435	71.760
q^1	5.000	5.658	5.315	5.489	5.777	6.331	4.101
q^2	4.000	5.486	5.486	6.070	6.070	6.524	5.547
IC^1	5.000	5.658	5.658	5.777	5.777	6.331	6.331
IC^2	4.000	5.486	5.486	6.070	6.070	6.524	6.524
I^1	.658	.119	.673				
I^2	1.486	.584	1.083				
$\pi^{1(*)}$	331.536	242.495	230.433	86.655	128.917	228.613	95.247
π^2	280.543	254.279	252.521	106.087	147.764	250.733	131.451

(*) Profits are in present values.

IV. Multilevel Market Equilibrium

In this section we define an equilibrium concept corresponding to a situation where the players have asymmetric information, due to the fact that they are associated with different transformation stages of the resource. Defining a sequential, or multilevel, game structure, where each level is a transformation stage, we propose an equilibrium concept, based on dynamic programming, which is closely related to *feedback Nash equilibria* in dynamical games [3]. We give a set of dynamic programming optimality conditions and we illustrate the concept on a simple numerical example.

A. Definition of Multilevel Market Equilibria

Let p^ℓ be the complete price system set forth by players at stage ℓ. A strategy for player $j \in J^\ell$, $\ell = 1, \ldots, L-1$, is a mapping

$$\tilde{\sigma}^j(p^{\ell-1}) \mapsto (q_{\cdot j}, p_{j\cdot}) \geq 0. \tag{24}$$

A strategy for player $j \in J^L$ is a mapping

$$\tilde{\sigma}^j(p^{\ell-1}) \mapsto (q_{\cdot j}, (q_{jn}, n \in M)) \geq 0. \tag{25}$$

$\tilde{\sigma}_{[0]}$ denotes the strategy vector of all players and $\tilde{\sigma}_{[\ell]} = (\tilde{\sigma}^j)_{j \in J^k, k=\ell+1,\ldots,L}$ is the strategy vector of all players located downstream of level ℓ.

Given an initial price system p^0 at level 0, a strategy vector $\tilde{\sigma}_{[0]}$ generates a set of prices and a set of flows throughout the network. They are determined recurrently from stage to stage by applying the mappings (24, 25).

Similarly, given a price system p^ℓ at level ℓ, the restricted strategy vector $\tilde{\sigma}_{[\ell]}$ of all players located downtream of level ℓ generates a set of prices and a set of flows through the downstream levels $\ell + 1, \ldots, L$ of the network. Thus, for any given price system p^ℓ at level ℓ, the profit of any player $j \in J^k, k = 1, \ldots, L$ can be expressed as a function of the restricted strategy vector $\tilde{\sigma}_{[\ell]}$:

$$\phi^j(\tilde{\sigma}_{[\ell]}, p^\ell) = \Pi^j(p_{\cdot j}, q_{\cdot j}, p_{j\cdot}, q_{j\cdot}) \tag{26}$$

where the prices $p_{\cdot j}, p_{j\cdot}$ and the flows $q_{\cdot j}, q_{j\cdot}$ are determined recurrently from (24, 25) by $\tilde{\sigma}_{[\ell]}$ and p^ℓ.

A strategy $\tilde{\sigma}_{[\ell]}$ is *admissible at* p^ℓ, $\ell = 0, \dots, L-1$, if it generates a set of flows satisfying the conditions

$$\sum_{i \in J^{k-1}} q_{ij} \geq \sum_{\iota \in J^{k+1}} q_{j\iota} \quad j \in J^k,\ k = \ell+1, \dots, L. \tag{27}$$

We can now define

Definition 2 *The strategy $\tilde{\sigma}_* = (\tilde{\sigma}_*^j)_{j \in J}$ is a sequential market equilibrium if*

- *$\tilde{\sigma}_{*[\ell]}$ is admissible for any price system $p^\ell \geq 0$, $\ell = 0, \dots, L-1$;*
- *for any $j \in J^\ell$, $\ell = 1, \dots, L$, and any price system $p^k \geq 0$, $k = 0, \dots, \ell$, and for any admissible $\tilde{\sigma}_{*[k]_{(j)}}$*

$$\phi^j(\tilde{\sigma}_{*[k]}, p^k) \geq \phi^j(\tilde{\sigma}_{*[k]_{(j)}}, p^k) \tag{28}$$

where

$$\tilde{\sigma}_{*[0]_{(j)}} = (\tilde{\sigma}_*^1, \dots, \tilde{\sigma}_*^{j-1}, \tilde{\sigma}^j, \tilde{\sigma}_*^{j+1}, \dots, \tilde{\sigma}_*^m) \tag{29}$$

is the strategy obtained via a unilateral strategy change of player j, and $\tilde{\sigma}_{*[k]_{(j)}}$ is the restriction of $\tilde{\sigma}_{*[0]_{(j)}}$ to level k.

Notice that, due to the sequential nature of the game, an equilibrium strategy vector is such that, after players of the lower levels have committed themselves by announcing their price systems, its restriction to the downstream levels must also define an equilibrium for the embedded game defined among the remaining players.

B. Dynamic Programming Conditions

Following [11], we will use a dynamic programming argument to characterize a multi-level market equilibrium. Thus, the embedded equilibrium problems will be characterized recurrently from level L backwards; Players active at a given level, knowing the rational reaction functions of the players located at the downstream level, will strive to maximize their profit as a function of the price system set forth by the players located at the upstream level.

For any $i \in J^{\ell-1}, j \in J^{\ell}, \ell = 1, \ldots, L$, we define the mapping

$$\tilde{g}_{ij} : p^{\ell-1} \mapsto q_{ij} \tag{30}$$

where $\tilde{g}_{ij}$ is the projection on arc (ij) of a given strategy vector $\tilde{\sigma}$.

This function can be related to the concept of *derived demand* law since it expresses the quantity of inflow q_{ij} purchased by player $j \in J^{\ell}$ from a player $i \in J^{\ell-1}$ as a function of the price system set forth by the players in level $\ell - 1$.

The proof of the following proposition can be found in [11]:

Proposition 2 *A strategy vector $\tilde{\sigma}_*$ is a sequential market equilibrium if and only if the following holds:*

- *At level L, for all $j \in J^L$ and for all $p^{L-1} \geq 0$,*

$$\phi^j(\tilde{\sigma}_{*[L-1]}, p^{L-1}) = \max_{q_{\cdot j}, q_{j \cdot} \geq 0} \sum_{n \in M} q_{jn} f_n(q_{jn} + \sum_{i \in J^L - \{j\}} q_{*in}) - \sum_{i \in J^{L-1}} q_{ij} p_{ij} - TC^j(q_{\cdot j}, q_{j \cdot}) \tag{31}$$

$$\text{s.t.} \quad \sum_{i \in J^{L-1}} q_{ij} \geq \sum_{n \in M} q_{jn} \tag{32}$$

where q_{*in} is determined from (25) by $\tilde{\sigma}_{*[L]}$ and p^{L-1}.

- *At any intermediate level $\ell = 1, \ldots, L-1$, for all $j \in J^{\ell}$ and for all $p^{\ell-1} \geq 0$,*

$$\phi^j(\tilde{\sigma}_{*[\ell-1]}, p^{\ell-1}) = \max_{q_{\cdot j}, p_{j \cdot} \geq 0} \sum_{k \in J^{\ell+1}} q_{*jk} p_{jk} - \sum_{i \in J^{\ell-1}} q_{ij} p_{ij} - TC^j(q_{\cdot j}, q_{*j \cdot}) \tag{33}$$

$$\text{s.t.} \quad \sum_{i \in J^{\ell-1}} q_{ij} \geq \sum_{k \in \ell+1} q_{jk} \tag{34}$$

where

$$q_{*jk} = \tilde{g}_{*jk}(p_{jk}; (p_{*ik})_{i \in J^{\ell} - \{j\}}) \tag{35}$$

is the projection on arc (j,k) *of the strategy* $\tilde{\sigma}_{*}^{k}, k \in J^{\ell+1}$, *and* p_{*ik} *is determined from* (24) *by* $\tilde{\sigma}_{*[\ell]}$ *and* $p^{\ell-1}$.

The above characterization by dynamic programming suggests an algorithmic method to compute an equilibrium strategy vector.

At level L, given the price system p^{L-1} put forth by the players located at the upstream level $L-1$, and given the demand laws $f_n, n \in M$ on the various markets for the finished good, the solution of problem (31, 32) is equivalent to finding a Nash-Cournot equilibrium in an oligopoly. An equilibrium strategy vector for the players at level L is obtained by expressing the solution of problem (31, 32) as a function of the price system p^{L-1}. The projection of this equilibrium strategy vector on every arc $(i,j), i \in J^{L-1}, j \in J^{L}$ determines the form of the derived demand laws for the upstream level $L-1$.

Similarly, at level $\ell < L$, given the price system put forth by the players at the upstream level $\ell - 1$, and given functions $(\tilde{g}_{ij}(p^{\ell}))$, $i \in J^{\ell}, j \in J^{\ell+1}$, the solution of problem (33, 34) is equivalent to finding a Nash-Cournot equilibrium vector for the players in level ℓ. An equilibrium strategy vector for the players at level ℓ thus requires the expression of the solution of problem (33, 34) as a function of the price system $p^{\ell-1}$.

Starting from the last level L, an equilibrium strategy vector, if it exists, can thus be defined recursively. Then, from the first level, given the initial prices or royalties p^0, the equilibrium prices and flows can be determined from (24, 25).

At each level, two questions are raised: first, of the existence of a solution to problem (31, 32) or (33, 34), and then, of its uniqueness, in order for the derived demand laws to be well-defined. The determination of conditions guaranteeing the existence and uniqueness of equilibria for oligopolies is still an area of research, even in the static, one stage case.

In order to illustrate the algorithm, we will consider here the case where the final demand laws are (piecewise) affine, non increasing with respect to the total quantity, and where the transformation cost functions are non decreasing, quadratic and convex.

Consider the equilibrium problem at stage L and suppose that each demand law f_n is linear. Since the profit function of each player is quadratic, and since all constraints are linear, it is apparent that an equilibrium strategy, if it exists, will be piecewise linear w.r.t. p^{L-1}. Indeed, the solution to problem (31, 32) amounts to the simultaneous solution of linear (in p and q) equations and inequations defining the Kuhn-Tucker optimality conditions for each player.

In a similar manner, if at an intermediate stage ℓ, all the derived demand laws were linear, the solution to problem (33, 34) would again amount to the simultaneous solution of systems of equations and inequations, yielding again (if a unique equilibrium exists) piecewise linear equilibrium strategies.

The existence of discontinuities in the slopes of equilibrium strategies proceeds from non-negativity constraints on profits, prices and flows. To illustrate, suppose that, for a given price system and strategy vector of all other players, the optimal strategy of a firm $j \in J^{\ell}$ dictates the purchase of a very small inflow from a firm $i \in J^{\ell-1}$. Then, a slight increase in the price of inflow q_{ij} may change the set of "active" arcs on which a positive flow is demanded by firm j in order to achieve maximum profit (for example, removing from it arc (i, j)). A sufficient increase of the prices of its inflows may force firm j to leave the industry altogether.

Discontinuities in the slopes of the reaction functions of the players can thus be associated with the addition and deletion of arcs on the network (where an arc (i, j) is deleted if, at equilibrium, the associated flow $q_{ij} = 0$).

The solution of even a one-stage equilibrium problem for an oligopoly facing piecewise linear derived demand laws can be a formidable task. We propose here a heuristic method for the multi-level, linear-quadratic problem, which consists in the systematic computation of what we call *pseudo-equilibria*. A pseudo-equilibrium is a set of flows and prices defined on a subset of the original network, such that

- all flows, prices and profits are non-negative,
- these flows and prices satisfy conditions (31–34) on the reduced network.

Thus, for a given subset of the original network, a pseudo-equilibrium would be generated by an equilibrium strategy vector for the problem defined on the *reduced* network. Moreover, these prices and flows do not necessarily represent an equilibrium solution to the original problem, hence the term "pseudo-equilibrium".

In fact, in order to show that a pseudo-equilibrium is a solution to the original problem, we have to show that, given the prices obtained for the upstream level, and given the actual strategies used by players at the same level and the derived demand laws, no player can improve its profit on a modified network (by adding or deleting arcs he controls). Moreover, the difficulty here comes from the fact that the strategies and derived demand laws obtained on the reduced network are not defined for other network configurations, and are thus only valid over limited subsets of their domain.

The algorithm proceeds as follows: A subset of the original network is chosen. A relaxation of the dynamic program (31–34) is then solved on this reduced network; This relaxation consists in ignoring non-negativity constraints.

Notice that, under the assumptions above, inequalities (32) and (34) may be replaced by equalities (conservation of flow). The solution of the relaxed problem (31)–(32) thus amounts to solving simultaneously a set of linear equations expressing the necessary Kuhn-Tucker optimality conditions for the optimization problem of each player at level L, yielding strategies linear in p^{L-1}. In the same manner, at each level $\ell < L$, the solution of the relaxed problem (33)–(34) is obtained by solving a set of linear equations w.r.t. p^{ℓ}. A first backward sweep, consisting in multiplication and inversion of matrices, is followed by a forward sweep, where the prices and flows are computed. Network configurations resulting in negative values for any price or flow are discarded.

A systematic search among all the subsets of the original network determines the class of all pseudo-equilibria. An equilibrium strategy vector, if it exists, necessarily belongs to this class. A pseudo-equilibrium is an equilibrium if no player can unilaterally improve its profit by changing the configuration of the network, adding or deleting arcs. In some cases, it may be possible to check this equilibrium property.

C. Numerical Illustration

We consider a three-stage commodity market whose network representation is given in Figure 7. Two players, called extractors, are located at the first stage of the transformation process; two "transformers" and two "producers" are located respectively at the second and third stage of the process. The producers sell a homogeneous product on a unique market, where the consumers are represented by an affine demand law.

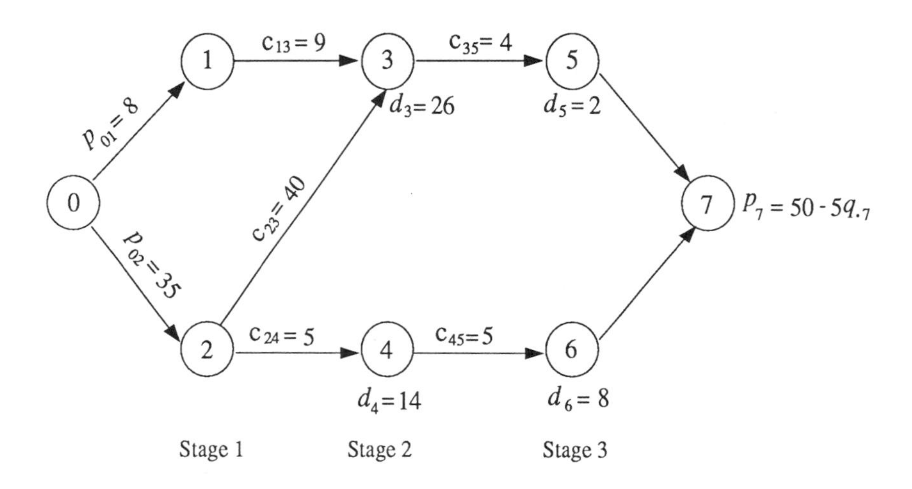

Figure 7:

The only cost incurred by extractor j is the royalty p_{0j} paid to have access to the primary resource. A transformer or producer $j \in J^{\ell}, \ell = 1, 2$, incurs pure quadratic transportation costs $c_{ij}q_{ij}^2$, $i \in J^{\ell-1}$, and transformation costs $d_j q_{.j}^2$. The various values for these parameters appear on Figure 7.

Applying the procedure outlined above on the complete network considered, we obtain the following three systems of linear equations:

Stage 3

$$\begin{bmatrix} q_{35} \\ q_{46} \end{bmatrix} = \begin{bmatrix} 1.91 \\ 1.55 \end{bmatrix} - \begin{bmatrix} .0475 & -.00914 \\ -.00914 & .0402 \end{bmatrix} \begin{bmatrix} p_{35} \\ p_{46} \end{bmatrix} \tag{36}$$

Stage 2

$$\begin{bmatrix} q_{13} \\ q_{23} \\ q_{24} \\ p_{35} \\ p_{46} \end{bmatrix} = \begin{bmatrix} .351 \\ .0791 \\ .538 \\ 37.8 \\ 33.8 \end{bmatrix} + \begin{bmatrix} -.0162 & .00883 & .000419 \\ .00883 & -.0105 & .0000944 \\ .000419 & .0000944 & -.0112 \\ .161 & .0364 & .0451 \\ .0263 & .00593 & .290 \end{bmatrix} \begin{bmatrix} p_{13} \\ p_{23} \\ p_{24} \end{bmatrix} \quad (37)$$

Stage 1

$$\begin{bmatrix} -.0325 & .00883 & .000419 \\ .00883 & -.0210 & .000188 \\ .000419 & .000188 & -.225 \end{bmatrix} \begin{bmatrix} p_{13} \\ p_{23} \\ p_{24} \end{bmatrix} = \begin{bmatrix} -.351 \\ -.0791 \\ -.538 \end{bmatrix} + \begin{bmatrix} -.0162 & 0 \\ 0 & -.0104 \\ 0 & -0111 \end{bmatrix} \begin{bmatrix} p_{01} \\ p_{02} \end{bmatrix} \quad (38)$$

Given the royalties p_{01} and p_{02}, we obtain by forward sweeping the values for all prices and flows indicated on Figure 8.

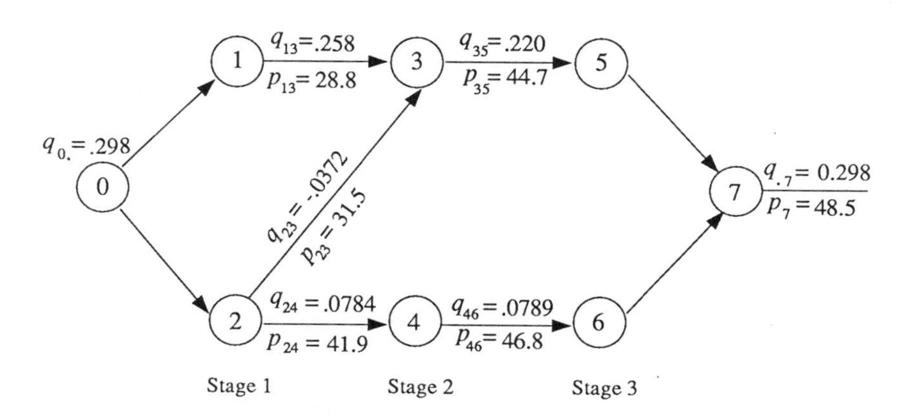

Figure 8:

Since the value of q_{23} is negative, this set of prices and flows cannot represent an equilibrium. Figure 9 represents the solution obtained after deleting arc (2, 3) from the original network and reapplying the optimization procedure.

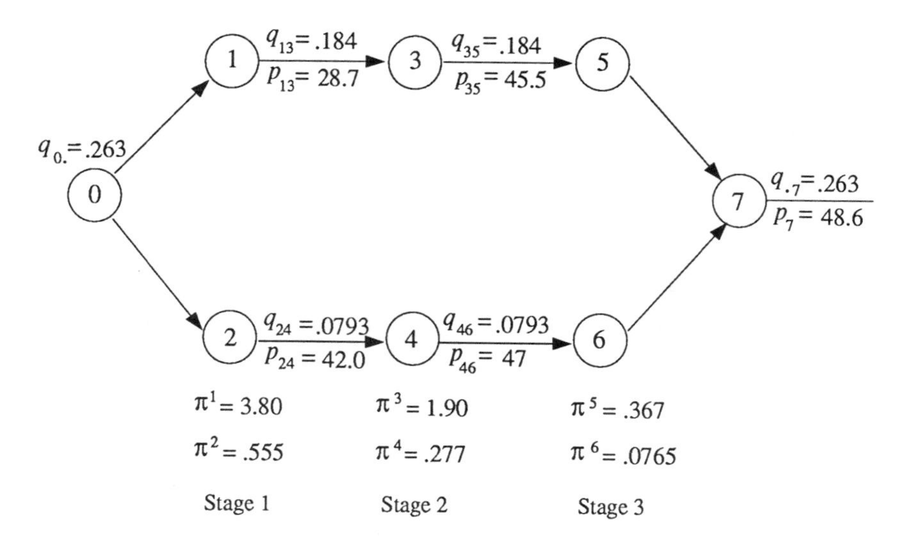

Figure 9:

All the prices, flows and profits are positive in this solution: this is a pseudo-equilibrium. To prove that this solution is also an equilibrium, we have to show that no other possible configuration unilaterally induced by either player improves his profit. Clearly, all configurations apart from the two studied exclude at least one of the players, therefore they are not optimal for him. It remains to verify that, given the prices and strategies obtained with the network configuration of Figure 9, player 2 has no incentive to sell his output to player 3. The profit function of player 2 is given by:

$$\pi^2 = -35(q_{23} + q_{24}) + p_{23}q_{23} + p_{24}q_{24} \tag{39}$$

where, using (37),

$$q_{23} = .0791 + .00883p_{13} - .0105p_{23} + .0000944p_{24} \tag{40}$$

$$q_{24} = .538 + .000419p_{13} + .0000944p_{23} - .0112p_{24}. \tag{41}$$

We can easily check that, at the current point ($p_{13} = 28.7$, $q_{23} = 0$, $p_{23} = 0$, $q_{24} = .0793$, $p_{24} = 42$), the partial derivative of π^2 w.r.t. q_{23} is negative, hence player 2 has no incentive of change unilaterally his strategy and the solution given in Figure 9 is indeed an equilibrium.

V. Conclusion

The two modeling frameworks considered in this paper correspond to different points of view for the analysis of commodity markets. In the first one we represent such an economic structure as a competition on a transportation network, subject to dynamic transformations and to stochastic perturbations. In the second point of view adopted, we give a special attention to the mechanism of derived demands generated in multilevel resource transformation processes.

These two approaches could be combined in a general modeling framework permitting a representation of dynamic, stochastic and multilevel network of oligopolies. We could then take into consideration, for example:

- entry and exit of firms,
- expansion of production capacities,
- shifts (deterministic or stochastic) in demand laws,
- effects of vertical and horizontal integration, etc.

Obviously the first element, when it occurs, modifies the topology of the network at each time period. Expansion or entry decisions constitute the sets of possible actions in the tree representation of the dynamic game. Stochastic changes in the demand laws can be modelled by an event tree as in the first example given above. Clearly if congestion effects on transportation links are added, we then obtain a very general model. In addition, the integration issue (vertical or horizontal) could be tackled with the help of a definition of a dynamic coalition structure. Some preliminary elements of this approach appeared in [11, 13].

References

[1] J.W. Friedman, "Oligopoly and the Theory of Games," North Holland, 1977.

[2] J.W. Friedman, "Game Theory with Applications to Economics," New York, Oxford University Press, 1986.

[3] T. Basar and G. Olsder, "Dynamic Noncooperative Game Theory," Academic Press, 1982.

[4] P.T. Harker, "Alternative Models of Spatial Competition," *Operations Research* **34**(3), 410–425 (1986).

[5] A. Haurie and P. Marcotte, "A Game-Theoretic Approach to Network Equilibrium," *Mathematical Programming Study* **26**, 252–255 (1986).

[6] P. Marcotte, "Algorithms for the Network Oligopoly Problem," Centre de recherche sur les transports, publication #426 (1985).

[7] A. Haurie, G. Zaccour and Y. Smeers, "Stochastic Equilibrium Programming for Dynamic Oligopolistic Markets," *Journal of Optimization Theory and Applications*, forthcoming.

[8] C. Van Duyne, "Commodity Cartels and the Theory of Derived Demand," *Kyklos* **28**, 597–612 (1975).

[9] M.L. Greenhut and M. Ohta, "Related Market Conditions and Interindustrial Mergers," *American Economic Review* June, 267–277 (1977).

[10] P.S. Armington, "A Theory of Demand for Products Dinstinguished by Place of Production," *IMF Staff Papers* **16**, 159–178 (1969).

[11] A. Haurie and M. Breton, "Market Equilibrium in a Multistage Commodity Network ," *Automatica* **21**(5), 585–597 (1986).

[12] G. Zaccour, "Calcul des pseudo-équilibres sur des réseaux représentant des marchés séquentiels," M. Sc. thesis, H.E.C., Montréal, 1983.

[13] H.D. Sherati and J.M. Lebano, "A Mathematical Programming Approach to a Nash-Cournot Equilibrium Analysis for a Two-Stage Network of Oligopolies," *Operations Research* **36**, 682–702 (1988).

[14] J.S. Pang and D. Chang, "Iterative Methods for Variational and Complementarity Problems," *Mathematical Programming* **24**, 284–313 (1982).

[15] S.C. Dafermos, "An Iterative Scheme for Variational Inequalities," *Mathematical Programming* **29**, 40–47 (1983).

[16] A. Haurie, G. Zaccour, J. Legrand and Y. Smeers, "A Dynamic Stochastic Nash-Cournot Model for the European Gas Market," Les cahiers du GERAD, G–87–24, H.E.C., Montréal, Submitted for publication, (1989).

[17] D. Gabay and H. Moulin, "On the Uniqueness and Stability of Nash-Equilibria in Noncooperative Games," *In*: A. Bensoussan, P. Kleidorfer and C.S. Tafiero, eds., " Applied Stochastic Control in Econometrics and Management Science," North-Holland, Amsterdam, 1980.

[18] P.T. Harker, "A Variational Inequality Approach for the Determination of Oligopolistic Market Equilibrium," *Mathematical Research* **30**, 105–111 (1984).

[19] P. Marcotte, "Quelques notes et résultats nouveaux sur le problème d'équilibre d'un oligopole," *R.A.I.R.O. Recherche opérationnelle* **18**(2), 147–171 (1984).

[20] G. Zaccour, "Théorie des jeux et marchés énergétiques: marché européen du gaz naturel et échanges d'électricité," Les cahiers du GERAD, G–87–37, H.E.C., Montréal, (1987).

A CONTROL PROBLEM FOR PROCESSING DISCRETE PARTS WITH RANDOM ARRIVALS[1]

Chih-Ping Lee
N. Harris McClamroch

Department of Electrical Engineering and Computer Science
The University of Michigan
Ann Arbor, Michigan 48109

ABSTRACT

An M/G/1 queue is introduced to model the productivity of a production system. Parts arrive according to to a Poisson process. The distribution of the production time is of general form, parameterized by a control variable - *the production rate*. The decision epochs for the control policy are restricted to the times when production is initiated for a part. A short-term cost is introduced as a weighted sum of penalties associated with the production rate control, penalties associated with part waiting time and penalties on the increase of parts in the queue, assessed during the time required to produce a single part. The major advantages of this formulation are generality, mathematical tractability of the problem and the properties of the closed-loop production system. A simple nonlinear search routine is required to determine the optimal production rate. For the special case of an exponential production time distribution and the special case where the production is modeled by a diffusion-threshold process, analytical expressions for the optimal control policy in terms of the queue length are obtained. A transportation example is presented to illustrate the design methodology; statistical data from computer simulation is obtained to evaluate the performance of the adaptive control policy.

[1]This paper is an expanded version of the paper by C. P. Lee and N. H. McClamroch, "Control of Service Rate for an M/G/1 Queue Using a Short-Term Cost Criterion", presented at the 27th Conference on Decision and Control, Dec. 1988

I. INTRODUCTION

In this paper, an M/G/1 queue is used to model a production system which produces or carries out processing on discrete parts. The terminology "production system" might refer to a single machine which performs an elementary operation on parts such as drilling or grinding; or it can refer to a material handling operation using an automated guided vehicle, a robot or a conveyor.

The part production times of the production system are assumed to be random and can be affected by an active control variable - the production rate. We assume that a constant production rate is selected at each production initiation epoch and is used throughout the period of producing one part. Our goal is to design a real-time control policy which has some desirable closed-loop characteristics, minimizes an expected long-term average cost criterion and reduces the mean and variance of the production queue length in the long run. Toward this end, we introduce a short-term cost measure, which involves only quantities determined during one production period. By minimizing this cost, we obtain the optimal control policy for the production rate which depends on the queue length observed at each decision epoch.

Contrary to what the name might suggest, this short-term cost criterion does aim at achieving a control policy which results in good long-term average cost per unit time for the controlled M/G/1 production system. Even though the resultant control policy is actually sub-optimal from the long-term point of view, our approach does possess some important features: namely generality and mathematical tractability. Furthermore, concrete results can be obtained for the special case of a controlled M/M/1 queueing system, and for the special case that the production process is modeled by an underlying diffusion-threshold process, for which the production times are inverse-Gaussian distributed. A detailed study of an example where the "production" time is due to a transportation delay is also presented.

This paper is organized as follows: in section II, the model for the production system as a controlled M/G/1 queueing system is briefly described; section III contains the short-term optimization problem formulation, where control decisions are allowed at production initialization epochs; section IV presents properties of the resultant optimal control policy; section V discusses the closed-loop behavior of the production system under the optimal control policy; section VI contains a numerical procedure to calculate the steady state probability mass function of the queue length under the optimal control policy; in section VII, results obtained in the previous sections are applied to two special cases: (i) a system with exponentially distributed production time (a controlled M/M/1 queue) and (ii) a system whose production process is modeled by a diffusion-threshold process. In section VIII, an example where the service time is due to a transportation delay is presented; section IX contains the conclusions.

II. MODEL OF THE PRODUCTION QUEUEING SYSTEM

The production system is a single server single channel system with Poisson arrivals and random service times with arbitrary probability distribution. The positive random variable for the service time T_B has distribution denoted by $F_B(t\,|u)$ with density function $f_B(t\,|u)$. The distribution and density are assumed to be functions of the production rate u. The interarrival time, represented by a random variable T_A, has the exponential distribution:

$$F_A(t) = P\,[T_A \le t] = 1 - e^{-\lambda t}, \quad t \ge 0, \tag{1}$$

where $\lambda > 0$ is the constant Poisson arrival intensity. When a part arrives, if the server is not busy, the part immediately receives a non-interrupted period of processing and leaves the system; otherwise, the part joins the end of a queue assumed to have an infinite capacity. The length of a production period obeys the probabilistic distribution law $F_B(t\,|u)$ which is a function of the *selected production rate* u. On completion of the processing of a part, processing of the first part waiting in the queue (first-come-first-serve queueing discipline) is immediately initiated if the queue is not empty; otherwise, the system is idle until a part arrives. This part-driven process is repeated *ad infinitum.* Our goal is to select the production rate according to on-line queue length observations to affect the operation of the production system in a desirable way from an economic standpoint.

III. SHORT-TERM OPTIMIZATION PROBLEM FORMULATION

We introduce the idea of a short-term cost for the controlled production queueing system, which is the *cost incurred during the period of producing a single part.* The cost is a weighted sum of a cost due to control effort, a cost due to part waiting times of all parts in the queue and a cost due to the increase of parts waiting in the queue during one production period. Then, an optimization problem is formulated to minimize this short-term cost.

In our model, control decisions are allowed only at epochs when production of a part is initiated. Such epochs are convenient since, under appropriate assumptions, they are *renewal times* in the sense that the distribution of the production time for the part being produced and the distribution for the arrival time of the next part are determined. The optimal production rate depends on observation of the number of parts in the queue but is independent of time; that is, the obtained control policy is a *feedback* control policy with *stationary* characteristics.

A. Short-Term Cost Criterion

Let T_P^k denote the random time when production for the k-th part is begun; let T_F^k denote the random time when production for the k-th part is completed. The short-term cost incurred during the time interval $[T_P^k, T_F^k]$ when the k-th part is being produced is given by:

$$J^k = E\{w_1 \int_{T_P^k}^{T_F^k} f(u^k)dt + w_2 \sum_{j=1}^{n^k} (T_j^k + \int_{T_P^k}^{T_F^k} dt) + w_2 \sum_{l=1}^{\infty} 1_{(T_P^k < t_l < T_F^k)} \int_{t_l}^{T_F^k} dt$$

$$+ w_3 \sum_{l=1}^{\infty} 1_{(T_P^k < t_l < T_F^k)} | n^k, u^k, T_1^k, T_2^k, \cdots, T_{n^k}^k\}, \qquad n^k \geq 1, \tag{2}$$

where n^k is the number of parts in the queue at time T_P^k, u^k is the production rate *assumed to be constant* during the period $[T_P^k, T_F^k]$. The random variable T_j^k denotes the length of time the j-th part has been in the queue at time T_P^k, $j = 1, 2, \cdots, n^k$. The random time t_l denotes the arrival time of the l-th part, $l = 1, 2, 3, \cdots$.

The function $f : \mathbf{R} \to \mathbf{R}$ is assumed to be continuous, positive and monotone increasing function satisfying $f(0) = 0$. The constants $w_1 > 0$, $w_2 > 0$ and $w_3 > 0$ are cost weighting factors. The indicator function is defined as:

$$1_{(a<t<b)} = \begin{cases} 1, & \text{if } a < t < b, \\ 0, & \text{otherwise} . \end{cases}$$

The first term in (2) is the cost, accumulated during $[T_P^k, T_F^k]$, associated with the production rate u^k; the second term in (2) is the cost of waiting, up to time T_F^k, of all the parts which are already in the queue at T_P^k; the third term in (2) represents the cost of waiting, up to time T_F^k, of all the parts which arrive during $[T_P^k, T_F^k]$; and the last term in (2) is the cost associated with the increase of parts joining the queue during $[T_P^k, T_F^k]$. Therefore, J^k in (2) is the weighted sum of penalties on control (the production rate), part waiting time and increase of parts.

The first three terms in eq. (2) are included in order to *strike a balance between costs due to part waiting time and control effort.* The last term in eq. (2) is included so that the resultant Q-optimal control policy is not *myopic.* By introducing a cost term associated with the number of parts joining the queue during one production period, we aim to control the number of parts waiting in the queue *in the long run.*

After some simplifications, (2) can be rewritten as:

$$J^k = E\{w_1 f(u^k) T_B^k(u^k) + w_2 \sum_{j=1}^{n^k} (T_j^k + T_B^k(u^k)) + w_2 \sum_{l=1}^{\infty} 1_{(T_P^k < t_l < T_F^k)} (T_F^k - t_l)$$

$$+ w_3 \sum_{l=1}^{\infty} 1_{(T_P^k < t_l < T_F^k)} | n^k, u^k, T_1^k, T_2^k, \cdots, T_{n^k}^k\}, \quad n^k \geq 1, \tag{3}$$

where $T_B^k(u^k) = T_F^k(u^k) - T_P^k$ is the production time for the k-th part, which we express as an explicit function of the production rate u^k.

Due to the memoryless property of the exponentially distributed interarrival times, we have:

$$\sum_{l=1}^{\infty} 1_{(T_P^k < t_l < T_F^k)} (T_F^k - t_l) = \sum_{l=1}^{\infty} 1_{(0 < \tau_l^k < T_B^k)} (T_B^k - \tau_l^k) \tag{4}$$

and

$$\sum_{l=1}^{\infty} 1_{(T_P^k < t_l < T_F^k)} = \sum_{l=1}^{\infty} 1_{(0 < \tau_l^k < T_B^k)} , \tag{5}$$

where $\tau_l^k = t_l - T_P^k$, $l = 1, 2, 3, \cdots$, are exponentially distributed. According to eqs. (4) and (5), the arrival process can be seen as being restarted at T_P^k, $k \geq 1$, and the elapsed time since the last arrival is irrelevant. Furthermore, it is observed that J^k in (3) has the same functional form for all $k \geq 1$. Therefore, the superscript k can be dropped from (3); we then have:

$$J(n, u) = E\{w_1 f(u) T_B + w_2 \sum_{j=1}^{n} (T_j + T_B) + w_2 \sum_{l=1}^{\infty} 1_{(0 < \tau_l < T_B)} (T_B - \tau_l)$$

$$+ w_3 \sum_{l=1}^{\infty} 1_{(0 < \tau_l < T_B)} \mid n, u, T_1, T_2, \cdots, T_n\}, \quad n \geq 1, \tag{6}$$

which is the short-term cost accrued during the production of a single part where n parts are in the queue initially with respective waiting times $T_1, T_2, T_3, \cdots, T_n$, and the production rate has value u. This cost expression is valid at each initialization of production of a part.

Now we evaluate each term in (6). It is straightforward to obtain:

$$E\{w_1 f(u) T_B | n, u, T_1, T_2, \cdots, T_n\} = w_1 f(u) E[T_B | u], \tag{7}$$

$$E\{w_2 \sum_{j=1}^{n} (T_j + T_B) | n, u, T_1, T_2, \cdots, T_n\} = w_2 \sum_{j=1}^{n} T_j + w_2 n E[T_B | u], \tag{8}$$

and

$$E\{w_3 \sum_{l=1}^{\infty} 1_{(0 < \tau_l < T_B)} \mid n, u, T_1, T_2, \cdots, T_n\} = w_3 \lambda E[T_B | u]. \tag{9}$$

Also, we have:

$$E\{w_2 \sum_{l=1}^{\infty} 1_{(0 < \tau_l < T_B)} (T_B - \tau_l) | n, u, T_1, T_2, \cdots, T_n\}$$

$$= E\{w_2 \sum_{l=1}^{\infty} 1_{(0 < \tau_l < T_B)} (T_B - \tau_l) | u\}$$

$$= E\{E[E(w_2 \sum_{l=1}^{\infty} 1_{(0 < \tau_l < T_B)} (T_B - \tau_l) | N_{T_B}, T_B, u) | T_B, u] | u\} \tag{10}$$

by the smoothing property of the conditional expectation, where N_{T_B} is a random variable representing the number of parts arriving during a period of length T_B. It is true that for a Poisson arrival process, if there are N_{T_B} arrivals in a period of length T_B, the arrival times of each of the N_{T_B} parts are uniformly distributed in $[0, T_B]$ [4]. Therefore, (10) can be rewritten as:

$$E\{w_2 \sum_{l=1}^{\infty} 1_{(0 < \tau_l < T_B)} (T_B - \tau_l) | n, u, T_1, T_2, \cdots, T_n\}$$

$$= w_2 E\{E[N_{T_B} \int_0^{T_B} (T_B - \tau) \frac{d\tau}{T_B} | T_B, u] | u\}$$

$$= \frac{w_2 \lambda}{2} E[T_B^2 | u], \tag{11}$$

where (11) is obtained by observing:

$$E[N_{T_B}|T_B] = \sum_{j=0}^{\infty} \frac{je^{-\lambda T_B}(\lambda T_B)^j}{j!} = \lambda T_B. \tag{12}$$

From eqs. (7)-(9) and (11), (6) can be rewritten as:

$$J(n, u) = w_2 \sum_{j=1}^{n} T_j + (w_1 f(u) + w_2 n + w_3 \lambda) E[T_B|u] + \frac{\lambda w_2}{2} E[T_B^2|u], \qquad n \geq 1. \tag{13}$$

Equation (13) is the final simplified form of the general short-term cost function from which the desired production rate can be derived by numerical search.

B. Optimal Control Problem

We now formulate the following one-dimensional optimization problem: given any positive integer n, choose a production rate u to minimize

$$J(n, u) = w_2 \sum_{j=1}^{n} T_j + (w_1 f(u) + w_2 n + w_3 \lambda) E[T_B|u] + \frac{\lambda w_2}{2} E[T_B^2|u], \qquad n \geq 1. \tag{14}$$

Let

$$u^*(n) = arg\{\min_u J(n, u)\}, \qquad n \geq 1. \tag{15}$$

We denote the policy expressed in eq. (15) as the Q-optimal control policy. It can be seen from (14) that the Q-optimal production rate is independent of the part waiting history T_j, $j \geq 1$, which is due to *linearity* of the assumed waiting cost. Furthermore, we assume that during periods when no parts are in the queue, i.e., the system is idle, the production rate is $u^*(0) = 0$.

In general, the Q-optimal production rate depends on the number of parts in the queue. The Q-optimal control policy obtained is therefore a *feedback control policy* where the observed queue length is utilized at each production initialization epoch to determine a *constant production rate* for the part beginning production. Furthermore, it is obvious that the resulting Q-optimal control policy is also *stationary* and *causal.*

In general, the Q-optimal production rate cannot be expressed in a simple analytic form. But a one-dimensional nonlinear search routine can be employed to determine the Q-optimal production rate. Also, the Q-optimal production rate can always be calculated *off-line* and stored *apriori* in a table look-up form. This reduces the complexity of control implementation because all that is required is to retrieve the Q-optimal production rate based on observation of the number of parts in the queue at each decision epoch.

Furthermore, we notice that only the *first and second moments* of the production times appear in eq. (14); the exact expression for the production time distribution is not required.

IV. PROPERTIES OF THE Q-OPTIMAL CONTROL POLICY

In this section, we present several qualitative results about the Q-optimal production rate as a function of the number of parts in the queue.

We first make the following mild assumptions:

<u>Assumption 1:</u> For any $u \in (0, \infty)$, $f(u) < \infty$, $E[T_B|u] < \infty$ and $E[T_B^2|u] < \infty$.

<u>Assumption 2:</u> $E[T_B|u]$ is a monotone decreasing function of u.

Assumption 1 is simply a statement of finiteness for all the functions involved; Assumption 2 says that the expected production time decreases as the production rate increases, which is intuitively true for a failure free production system.

For simplification, let

$$g_1(u) = w_2\sum_{j=1}^{n} T_j + (w_1 f(u) + w_3\lambda)E[T_B|u] + \frac{\lambda w_2}{2}E[T_B^2|u] \tag{16}$$

and

$$g_2(u) = w_2 E[T_B | u]. \tag{17}$$

Then, eq. (14) can be rewritten as:

$$J(n, u) = g_1(u) + n g_2(u). \tag{18}$$

Theorem 1: The Q-optimal control policy u^* is a monotone non-decreasing function of n.

Proof: Let u_0^* be the Q-optimal production rate corresponding to n_0, then for *any* $n_1 > n_0$ and for *all* $u < u_0^*$, we have from eq. (18):

$$\begin{aligned}
J(n_1, u) - J(n_1, u_0^*) &= [g_1(u) + n_1 g_2(u)] - [g_1(u_0^*) + n_1 g_2(u_0^*)] \\
&= [g_1(u) + n_0 g_2(u) + (n_1 - n_0) g_2(u)] \\
&\quad - [g_1(u_0^*) + n_0 g_2(u_0^*) + (n_1 - n_0) g_2(u_0^*)] \\
&= [(g_1(u) + n_0 g_2(u)) - (g_1(u_0^*) + n_0 g_2(u_0^*))] \\
&\quad + (n_1 - n_0)[g_2(u) - g_2(u_0^*)] \\
&> 0,
\end{aligned}$$

which follows from the fact that u_0^* is a minimizer of $J(n_0, u)$ and that $g_2(u)$ is a monotone decreasing function of u by Assumption 2. Therefore, the minimizer corresponding to n_1 must lie in $[u_0^*, \infty)$.

Q.E.D.

It can easily be shown that Theorem 1 still holds if $E[T_B | u]$ is only a monotone non-increasing function of u.

The next theorem states that if the number of parts waiting in the queue increases without limit, the Q-optimal production rate increases without limit.

Theorem 2: The Q-optimal control policy satisfies $u^* \to \infty$ as $n \to \infty$.

Proof: We want to show that for *any* positive real number r, the Q-optimal production rate $u^*(n)$ corresponding to state n will exceed r for n sufficiently large.

Given $r \in (0, \infty)$, for *a fixed* $u_2 \in (r, \infty)$, and for *any* $u_1 \in [0, r]$ then

$$
\begin{aligned}
J(n, u_1) - J(n, u_2) &= [g_1(u_1) - g_1(u_2)] + n[g_2(u_1) - g_2(u_2)] \\
&\geq -g_1(u_2) + n[g_2(u_1) - g_2(u_2)] \\
&> 0
\end{aligned}
$$

when n is sufficiently large. This means that for any bounded interval $[0, r]$, we can always find an integer n_r so that $u^*(n_r) > r$. But from Theorem 1, we know that $u^*(n) > r$ for all $n > n_r$, which proves the theorem.

Q.E.D.

V. ERGODICITY OF THE CLOSED-LOOP PRODUCTION SYSTEM UNDER THE Q-OPTIMAL CONTROL POLICY

For an M/G/1 queue, the asymptotic behavior of the system is mainly determined by the traffic intensity, which is the ratio between the expected service time and the expected interarrival time [4]. Our production queueing system can be viewed as a generalized M/G/1 queueing system, where the mean Q-optimal production time $E[T_B | u^*(n)]$ implicitly depends on the number of parts n in the queue. We define the mean number of arrivals between production completions, given that there are n parts waiting in queue initially, as:

$$
\rho_n^* = \begin{cases} 1 + \lambda E[T_B | u^*(1)], & n = 0, \\ \lambda E[T_B | u^*(n)], & n \geq 1; \end{cases} \tag{19}
$$

this quantity is referred to as the *state dependent traffic intensity* at each *production completion* (part departure) epoch. It is well known that the numbers of parts waiting in the queue observed at all production completion epochs of an generalized M/G/1 queueing system constitute the states of an embedded semi-Markov chain. The following theorem guarantees ergodicity of the embedded semi-Markov chain under the Q-optimal control policy.

Theorem 3: If $\lim_{u \to \infty} E[T_B | u] < \frac{1}{\lambda}$, then the embedded semi-Markov chain of the closed-loop Q-optimal queue is ergodic.

Proof: From Theorem 2, we have:

$$\limsup_{n\to\infty} \rho_n^* = \limsup_{n\to\infty} \lambda E[T_B | u^*(n)] = \lambda \lim_{u\to\infty} E[T_B | u], \tag{20}$$

where the existence of the limit is ensured by the monotonicity assumption of $E[T_B | u]$. Then we know $\limsup_{n\to\infty} \rho_n^* < 1$. Furthermore, from Assumption 1, 2 and Theorem 1, we have:

$$\rho_n^* = \lambda E[T_B | u^*(n)] \le \lambda E[T_B | u^*(1)] < \infty, \quad \text{for all } n \ge 1, \tag{21}$$

and $\rho_0^* = 1 + \lambda E[T_B | u^*(1)] < \infty$. Therefore, the queue length, as an embedded Markov chain, is ergodic [2] and has a non-degenerative stationary distribution: $\{p_n\}_{n=0}^{\infty}$ with $p_n > 0$ for all $n \ge 0$ and $\sum_{n=0}^{\infty} p_n = 1$. Here p_n is the probability that there are n parts waiting in the queue at a production completion epoch. The embedded semi-Markov chain is recurrent and the mean recurrence time for queue length j, μ_j, can be expressed as [8]:

$$\mu_j = \frac{p_0(\frac{1}{\lambda} + E[T_B | u^*(1)]) + \sum_{n=1}^{\infty} p_n E[T_B | u^*(n)]}{p_j} \tag{22}$$

$$\le \frac{E[T_B | u^*(1)] + \frac{p_0}{\lambda}}{p_j} < \infty, \qquad \text{for all } j \ge 0, \tag{23}$$

which shows that the embedded semi-Markov chain is positive recurrent, hence ergodic [8].

Q.E.D.

According to Theorem 3, if the limiting expected production time is smaller than the expected interarrival time, then the closed loop Q-optimal queue is ergodic.

Theorem 3 reveals that the Q-optimal closed-loop production system is stable, i.e., the number of parts waiting in the queue does not explode, almost surely. This is because for any time t such that $T_F^k < t < T_F^{k+1}$, we have $n(t) \le n(T_F^{k+1}) + 1$, where $n(t)$ is the number of parts in the queue at time t; hence if the embedded semi-Markov chain is non-explosive, so is the *complete* process associated with the number of parts in the queue.

Also, a *robustness property* of the closed-loop queue is achieved under the Q-optimal control policy. Even if an incorrect value of the arrival intensity is used in the control design process, as long as the actual value of the arrival intensity λ is bounded and satisfies $\lim_{u \to \infty} E[T_B | u] < \frac{1}{\lambda}$, the stability of the resultant closed-loop queue is guaranteed.

Finally, the Q-optimal closed-loop production system reaches a steady state; the steady state probability distribution of the number of parts in the queue can be calculated and used for system design and performance evaluation. Section 6 discusses a recursive procedure to obtain the steady state probability distribution for the queue length process, as an embedded Markov chain, of the closed-loop Q-optimal production system.

VI. STEADY-STATE STATISTICS OF THE QUEUE LENGTH UNDER THE Q-OPTIMAL CONTROL POLICY

It is obvious from the results of the last section that the Q-optimal control strategy obtained by minimizing the short-term cost in eq. (14) is stationary. Furthermore, from Theorem 3, we know that if $E[T_B | u]$ is a monotone decreasing function of u and $\lim_{u \to \infty} E[T_B | u] < \frac{1}{\lambda}$, then the embedded semi-Markov chain is *ergodic*. Therefore, there is a steady state probability distribution for the number of parts in the queue for the embedded Markov chain.

As in the proof of Theorem 3, let p_n denote the probability that there are n parts waiting in the queue at a *production completion* epoch; then $\{p_n\}_{n=0}^{\infty}$ is the steady state probability distribution for the embedded Markov chain under the Q-optimal control policy. Hence $\mathbf{p} = (p_0, p_1, p_2, \cdots)^T$ satisfies the system of linear equations:

$$\mathbf{p} = \mathbf{pP} \tag{24}$$

and $\sum_{n=0}^{\infty} p_n = 1$, where $\mathbf{P} = \{P_{n,j}, n = 0, 1, 2, \cdots ; j = 0, 1, 2, \cdots\}$ is the state transition matrix of the embedded Markov chain. An entry $P_{n,j}$ is the probability that the number of parts in the queue is j at the next production completion epoch, given that there are n parts in the queue at the present production completion epoch. Therefore, $P_{n,j}$ is the probability that $(j-n+1)$ parts arrive during one production period if $n \geq 1$ and if $j \geq n-1$; and $P_{0,j}$ is the probability that j parts arrive during the first production period after the server is idle; otherwise $P_{n,j} = 0$. That is,

$$P_{n,j} = \begin{cases} q_j(1), & n = 0, j \geq 0, \\ q_{j-n+1}(n), & n \geq 1, j \geq n-1, \\ 0, & \text{otherwise,} \end{cases} \tag{25}$$

where

$$q_k(n) = \int_0^\infty \frac{e^{-\lambda t}(\lambda t)^k}{k!} dF_B(t \,|u^*(n)) \tag{26}$$

is the probability that k parts arrive during a production period for which the constant production rate is $u^*(n)$. In general, the transition probabilities $P_{n,j}$ do not have a closed-form expression but they can be obtained by numerical integration.

If the transition probabilities $P_{n,j}$ are computed (and are all nonzero for $j \geq n-1$, $n \geq 0$, $j \geq 0$, as is usually the case), then eq. (24) can be explicitly solved for the embedded steady state probabilities p_n, $n = 0, 1, 2, \cdots$. Let $\{x_n\}_{n=0}^\infty$ be a sequence generated by assuming $x_0 = 1$ and by solving $\mathbf{x} = \mathbf{xP}$ recursively, where $\mathbf{x} = (x_0, x_1, x_2, \cdots)$:

$$x_{n+1} = \frac{1}{P_{n+1,n}}(x_n - \sum_{k=0}^{n} P_{k,n} x_k), \qquad n \geq 0. \tag{27}$$

Then $\{p_n\}_{n=0}^\infty$ is obtained by normalizing $\{x_n\}_{n=0}^\infty$ to be a probability distribution:

$$p_n = \frac{x_n}{\sum_{k=0}^{\infty} x_k}. \tag{28}$$

Note that $\infty > \sum_{k=0}^{\infty} x_k = \frac{1}{p_0} > 0$ because the embedded Markov chain is ergodic. Hence, existence of the infinite sum in eq. (28) is ensured.

Most importantly, the significance of $\{p_n\}_{n=0}^\infty$ is that not only is p_n the probability that there are n parts waiting in the Q-optimal queue at any *production completion* epoch in steady state, it is also the probability of having n parts in the Q-optimal queue *at any time* in steady state [11,12]; i.e., $p_n = \lim_{t \to \infty} P\{n(t) = n\}$, where $n(t)$ is the number of parts in the queue at time

t.

Once the steady-state probability distribution of the queue length is obtained, it is easy to determine the first and second moments of the queue length, which can be used for evaluating performance of the Q-optimal control policy:

$$\overline{N}^{\pi^*} = \sum_{n=0}^{\infty} n p_n , \tag{29}$$

$$\sigma_N^{\pi^*} = \sqrt{\sum_{n=0}^{\infty} p_n (n - \overline{N}^{\pi^*})^2} . \tag{30}$$

VII. SPECIAL CASES

In this section, two special cases of a controlled production queueing system are presented. In the first subsection, the production time is assumed to be exponentially distributed (a controlled generalized M/M/1 queue) with intensity proportional to the production rate u. In the second subsection, the production process is modeled by a diffusion-threshold process, so that the production time has an inverse-Gaussian distribution parameterized by the production rate u. For these two cases, problems with linear and quadratic control costs are investigated. Closed form expressions for the Q-optimal production rates are obtained. Properties of the Q-optimal control policy and long-term system behavior are derived using previous results. Similarities are observed between these two cases.

A. Exponential Production Time Distribution

In this subsection, we assume that the production time has an exponential distribution, i.e.,

$$F_B(t \mid u) = 1 - e^{\frac{-u}{T} t} , \tag{31}$$

where $T > 0$ is a production time scaling factor so that the production rate u is dimensionless. Equation (31) indicates that the instantaneous part production rate is proportional to the production rate u. Then it is easy to show that

$$E[T_B \mid u] = \frac{T}{u} , \tag{32}$$

and

$$E[T_B^2|u] = \frac{2T^2}{u^2}. \tag{33}$$

By eq. (14), the short-term cost for this case can be written as:

$$J(n, u) = w_2 \sum_{j=1}^{n} T_j + \frac{w_1 T f(u)}{u} + (w_2 n + w_3 \lambda)\frac{T}{u} + \frac{w_2 \lambda T^2}{u^2}, \qquad n \geq 1. \tag{34}$$

From eq. (34), it is obvious that if $f(u) = u$, then the *full service policy* (maximal production rate) is Q-optimal. This result coincides with that obtained in [14], where an optimal state-dependent control policy is sought to minimize the *long-term average cost per part.*

If the production rate penalty function is quadratic, i.e., $f(u) = u^2$, then:

$$J(n, u) = w_2 \sum_{j=1}^{n} T_j + w_1 T u + (w_2 n + w_3 \lambda)\frac{T}{u} + \frac{w_2 \lambda}{2} \frac{1}{u^2}, \qquad n \geq 1. \tag{35}$$

From eq. (35), we can obtain an analytic expression for the Q-optimal production rate.

Theorem 4: The cost function in (35) is a strictly convex function of the production rate u for $u \in (0, \infty)$ and has a unique minimum on $(0, \infty)$ given by:

$$u^*(n) = \begin{cases} \left(\frac{-b}{2} + \sqrt{K}\right)^{\frac{1}{3}} - \left(\frac{b}{2} + \sqrt{K}\right)^{\frac{1}{3}}, & \text{if } K \geq 0, \\ 2\sqrt{\frac{-a}{3}}\cos\theta, & \text{if } K < 0, \end{cases} \tag{36}$$

where

$$K = \frac{b^2}{4} + \frac{a^3}{27}, \qquad a = -\frac{w_2 n + w_3 \lambda}{w_1},$$

$$b = -\frac{w_2 \lambda T}{w_1}, \qquad \theta = \frac{1}{3}\tan^{-1}\left(\frac{2\sqrt{-K}}{-b}\right). \tag{37}$$

Proof: A calculation gives

$$\frac{\partial^2 J(n, u)}{\partial u^2} = 2(w_2 n + w_3 \lambda)\frac{T}{u^3} + \frac{6\lambda w_2 T^2}{u^4} > 0 \tag{38}$$

for all $u > 0$. Hence $J(n, u)$ is a strictly convex function of u. From eq. (35), we have

$$\lim_{u \to 0} J(n, u) = \lim_{u \to \infty} J(n, u) = \infty \tag{39}$$

for any positive integer n. Hence $u^*(n)$ exists and is uniquely determined. The expression (36) is obtained by solving

$$\frac{\partial J(n, u)}{\partial u}\Big|_{u = u^*(n)} = 0. \tag{40}$$

Q.E.D.

Also, we notice from eq. (32) that $E[T_B | u]$ is monotone decreasing as a function of $u \in (0, \infty)$; and

$$\lim_{u \to \infty} E[T_B | u] = \lim_{u \to \infty} \frac{1}{u} = 0 < \frac{1}{\lambda}. \tag{41}$$

Then from the theorems in sections IV and V, we have:

Corollary 5: For the controlled production (generalized M/M/1) queueing system:

(i) The Q-optimal control policy $u^*(n)$ is a monotone non-decreasing function of n, and $u^*(n) \to \infty$ as $n \to \infty$.

(ii) the embedded semi-Markov chain of the closed-loop Q-optimal queue is ergodic.

Since the production queueing system, under the Q-optimal control policy $u^*(n)$, $n \geq 0$, is an ergodic birth and death process, the steady state probability distribution of the number of parts waiting in the queue can be obtained by the following expression [5]:

$$p_n = \frac{(\lambda T)^n}{\prod_{k=1}^{n} u^*(k)} p_0, \qquad n \geq 0, \tag{42}$$

where

$$p_0 = \frac{1}{1 + \sum_{k=1}^{\infty} \prod_{i=0}^{k-1} \frac{\lambda T}{u^*(i+1)}}. \tag{43}$$

Furthermore, from eqs. (25) (26) and (31), we can derive a closed-form expression for $P_{n,j}$:

$$P_{n,j} = \begin{cases} \dfrac{[\rho_1^*]^j}{[1+\rho_1^*]^{j+1}}, & n = 0,\ j \geq 0, \\ \dfrac{[\rho_n^*]^{j-n+1}}{[1+\rho_n^*]^{j-n+2}}, & n \geq 1,\ j \geq n-1, \\ 0, & \text{otherwise}, \end{cases} \tag{44}$$

where $\rho_n^* = \dfrac{\lambda T}{u^*(n)}$, $n \geq 1$.

The following theorem summarizes certain properties of $P_{n,j}$ defined in eq. (44).

Theorem 6: The transition probabilities $P_{n,j}$ of the Q-optimal production system (generalized M/M/1 queue) has the following properties:

(i) For any positive n, $P_{n,j}$ is a monotone decreasing function of j.

(ii) $P_{n,n-1} \to 1$ and $P_{n,j} \to 0$, $j > n-1$ as $n \to \infty$.

B. A Production Process Described by a Diffusion-Threshold Model

In this section, we introduce another model for a production process. The extent of processing a part receives is described by a controlled diffusion process where a threshold crossing signifies the completion of the part [1,7]. The part production time is found to be an inverse-Gaussian (IG) random variable parameterized by the production rate.

Let x_t be a diffusion process representing the *extent of processing* for a part up to time t. Then x_t is assumed to satisfy:

$$x_t = k_1 u t + k_2 u W_t, \qquad x_0 = 0, \tag{45}$$

where u is the controlled production rate, $k_1 u$ is the drift rate, $k_2 u$ is the diffusion coefficient and W_t is a Wiener process. Then the diffusion process x_t has mean and variance that increase linearly with time t and linearly with u.

The time elapsed before completing a part is determined by

$$T_B = \inf \{t > 0 : x_t = L\}, \tag{46}$$

so that a part is produced when the extent of processing achieves the threshold value $L > 0$. Notice that T_B is the first passage time for the random process x_t.

The probability distribution for T_B, with a constant production rate u, has been determined [1]:

$$F_B(t \mid u) = \int_0^t \frac{L}{\sqrt{2\pi} k_2 u \tau^{3/2}} \exp\{\frac{-(L - k_1 u \tau)^2}{2k_2^2 u^2 \tau}\} d\tau . \tag{47}$$

This resultant distribution function in eq. (47) is often referred to as the inverse-Gaussian or the Wald distribution.

From eq. (47), it is easy to obtain:

$$E[T_B \mid u] = \frac{L}{k_1 u}, \tag{48}$$

and

$$\mathrm{var}[T_B \,|u] = \frac{k_2^2 L}{k_1^3 u}. \tag{49}$$

From eqs. (48) and (49), the optimization problem in (14) can be specialized as: minimize

$$J(n, u) = w_2 \sum_{j=1}^{n} T_j + \frac{w_1 L}{k_1 u} f(u) + \left(\frac{w_2 L n}{k_1} + \frac{\lambda w_2 L k_2^2}{2k_1^3}\right)\frac{1}{u} + \frac{\lambda w_2 L^2}{2k_1^2}\frac{1}{u^2}, \qquad n \geq 1. \tag{50}$$

The close similarity between eq. (33) (for the controlled M/M/1 queue) and eq. (50) can be clearly observed. Therefore, the results stated in the rest of this subsection can be easily obtained.

First of all, the full service policy is S-optimal if $f(u) = u$. This result coincides with that obtained in [7], where an *open-loop constant control policy* is sought to minimize the *long-term average cost per part.*

Furthermore, if the production rate penalty function is quadratic, i.e., $f(u) = u^2$, then eq. (50) is:

$$J(n, u) = w_2 \sum_{j=1}^{n} T_j + \frac{w_1 L}{k_1} u + \left(\frac{w_2 L n}{k_1} + \frac{\lambda w_2 L k_2^2}{2k_1^3}\right)\frac{1}{u} + \frac{\lambda w_2 L^2}{2k_1^2}\frac{1}{u^2}, \qquad n \geq 1; \tag{51}$$

and we have the following theorem, whose proof is exactly the same as that for Theorem 4.

Theorem 7: The cost function in (51) is a strictly convex function of the production rate u for $u \in (0, \infty)$ and has a unique minimum on $(0, \infty)$. The Q-optimal control can be expressed as:

$$u^*(n) = \begin{cases} (\frac{-b}{2} + \sqrt{K})^{\frac{1}{3}} - (\frac{b}{2} + \sqrt{K})^{\frac{1}{3}}, & \text{if } K \geq 0, \\ 2\sqrt{\frac{-a}{3}}\cos\theta, & \text{if } K < 0, \end{cases} \qquad \textbf{(52)}$$

with parameters defined as:

$$K = \frac{b^2}{4} + \frac{a^3}{27}, \qquad a = -(\frac{w_2 n}{w_1} + \frac{\lambda w_2 k_2^2}{2k_1^2 w_1}),$$

$$b = -\frac{\lambda w_2 L}{k_1 w_1}, \qquad \theta = \frac{1}{3}\tan^{-1}(\frac{2\sqrt{-K}}{-b}). \qquad (53)$$

Also, we notice from eq. (48) that $E[T_B|u]$ is a monotone decreasing function of $u \in (0, \infty)$; and

$$\lim_{u \to \infty} E[T_B|u] = \lim_{u \to \infty} \frac{L}{k_1 u} = 0 < \frac{1}{\lambda}. \qquad (54)$$

Thus, we have:

Corollary 8: For the controlled production (M/IG/1) queueing system:

(i) The Q-optimal policy u^* is a monotone non-decreasing function of n, and $u^* \to \infty$ as $n \to \infty$.

(ii) The embedded semi-Markov chain of the closed-loop Q-optimal queue is ergodic.

Furthermore, $q_k(n)$ in eq. (26) can be rewritten as:

$$q_k(n) = \int_0^\infty \frac{e^{\lambda\tau}(\lambda\tau)^k L}{k!\sqrt{2\pi}k_2 u^*(n)\tau^{3/2}} \exp\{\frac{-(L - k_1 u^*(n)\tau)^2}{2k_2^2 u^*(n)^2 \tau}\} d\tau, \qquad (55)$$

The values for $q_k(n)$ and hence for the steady state probabilities $(p_0, p_1, p_2, \cdots)$ can only be obtained numerically.

VIII. A TRANSPORTATION DELAY EXAMPLE

In this section, we consider an engineering example which illustrates the approach presented in the previous sections. Parts to be transported over a fixed distance arrive at a station according to a homogeneous Poisson process. Transportation of parts is provided on a first come first serve basis by a single server. Any part arriving at an instant when the server is busy in transporting a part will have to join the end of a queue with infinite capacity. The mass of a part is assumed to be a random variable with given probability distribution function; hence the transportation time is a random variable depending on the mass of the transported part and the control force required to move the part. An economic problem is to minimize a weighted cost associated with the part waiting time and the control force.

The rest of this section is organized as follows: in section A., the mathematical model for the transportation system is formally presented; in section B., the *Q-optimal* control policy is obtained; section C. contains the derivation of the *best constant control policy* and section D. discusses and compares simulation results under both policies based on several long-term performance measures.

A. Mathematical Model

In Fig. 1, parts to be transported over a distance L arrive at a station according to a Poisson process with intensity λ. Let $x(t)$ be the distance traveled by a transported part up to time t and let T_B be the total transportation time required to move a part from $x = 0$ to $x = L$. Let $F(x)$ be the force exerted on the part being transported at location x.

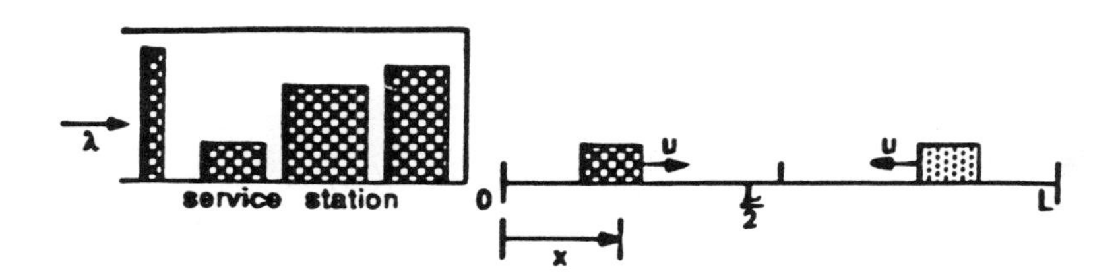

Fig. 1 A Transportation System with Queue

We assume the force on the part is given by the bang-bang function:

$$F(x) = \begin{cases} u, & 0 \le x < \dfrac{L}{2}, \\ -u, & \dfrac{L}{2} \le x < L, \end{cases} \tag{56}$$

where $u > 0$ is the control parameter. Therefore, the total transportation time $T_B(u)$ for a part is:

$$T_B = 2\sqrt{\frac{LM}{u}}, \tag{57}$$

where M is the random variable representing the mass of a transported part with distribution function $F_M(m)$. Notice that, given the mass M of a part, the transport time is *deterministic;* it is the *uncertainty of the mass of the* arriving parts that makes the "production time" random.

It is easy to show that the following boundary conditions for each transported part are satisfied:

$$x(0) = 0, \tag{58}$$

$$x(T_B) = L, \tag{59}$$

$$\text{v}(0) = 0, \tag{60}$$

$$v(L) = 0. \tag{61}$$

where v(x) is the velocity of the part at location x. Using eq. (57), the transportation time T_B is a random variable with distribution function $F_B(t\,|u)$ expressed as:

$$F_B(t\,|u) = F_M(\frac{t^2u}{4L}). \tag{62}$$

Then the first and second moments of T_B can be expressed as:

$$E[T_B \mid u] = \int_0^\infty t dF_B(t\,|u)$$

$$= 2\sqrt{\frac{L}{u}} \int_0^\infty \sqrt{m}\, dF_M(m), \tag{63}$$

and

$$E[T_B^2 \mid u] = \int_0^\infty t^2 dF_B(t \mid u)$$

$$= \frac{4L}{u} \int_0^\infty m dF_M(m). \tag{64}$$

Our objective is to evaluate various control policies on the basis of long-term statistical performance measures. One such measure is given by the *expected long term average cost per unit time* [8,9] defined as:

$$\Phi^\pi = \overline{\lim_{t \to \infty}} E^\pi \left[\frac{Z^\pi(t)}{t} \right], \tag{65}$$

where π is a control policy and

$$Z^\pi(t) = \int_0^t (w_1 f(u^\pi(\tau)) + w_2 N(\tau)) d\tau \tag{66}$$

is the weighted cost associated with control and part waiting time up to time t; $N(\tau)$ is the number of parts waiting in the queue at time τ and $u^\pi(\tau) > 0$ is the value of control applied at time τ. Note that Φ^π is independent of the initial number of parts waiting in the queue [8].

Two other interesting long-term measures are the *mean queue length,* $\bar{N}^\pi$, and the *standard deviation of queue length,* σ_N^π, in the steady state under a control policy π which makes the closed-loop production system ergodic. (Refer to eqs. (29) and (30).)

B. Q-Optimal Control Policy

In the subsequent analysis, we assume that $F_M(m)$ is uniformly distributed between values 0.0 and 30.0 and that $L = 100.0$, $\lambda = 0.1$. We then have from eqs. (63) and (64):

$$E[T_B \mid u] = \frac{73.030}{\sqrt{u}} \tag{67}$$

and

$$E[T_B^2 \mid u] = \frac{6000}{u}. \tag{68}$$

Let the weighting factors $w_1 = 3.652$, $w_2 = 1.0$, $w_3 = 39700.0$; the short-term cost from equation (14) is:

$$J(n, u) = \sum_{j=1}^{n} T_j + \frac{266.706 f(u)}{\sqrt{u}} + \frac{(73.03n + 289929.1)}{\sqrt{u}} + \frac{300.0}{u}, \qquad n \geq 1. \tag{69}$$

Assume that the cost for transporting one part is proportional to the total energy W spent in transporting this part. From physical arguments, it follows that:

$$W = uL. \tag{70}$$

In terms of the control cost rate function $f(u)$, we have:

$$W = f(u) E[T_B \mid u] = \frac{73.03 f(u)}{\sqrt{u}}. \tag{71}$$

Thus,

$$f(u) = \frac{uL}{E[T_B \mid u]} = 1.369\sqrt{u^3}. \tag{72}$$

Consequently, the short-term cost in eq. (69) is:

$$J(n, u) = \sum_{j=1}^{n} T_j + 365.2u + \frac{(73.03n + 289929.1)}{\sqrt{u}} + \frac{300.0}{u}, \qquad n \geq 1. \tag{73}$$

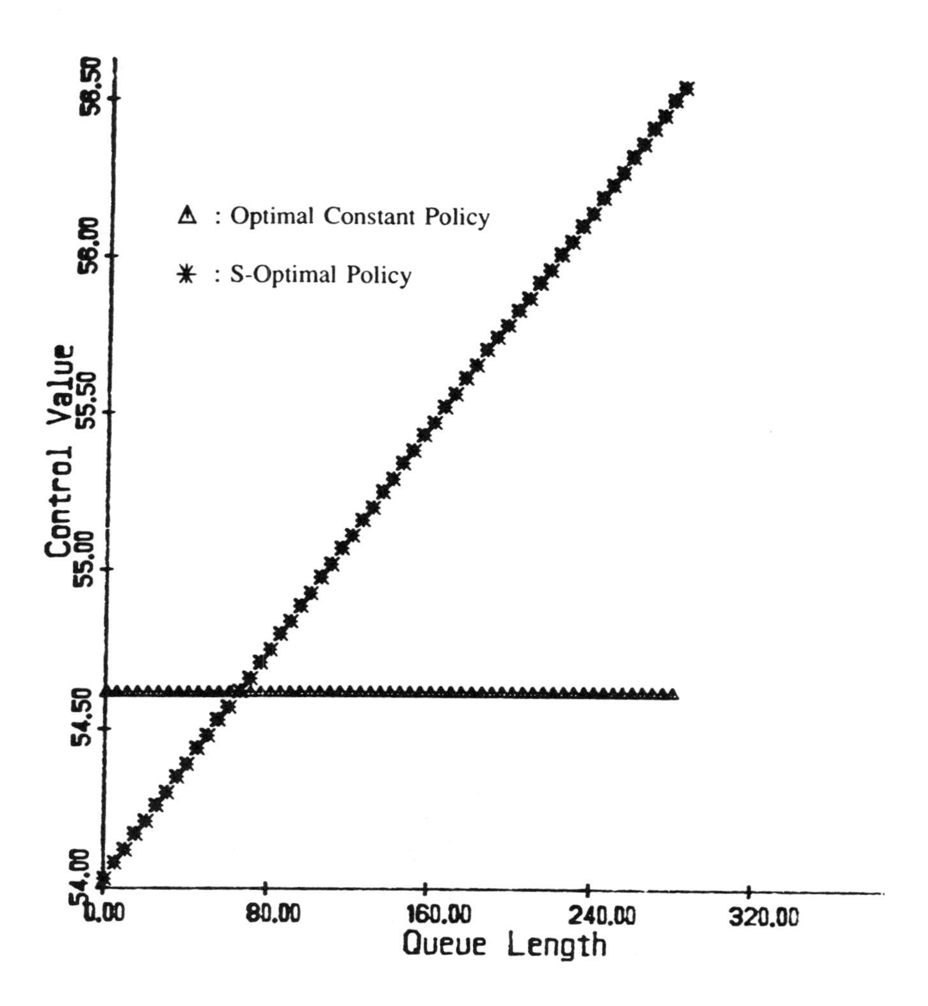

Fig. 2 Plot of State-Dependent Optimal Policy

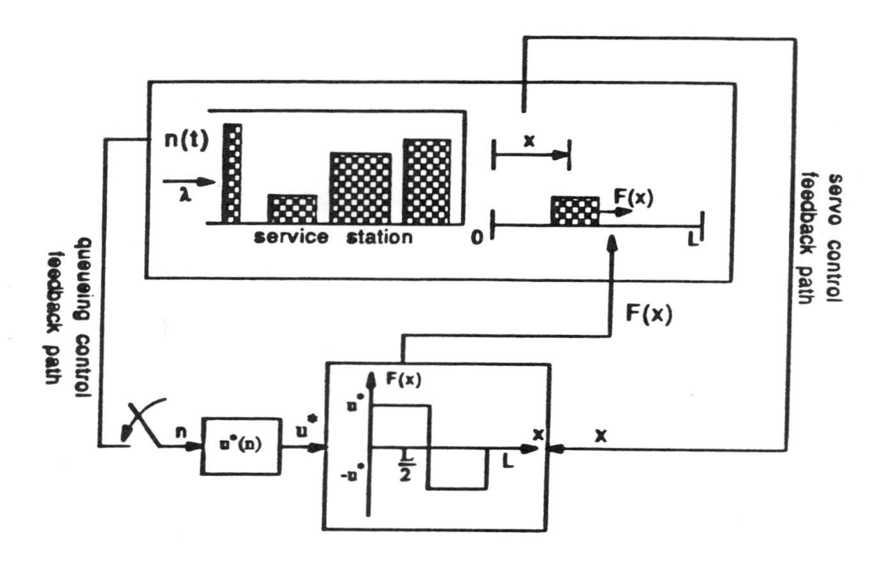

Fig. 3 Control Structure under π^*

It is easy to show that, for fixed queue length n, $J(n, u)$ is a *strictly convex function* of u for $u > 0$. Therefore, the Q-optimal control policy $\pi^* = (u^*(1), u^*(2), u^*(3), \cdots)$ under the short-term cost criterion (73) can be found by a simple one dimensional search; values for $u^*(n)$ are shown in Fig. 2.

Fig. 3 shows the closed-loop structure of the controlled transportation system using this control policy. There are two feedback loops: a continuous time servo loop for control of the motion of each part being transported and a discrete time queueing control feedback loop. In the queueing control feedback, information about the number of parts waiting in the queue is observed at each service initialization epoch, and this information is used to determine the corresponding Q-optimal control value. The selected Q-optimal control parameter is then used, according to eq. (56), in transporting the next part in the queue.

We anticipate that the control policy obtained by minimizing the short-term cost will perform well in the long run. Evaluation of this control policy is carried out by computer simulation and is presented in the last subsection.

C. Long-Term Optimal Constant Control Policy

In this section, an *optimal constant control policy* is determined which can be compared with the Q-optimal control policy obtained in section IV.C. A similar development has been reported in [7].

A constant control policy π_U is defined as:

$$u(n) = \begin{cases} U, & n > 0, \\ 0, & n = 0. \end{cases} \tag{74}$$

The queueing system under a constant control policy π_U is ergodic if and only if [5]:

$$\lambda E[T_B \mid U] < 1, \tag{75}$$

which results in the condition

$$U > 53.33. \tag{76}$$

Furthermore, if the system is ergodic, the expected long term average cost is characterized by the following theorem.

Theorem 9: If U satisfies condition (76), then

$$\Phi^{\pi_U} = w_1 \lambda f(U) E[T_B \mid U] + w_2 \overline{N}^{\pi_U}, \tag{77}$$

where

$$\overline{N}^{\pi_U} = \lambda E[T_B \mid U] + \frac{\lambda^2 E[T_B^2 \mid U]}{2(1 - \lambda E[T_B \mid U])}. \tag{78}$$

Proof: From eq. (66), we have:

$$E^{\pi_U}[Z^{\pi_U}(t)] = w_1\{\int_0^t f(U)1_{(N(\tau)>0)}d\tau\} + w_2\{\int_0^t N(\tau)d\tau\}. \tag{79}$$

where $1_{(N(\tau)>0)}$ is the indicator function which takes value 1 when $N(\tau) > 0$ and 0 otherwise. If U satisfies $\lambda E[T_B \mid U] < 1$, then the system is ergodic, and, consequently, $N(t)$ is bounded with probability one for any $t > 0$. By the Fubini Theorem [10], we can interchange expectation and integration to obtain:

$$E^{\pi_U}[Z^{\pi_U}(t)] = w_1\int_0^t E^{\pi_U}E\{1_{(N(\tau)>0)}\}d\tau + w_2\int_0^t E\{N(\tau)\}d\tau. \tag{80}$$

Also, with the ergodicity property of an M/G/1 queue, we know [5]:

$$E\{1_{(N(\tau)>0)}\} \rightarrow \lambda E[T_B \mid U] \qquad \text{as } \tau \rightarrow \infty; \tag{81}$$

and

$$E\{N(\tau)\} \rightarrow \overline{N} \qquad \text{as } \tau \rightarrow \infty, \tag{83}$$

where $\overline{N}$ is the expected number of parts waiting in the queue *at any time* in the long run; this is also the steady state expected number of parts waiting in the queue of the *embedded Markov chain* observed at all part departure epochs [5]. Hence [5],

$$\overline{N}^{\pi_U} = \lambda E[T_B \mid U] + \frac{\lambda^2 E[T_B^2 \mid U]}{2(1 - \lambda E[T_B \mid U])}. \tag{83}$$

From eqs. (65) and (80)-(82), we obtain eq. (77) and the theorem is proved.

Q.E.D.

By substituting numerical values into eq. (77), we have:

$$\Phi^{\pi_U} = 36.2U + \frac{7.302967}{\sqrt{U}} + \frac{30}{U - 7.302967\sqrt{U}} \qquad \text{if } U > 53.33. \tag{84}$$

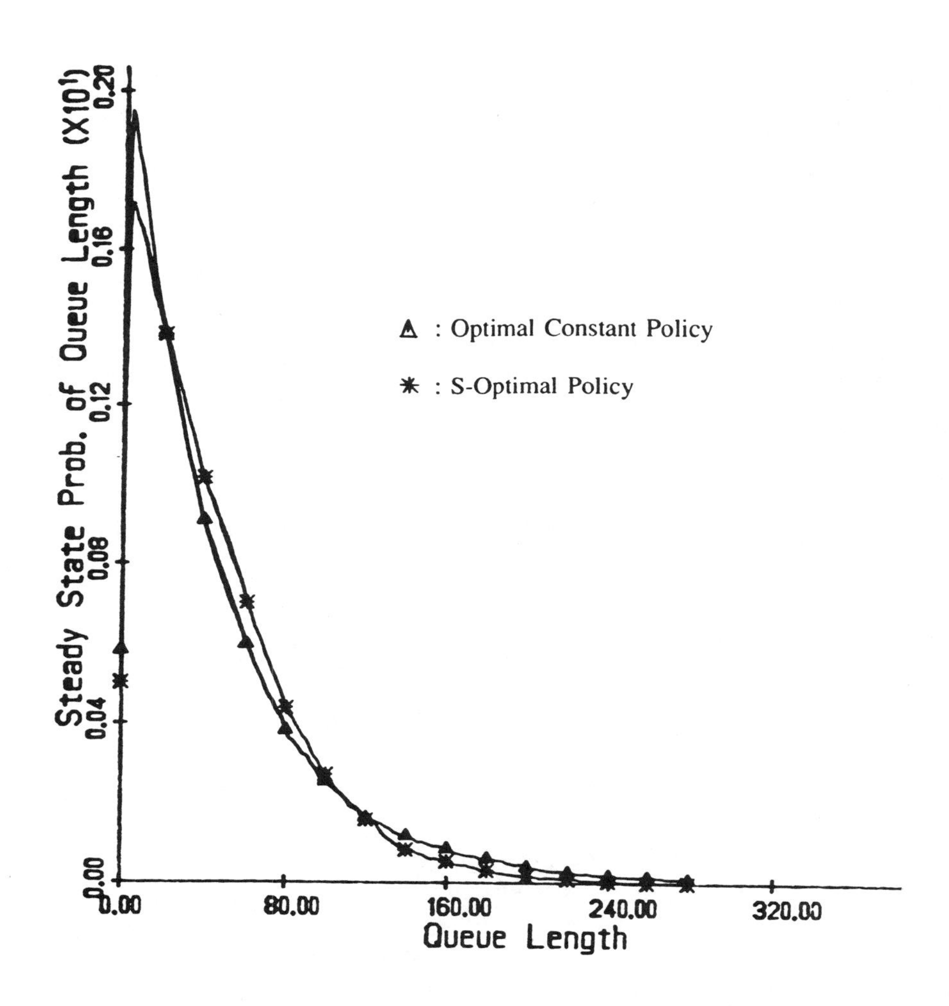

Fig. 4 Plot of Probability of Queue Length in the Steady State

It is easy to show that Φ^{π_U} is a *strictly convex function* of U for $U > 53.33$. Using a simple one dimensional search routine, we find the optimal control to be:

$$U^* = 54.62. \tag{85}$$

The long-term expected average cost per unit time and the mean queue length under this optimal constant control are:

$$\Phi^{\pi_{U^*}} = 2041.78, \tag{86}$$

$$\overline{N}^{\pi_{U^*}} = 47.52. \tag{87}$$

Fig. 2 also shows the best constant control U^* on the same graph as the Q-optimal control policy obtained in the last subsection.

D. Simulation Results

It is easy to verify that the closed-loop queue under the Q-optimal control policy π^* and the long-term optimal constant control policy π_{U^*} are both ergodic. By the regenerative reward theorem [8,9], we have:

$$\Phi^{\pi} = \lim_{t \to \infty} \frac{Z^{\pi}(t)}{t} \qquad \text{with probability 1,} \tag{88}$$

where π is either π^* or π_{U^*}. This indicates that a good estimate for Φ^{π^*} or $\Phi^{\pi_{U^*}}$ can be obtained by using only one computer simulation run with a large simulation time.

The queue length probability p_n^{π} in the steady state is also the probability that a departing part sees n parts waiting in the queue in the steady state [12]. Then, by the renewal reward theorem [8,9], we have:

$$p_n^{\pi} = \lim_{m \to \infty} \frac{\sum_{k=1}^{m} 1_{(N(T_F^k)=n)}}{m}, \tag{89}$$

where T_F^k is the k-th departing epoch.

Therefore, estimates of the steady state probability distribution $\{p_n^{\pi}\}_{n=0}^{\infty}$, the steady state mean queue length $\bar{N}^{\pi}$ and the standard deviation of queue length in the steady state σ_N^{π}, where π is either π^* or π_{U^*}, can all be calculated by information obtained in one single simulation run.

Table I summarizes the relevant statistics from a simulation with simulation time equal to 10^8. It can be seen that the average long-term cost per unit time is smaller under the Q-optimal control policy π^*. More significantly, the mean and the standard variation of the queue length in the steady state are much smaller under the optimal adaptive control policy π^*. The relative values for the latter two statistics are consistent with the steady state probability distribution $\{p_j\}_{j=0}^{\infty}$ which is indicated for both policies in Fig. 5 and Fig. 6. The smaller values of these two statistics under the Q-optimal control policy π^* are due to the shorter tail of the resultant probability distribution, which in turn is a consequence of the *adaptivity* of the queueing control scheme.

Fig. 5 (Fig. 6) compares the steady state queue length probabilities $\{p_n^{\pi^*}\}_{n=0}^{\infty}$ ($\{p_n^{\pi_{U^*}}\}_{n=0}^{\infty}$) obtained from simulation and from computations using eqs. (27) and (28). We observe a close match of these two results. Also, we see a close match of $\Phi^{\pi_{U^*}}$ and $\bar{N}^{\pi_{U^*}}$ between the simulation outcomes and those predicted by eqs. (86) and (87).

Table I. Simulation Statistics

Policy π	ϕ^{π}	$\bar{N}^{\pi}$	σ_N^{π}	$N_{\max}^{\pi}$
π^*	2032.642	45.510	38.766	284
π_{U^*}	2043.852	49.715	49.691	390

Simulation time = 10^8

ϕ^{π}: (expected) long-term average cost per unit time
$\bar{N}^{\pi}$: mean queue length in the steady state
σ_N^{π}: standard deviation of queue length in the steady state
$N_{\max}^{\pi}$: maximal number of parts waiting in the queue

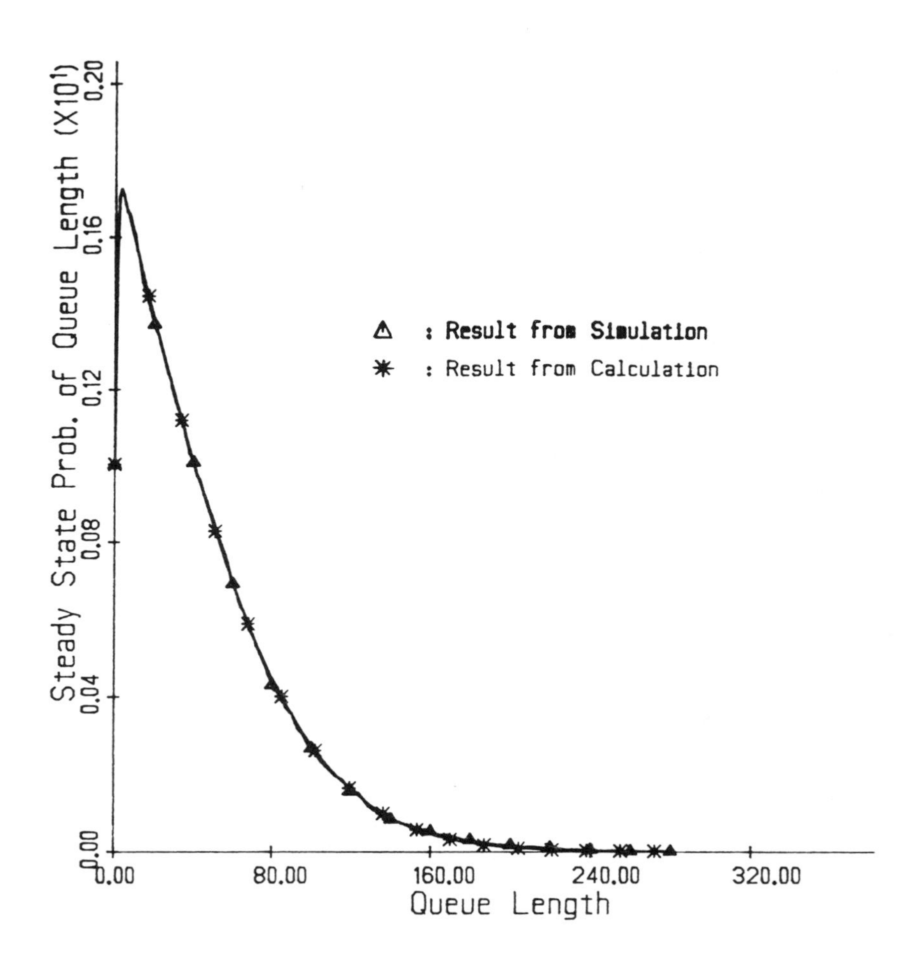

Fig. 5 Comparison of Steady State Queue Length Probabilities under π^*

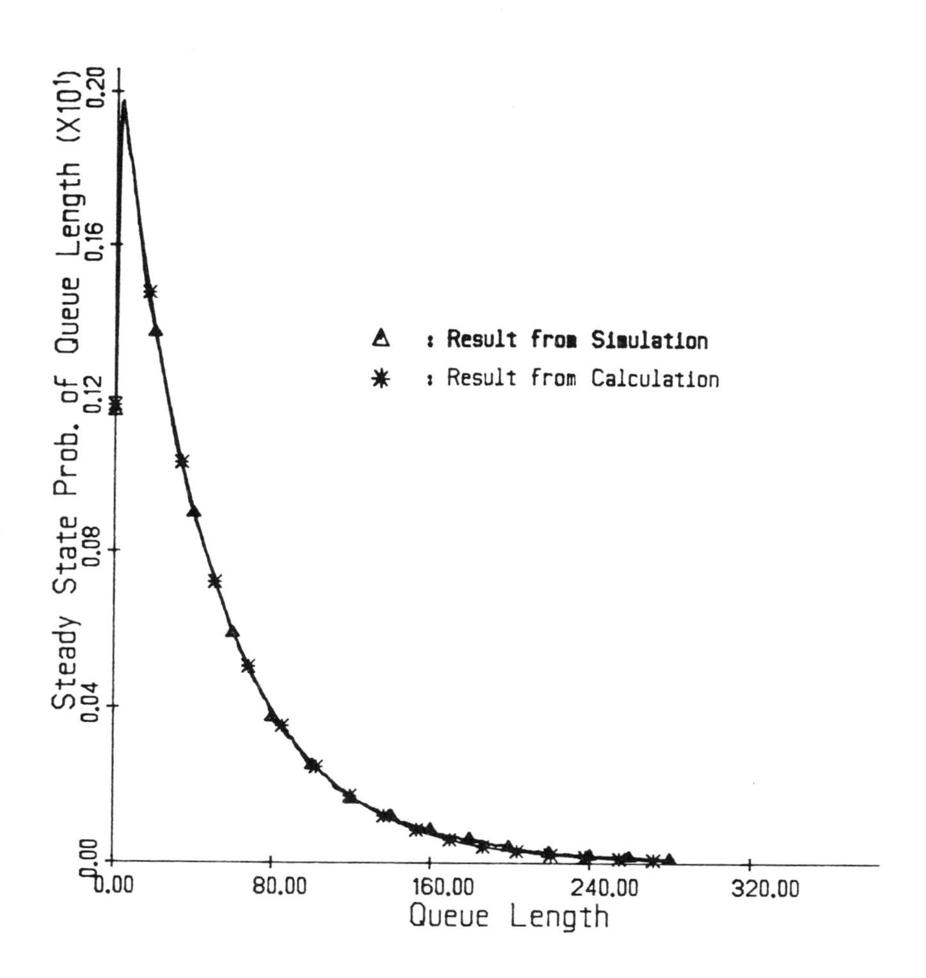

Fig. 6 Comparison of Steady State Queue Length Probabilities under π_U.

IX. CONCLUSIONS

The optimal control of a production (M/G/1) queueing system by selection of the production rate to achieve a *minimal expected long-term average cost* has been studied by other authors [3,12,13]. But the solution of a set of dynamic programming equations is required in all cases, which is very restrictive and numerically complex. By contrast, our proposed Q-optimal control policy based on a short-term cost criterion has some desirable properties: (i) Relatively non-restrictive assumptions are imposed on the problem formulation. (ii) Only the first several statistical moments of the production time, rather than complete knowledge of the production time distribution, are required to obtain the Q-optimal control policy. (iii) For special cases, such as those presented in section VII, an analytic expression for the Q-optimal control policy can be obtained. (iv) Ergodicity of the closed-loop Q-optimal production system can be guaranteed. (v) Stability robustness is ensured.

It is straightforward to make extensions to cases such as compound Poisson arrivals and state-dependent arrival intensity, which allows balking and finite queueing capacity. Due to the "short-term" nature of our cost criterion, it is also easy to accommodate other generalizations such as time-varying system parameters, e.g., arrival intensity, part load variations and cost weighting factors. Another interesting extension considers production systems that can fail as a result of continual usage [6].

REFERENCES

[1] C. J. Conrad and N. H. McClamroch, "The Drilling Problem, A Stochastic Modeling and Control Example in Manufacturing," IEEE Trans. Automat. Contr. AC-32, 947-958 (1987).

[2] T. B. Crabill, "Sufficient Conditions for Positive Recurrent and Recurrence of Specially Structured Markov Chains," Oper. Res. 16, 858-867 (1968).

[3] E. Gallisch, "On Monotone Optimal Policies in a Queueing Model of M/G/1 Type with Controllable Service Time Distribution," J. Appl. Prob. 16, 871-887 (1979).

[4] S. Karlin and H. Taylor, "A First Course in Stochastic Processes," Academic Press, New York, 1981.

[5] L. Kleinrock, "Queueing Theory Vol. I: Theory," John Wiley & Sons, New York, 1975.

[6] C. P. Lee, "Stochastic Modeling and Control of a Production System with Random Part Production Times," Computer, Information and Control Engineering, The University of Michigan, Ph.D. Dissertation (1989).

[7] C. P. Lee and N. H. McClamroch, "Control of an M/G/1 Queue with Service Times Modeled by a Diffusion-Threshold Process," in "Proc. 26th Conf. Decision Contr.," Los Angeles, 1987.

[8] S. M. Ross, "Applied Probability Model with Optimization Applications," Holden-Day, San Francisco, 1970.

[9] S. M. Ross, "Stochastic Processes," Wiley, New York, 1983.

[10] H. L. Royden, "Real Analysis," The Macmillan Company, New York, 1968.

[11] M. Schal, "The Analysis of Queues with State-Dependent Parameters by Markov Renewal Processes," J. Appl. Prob. 8, 155-175 (1971).

[12] M. Schal, "On the M/G/1 Queue with Controlled Service Rate," in "Optimization and Operations Research," (R. Henn and B. Korte, ed.), Spring-Verlag, New York, 1978.

[13] R. Schassberger, "A Note on Optimal Service Selection in a Single Server Queue," Management Science, Vol. 21, No. 11, 1326-1331 (1975).

[14] M. J. Sobel, "The Optimality of Full Service Policies," Oper. Res. 30, 636-649 (1982).

MODEL REDUCTION FOR FLEXIBLE STRUCTURES

WODEK GAWRONSKI
Jet Propulsion Laboratory
California Institute of Technology
Pasadena, California 91109

JER-NAN JUANG
NASA Langley Research Center
Hampton, Virginia 23665

I. INTRODUCTION

Model reduction is one of the important issues in the dynamical analysis of flexible structures. A model with a large number of degrees of freedom, while useful in a static analysis, can cause numerical difficulties, uncertainities, and high computational costs if used to study the dynamics of a system. The problem of model reduction has a long history, and among many techniques we mention balanced and modal truncation [1 – 8]. While these techniques are all comparatively simple, they do not necessarily give optimal results. On the other hand, the optimal reduction methods of Wilson, [9 – 10] and Hyland and Bernstein [11] are computationally expensive. Furthermore, existing literature are mainly concerned with the reduction of general linear systems. Model reduction for the specific case of flexible structures is studied by Gregory [5] Jonckheere [6] and Skelton, et al.[1, 2, 12]. Gregory [5] and Jonckheere [6] have shown that, in the case of small damping and widely separated natural frequencies, the balanced and modal representations of a flexible structure are almost identical.

Based on the papers [13 – 16], this study begins with the presentation of several conditions for a near-optimal reduction (with a small reduction error) of general dynamic systems. Since the near-optimal solution is a fuzzy concept (the solution is non-unique), different conditions can be used to determine it. Our attention concentrates on the reduction in balanced and modal coordinates. The conditions under which the modal truncation of a flexible structure yields a near-optimal reduced-order model, are presented. In particular, it is shown that modal and balanced reductions give very different results for the flexible structure with closely-spaced natural frequencies. The balanced reduction generally gives better results. Another question addressed is the sensitivity of modeling error to variations in the damping of a structure. A robust model reduction technique is developed to cope with this question. The practical importance of this problem is a consequence of the fact that the damping level present in a flexible structure is usually poorly defined, and difficult to identify in ground tests.

A reduced model is obtained by truncating part of the state variables. The reduction indices determine which component is deleted or retained in the reduced model. It is shown in this study how to obtain the reduction indices from the system transfer function, rather than from the system matrices. Test data, besides system dynamics, include also actuator and sensor dynamics. The reduced model obtained from these data can be far from the optimal because actuator and sensor dynamics are involved. The reconstruction of the flexible structure indices from the joint actuator-sensor-flexible structure indices is discussed and illustrated.

The reduction error (norm of the difference between the full and reduced system outputs) is determined in infinite time or frequency interval. While such approach is of great theoretical interest and useful in many applications, it is important to note that no realistic sensor or actuator can operate over an infinite frequency interval; nor can infinite time histories ever be obtained. Consequently, the classical infinite time and frequency grammians are only approximations to the quantities that actually describe any physical system.

New concepts of grammians defined over a finite time and/or a frequency interval are introduced including computational procedures to evaluate them.

Applying the presented model reduction technique to these grammians rather than to the classical ones, leads to a near-optimal reduced model which closely reproduces the full system output in the time and/or frequency interval of interest. This approach has clear connections with filter design, and can be applied to unstable systems without encountering the divergence problems inevitable with an infinite interval.

Extensive examples are given to illustrate all the concepts developed. It should be noted that, while these examples and the original motivation for this work are drawn from structural dynamics, the concepts derived are in fact applicable to any linear system.

II. MODEL REDUCTION

A reduced order model of a system is obtained by truncating appropriate state components of the system. In order to make the right choice of components, the system controllability and observability grammians are analyzed. In this section, grammians and its relation with the balanced representation are discussed and a reduction technique is introduced.

A. GRAMMIANS AND BALANCED REPRESENTATION

The triple $(A, B, C,)$ represents the continuous time linear system

$$\dot{x} = Ax + Bu, \quad y = Cx, \tag{1}$$

The system has p inputs, q outputs, n states, and is assumed stable, unless otherwise stated. Let λ_i be the i-th eigenvalue of A and impose the condition $\lambda_i + \lambda_k^* \neq 0$ for every $i, k = 1, ..., n$. The controllability and observability grammians are defined as follows [4], [17]

$$W_c = \int_0^t \exp(A\tau)BB^* \exp(A^*\tau)d\tau, \quad W_o = \int_0^t \exp(A^*\tau)C^*C \exp(A\tau)d\tau. \tag{2}$$

They can be determined from the equations

$$\dot{W}_c = AW_c + W_cA^* + BB^*, \quad \dot{W}_o = A^*W_o + W_oA + C^*C, \tag{3}$$

The stationary solutions of (2) such that $\dot{W}_c = \dot{W}_o = 0$ are determined from the Lyapunov equations

$$AW_c + W_cA^* + BB^* = 0, \quad A^*W_o + W_oA + C^*C = 0, \tag{4}$$

Note that the stationary solutions exist for stable as well as unstable systems; for stable systems they are positive definite, however.

The system triple is balanced, if its controllability and observability grammians are equal and diagonal [4]. In order to find the balanced representation, the semigrammians P, Q are first determined from the decomposition of controllability and observability grammians (for example, singular value, or Cholesky decomposition)

$$W_c = PP^T, \quad W_o = Q^TQ, \tag{5}$$

and next the matrix $H = QP$ is obtained. The singular value decomposition of H gives

$$H = V\Gamma^2U^T, \tag{6}$$

where $V^TV = I$, $U^TU = I$, and Γ is a positive semidefinite diagonal matrix. Denoting

$$R = PU\Gamma^{-1} = Q^{-1}V\Gamma, \; R^{-1} = \Gamma^{-1}V^TQ = \Gamma U^TP^{-1}, \tag{7}$$

the balanced representation is obtained from

$$A_b = R^{-1}AR, \quad B_b = R^{-1}B, \quad C_b = CR, \tag{8}$$

since the grammians in balanced coordinates are $W_{cb} = R^{-1}W_cR^{-T} = \Gamma^2 = R^TW_oR = W_{ob}$. This indicates that the controllability and observability grammians, after transformation, are equal and diagonal. This is the conventional definition of the balanced representation.

From now on, define that a system is balanced if its observability and controllability grammians are equal, but not necessarily diagonal. In this case the balanced representation is not unique, and is invariant under an orthogonal transformation θ. Let $\bar{x} = \theta x$, and $W_c = W_o$, then also $\bar{W}_c = \bar{W}_o$, since $\bar{W}_c = \theta^{-1}W_c\theta^{-T} = \theta^TW_o\theta$, $\bar{W}_o = \theta^TW_o\theta$. This allows one to chose different forms of the balanced representation. Among many are Moore form, where the grammians are diagonal, or block-diagonal; lower- or upper-Schur form, where matrix A is lower- or upper-triangular, or block-triangular; Hessenberg forms (not unique), where matrix A is in Hessenberg, or block-Hessenberg form. The application of different balanced forms to model reduction will be presented later.

B. REDUCTION TECHNIQUE

Let the state vector x of the representation (A, B, C) be partitioned as $x^T = \left[x_R^T \; x_T^T\right]$, and (A, B, C) be partitioned accordingly

$$A = \begin{bmatrix} A_R & A_{RT} \\ A_{TR} & A_T \end{bmatrix}, \quad B = \begin{bmatrix} B_R \\ B_T \end{bmatrix}, \quad C = [C_R \; C_T] \tag{9}$$

where x_R is a k by 1 vector and x_T is an $(n-k)$ by 1 vector. The reduced model is obtained by deleting the last n–k rows of A, B, and the last n–k columns of A, C, i.e.

$$A_R = LAL^T, \quad B_R = LB, \quad C_R = CL^T, \quad L = [I_k \; 0]\,. \tag{10}$$

This approach is considered a general case of reduction via truncation, since the system before reduction can be transformed to a suitable representation through an appropriate similarity transformation.

The reduction index, defined as follows

$$J = \|y - y_R\|, \tag{11}$$

is determined from the formula

$$J^2 = \|y\|^2 + \|y_R\|^2 - 2tr(C^* C_R W_{cRO}) = \|y\|^2 + \|y_R\|^2 - 2tr\left(BB_R^* W_{oRO}\right). \tag{12}$$

where $W_{cRO} = E\left(x_R x^*\right)$ is the covariance between the states of the original and the reduced model. A normalized ***reduction index*** $\boldsymbol{\delta}$, also called a ***relative error***, is defined

$$\delta = \|y - y_R\| / \|y\| = J / \|y\|. \tag{13}$$

For the optimal reduction, the vectors $y - y_R$ and y_R are orthogonal, in consequence $\|y_R\|^2 + \|y - y_R\|^2 = \|y\|^2$, so that $\|y - y_R\|^2 = \|y\|^2 - \|y_R\|^2$.

For the near-optimal reduction $\|y - y_R\|^2 \approx \|y\|^2 - \|y_R\|^2$, so that the index

$$J_p \approx | \ \|y\|^2 - \|y_R\|^2 \ | \tag{14}$$

is an approximation of the index J, and is used as a ***prediction error***, since it can be determined before reduction.

III. NEAR OPTIMAL REDUCTION

Here, the global near-optimal reduction is considered such that the reduction error is almost zero. Given a system representation, the reduced model is obtained by truncating a set of selected system coordinates. The near-optimal reduction can be successful only for specific representations, and balanced and modal representations are considered for this purpose. Several conditions for the reduced model to be near the global optimum are presented.

A. NEAR-OPTIMAL REDUCTION IN BALANCED COORDINATES

In this section, conditions for the near-optimal reduction in balanced coordinates are presented. Proofs of the conditions can be found in [13].

In the balanced coordinates, the reduction index J determined from (12) is

$$J^2 = tr\,(C^*CW_{cO}) + tr\,(C_R^*C_RW_{cR}) - 2tr\,(C^*C_RW_{cRO}), \tag{15}$$

where $W_{cRO} = E\left(x_R x^*\right)$ is the covariance between the states, of the original, x, and the reduced model, x_R.

As defined previously, the balanced representation is not unique, and its different forms may be used for model reduction. For any balanced representation the following statement holds.

Condition 1. The following

$$\|W_{cT}\| = \|W_{oT}\| \ll \|W_{cR}\| = \|W_{oR}\| \tag{16}$$

is a necessary and sufficient condition for near-optimal reduction in any balanced coordinates. The quantities J, δ are predicted as J_p and δ_p

$$\begin{aligned} J^2 \approx J_p^2 &= | \ \|y\|^2 - \|y_R\|^2 | = tr\,(C_T^*C_TW_{cT}) = tr\,(B_TB_T^*W_{oT}) \\ &= -2tr\,(A_TW_{cT}W_{oT}), \end{aligned} \tag{17}$$

as well as

$$\delta^2 \approx \delta_p^2 = |1 - \|y_R\|^2/\|y\|^2|. \tag{18}$$

Corollary 1. In Moore balanced representation, with the diagonal entries of Γ in decreasing order, the condition

$$\sum_{i=k+1}^{n} \gamma_i^4 \ll \sum_{i=1}^{k} \gamma_i^4 \tag{19}$$

is necessary and sufficient for near-optimal reduction. However, the condition

$$\gamma_{k+1} \ll \gamma_k. \tag{20}$$

is sufficient but not necessary, for near-optimal reduction. Condition (20) is well-known from Moore [4] and is stronger than Condition 1, or Corollary 1.

Now, denote c_i the i-th column of C, b_i the i-th row of B, a_{ii} the i-th diagonal element of A, and define matrix Σ

$$\Sigma = \operatorname{diag}(\sigma_i), \quad \sigma_i = \alpha_i \gamma_i, \ i = 1, \ldots, n, \tag{21}$$

where $\alpha_i^2 = c_i^* c_i = b_i b_i^* = -2\gamma_i^2 a_{ii}$ (this equality follows from the Lyapunov equations for the controllability and observability grammians in Moore balanced representation). Denote $\Sigma = \operatorname{diag}(\Sigma_R, \Sigma_T)$, then in Moore balanced representation $\|y\|^2 = tr(\Sigma^2)$, $\|y_R\|^2 = tr\left(\Sigma_R^2\right)$, and $\|y\|^2 - \|y_R\|^2 = tr\left(\Sigma_T^2\right)$. Therefore, for the optimal reduction, according to (14), $\|y - y_R\|^2 = tr\left(\Sigma_T^2\right)$. If the solution is near-optimal, the latter equation holds approximately, and since Σ_T is known before reduction, it can serve for an *a priori* determination of the reduction error. This error is called the ***prediction error***, and denoted J_p or δ_p

$$J_p^2 = tr\left(\Sigma_T^2\right), \quad \delta_p^2 = tr\left(\Sigma_T^2\right)/tr(\Sigma^2). \tag{22}$$

In order to obtain a near-minimal reduction error, rearrange the coordinates such that the matrix Σ has its diagonal entries in decreasing order, i.e. $\sigma_i \geq \sigma_{i+1} > 0$, $i = 1, \ldots, n-1$. Thus the approximate value of J obtained from (22) is the smallest possible, since Σ_T contains the smallest diagonal entries of Σ. The above considerations are formalized as follows:

Condition 2. Let the system be in Moore balanced representation, and Σ as defined in (2) with its diagonal entries in decreasing order, then

$$\|\Sigma_T\| \ll \|\Sigma_R\| \tag{23}$$

is a necessary and sufficient condition for the near-optimal reduction.

Condition 2 is weaker than Condition 1, and therefore more precisely determines the near-optimal model. The fact that σ-evaluation of the system components for a model reduction gives better results than γ-evaluation was noted by Kabamba [18] and Skelton and Kabamba [19]. The following example shows the case where the reduction with the coordinate ranking, based on the Hankel singular values, γ_i, leads to a larger error than the ranking based on the reduction costs σ_i.

In the following example, the distance of the reduced model from the hypothetical optimal one is evaluated from the necessary condition of optimality [20], i.e. its output y_R is orthogonal to the error $y - y_R$. This condition can be written $< y_R, y - y_R >= 0$. If ϕ is the angle between the vectors y_R and $y - y_R$, then

$$\varepsilon = |\cos\phi| \tag{24}$$

is a measure of the distance of the reduced model from the optimal one. Surely, $\varepsilon = 0$ for the optimal reduced model, and $\varepsilon \ll 1$ for a near-optimal model. The measure ε is called the ***optimality index***, which is determined by

$$\begin{aligned}\cos\phi &=< y - y_R, y_R > /(\|y - y_R\| \, \|y_R\|) \\ &= \left(tr\,(C^* C_R W_{cRO}) - \|y_R\|^2\right)/(J\|y_R\|),\end{aligned} \tag{25}$$

where

$$\|y\|^2 = tr\,(C^* C W_c)\,, \quad \|y_R\|^2 = tr\,(C_R^* C_R W_{cR})\,. \tag{26}$$

Example 1. Consider the reduction in Moore balanced representation for the system given by the following triple

$$A = \begin{bmatrix} 0 & 0 & 0 & 1 & 0 & 0 \\ 0 & 0 & 0 & 0 & 1 & 0 \\ 0 & 0 & 0 & 0 & 0 & 1 \\ -5.4545 & 4.5454 & 0 & -0.0273 & 0.0227 & 0 \\ 10 & -21 & 11 & 0.05 & -0.105 & 0.055 \\ 0 & 5.5 & -6.5 & 0 & 0.0275 & -0.0325 \end{bmatrix},$$

$$B^T = [0\ 0\ 0\ 0.0909\ 0.4\ \ -0.5],\ C = [2\ \ -2\ 3\ 0\ 0\ 0],$$

Its reduction costs are $\sigma_1 = 1.9488$, $\sigma_2 = 1.9305$, $\sigma_3 = 0.4494$, $\sigma_4 = 0.4369$, $\sigma_5 = 0.3282$, $\sigma_6 = 0.2932$, and the related Hankel singular values are $\gamma_1 = 5.6089$, $\gamma_2 = 5.5967$, $\gamma_3 = 1.0998$, $\gamma_4 = 1.1139$, $\gamma_5 = 1.3447$, $\gamma_6 = 1.3365$. It is seen that the Hankel singular values are not in decreasing order. The reduction of the system by deleting the two state variables corresponding to the smallest two Hankel singular values (γ_3 and γ_4) gives the reduction index $\delta = 0.2208$, with the optimality index $\varepsilon = 0.7708\ 10^{-3}$, and the prediction index $\delta_p = 0.2201$. The reduction by deleting the two state variables corresponding to the two smallest reduction costs (σ_5 and σ_6) gives the reduction index $\delta = 0.1544$, the optimality index in this case $\varepsilon = 0.1267\ 10^{-3}$, and the prediction index $\delta_p = 0.1545$. The index ε and the index δ show that the latter solution is closer to the optimal one. Table 1 shows the reduction index δ and the optimality index ε versus the number k of state variables of the reduced model. The comparatively large values of ε for the odd k are observed. This fact can be explained by the following near-optimality condition.

Table 1

k	δ	ε
1	1.2070	0.5601
2	0.2689	$0.7116\ 10^{-6}$
3	0.3187	0.08707
4	0.1544	$0.1267\ 10^{-3}$
5	0.1865	0.06519

Conditions 3. Let the system be in Moore balanced representation, then one of the following

$$\|A_{RT}\|^2 \ll \|A_T\| \, \|A_R\|, \text{ and } \|A_{TR}\|^2 \ll \|A_T\| \, \|A_R\|, \tag{27a}$$

$$\|B_T\| \ll \|B_R\|, \tag{27b}$$

$$\|C_T\| \ll \|C_R\|, \tag{27c}$$

is a necessary and sufficient condition for a near optimal reduction.

Condition 3 shows that a block-diagonal matrix A, or zero truncated parts of B and C, make the reduction optimal in Moore balanced coordinates. It gives a clue for the near-optimal reduction of systems with complex poles. For a system with complex poles, the ranking of every single coordinate based on the value of σ_i (such that $\sigma_i \geq \sigma_{i+1}$) does not lead to a near-optimal result, since its corresponding matrix A is block-diagonally dominant (with 2×2 blocks). Every pair of coordinates, rather than every single coordinate, must be evaluated simultaneously. If absolute values of the real parts of the poles are small, the following evaluation is applied. In this case, note that the singular values γ_i related to every 2×2 block (i.e., to two complex conjugate poles) are almost equal (for $\lambda_i = \lambda_j^*$, one obtains $\gamma_i \approx \gamma_j$). First, put γ_i in Γ into decreasing order, and evaluate σ_i for each γ_i from (22). Then compute

$$\sigma_i = 0.5\,(\sigma_{2i-1} + \sigma_{2i}), \quad \sigma_{i+1} = \sigma_i, \quad i = 1, \ldots, n/2 \tag{28}$$

and finally rearrange σ_i into the decreasing order.

Example 2. Consider the system from Example 1 with $C = B^T$. For this case one obtains Σ = diag (0.6457, 0.5104, 0.007405, 0.005630, 0.005485, 0.000150), Γ = diag (1.6269, 0.9982, 0.2911, 1.6270, 0.9980, 0.2912). The reduction of the two state variables corresponding to the two smallest σ_i, namely σ_5 and σ_6 leads to comparatively large indices $\varepsilon = 0.4379$, and $\delta = 0.8758$. This solution cannot be considered near-optimal. The reason is that the matrix A in the balanced representation is not diagonally dominant, but block-diagonally dominant:

$$A = \begin{bmatrix} 0.0000 & -2.4374 & 0.0011 & 0.0000 & 0.0000 & 0.0000 \\ 2.4374 & -0.0298 & 0.0557 & 0.0013 & 0.0001 & -0.0039 \\ -0.0011 & 0.0557 & -0.1312 & -5.1238 & -0.0002 & 0.0143 \\ 0.0000 & -0.0013 & 5.1238 & 0.0000 & 0.0000 & 0.0001 \\ 0.0000 & 0.0001 & -0.0002 & 0.0000 & 0.0000 & 0.8738 \\ 0.0000 & 0.0039 & -0.0143 & 0.0001 & -0.8738 & -0.0038 \end{bmatrix}.$$

At the next step, instead of ranking every single coordinate, we rank every pair of variables, according to the procedure given above. This requires an rearrangement of the state variables in the above representation according to the permutation matrix [4 1 2 5 6 3]. Deleting now the last two variables, (related to σ_6 and σ_3) the optimality indices $\varepsilon = 0.9739\ 10^{-6}$, and $\delta = 0.0090$ are obtained. The obtained solution is considered close to the optimal one. The prediction error is $\delta_p = 0.008998$.

In Schur balanced representation matrix A is in block-triangular form. Lower and upper Schur balanced forms are distinguished. In the first case, with matrix A lower-triangular, $A_{RT} = 0$ so that (12) may be simplified to

$$J^2 = tr\left(C_T^* C_T W_{cT}\right). \tag{29a}$$

In the second case with A upper block-triangular, $A_{TR} = 0$, consequently (12) becomes

$$J^2 = tr\left(B_T B_T^* W_{oT}\right). \tag{29b}$$

and

$$J^2 = -tr\left(\left(A_T + A_T^*\right) W_{oT} W_{cT}\right) = -tr\left(\left(A_T + A_T^*\right) W_{cT} W_{oT}\right). \tag{29c}$$

The importance of Schur balanced representation is readily seen from (29), since the indices ε and J are evaluated before reduction. For other balanced representations J is predicted as J_p, and the exact values of ε and J are determined *a posteriori* after reduction. Furthermore, since $A_{RT} = 0$ (or $A_{TR} = 0$), there is no pole shifting in the reduced model, i.e., the set of poles of the reduced model is a subset of the poles of the original one. For Schur balanced representation the following condition applies:

Condition 4. In lower Schur form, the conditions

$$\|A_{TR}\|^2 \ll \|A_T\|\ \|A_R\|, \text{ or } \|C_T\| \ll \|C\|, \tag{30a}$$

or in upper Schur form, the conditions

$$\|A_{RT}\|^2 \ll \|A_T\| \; \|A_R\|, \text{ or } \|B_T\| \ll \|B\|, \tag{30b}$$

are necessary and sufficient for near-optimal reduction.

In lower Schur form, each zero column of C indicates unobservable part of the system, whereas, in upper Schur form, each zero row of B indicates uncontrollable part of the system. Define $\varepsilon_c = \|C_T\|/\|C\|$, $\varepsilon_b = \|B_T\|/\|B\|$. Then ε_c, ε_b are indicators of near-optimality.

Consider now the upper-Hessenberg balanced form,

$$A = \begin{bmatrix} a_{11} & a_{12} & a_{13} & \dots & a_{1n} \\ a_{21} & a_{22} & a_{23} & \dots & a_{2n} \\ 0 & a_{32} & a_{33} & \dots & a_{3n} \\ 0 & 0 & a_{43} & \dots & a_{4n} \\ \dots & \dots & \dots & \dots & \dots \\ 0 & 0 & 0 & \dots & a_{nn} \end{bmatrix},$$

with the matrix A partitioned according to (9) so that $A_{TR} = \begin{bmatrix} 0 & a_{k+1,k} \\ 0 & 0 \end{bmatrix}$. Then from Condition 4, the following corollary is obtained.

Corollary 2. In upper-Hessenberg form the conditions

$$|a_{k+1,k}|^2 \ll \|A_R\| \; \|A_T\|, \tag{31a}$$

and, either

$$\|A_{RT}\|^2 \ll \|A_R\| \; \|A_T\| \tag{31b}$$

or

$$\|B_T\| \ll \|B\| \tag{31c}$$

are necessary and sufficient for the near-optimal reduction. Similar conclusions can be given for the lower Hessenberg form. The above corollary makes the reduction clues from [21] more specific, namely, not only $\|a_{k+1,k}\|$ must be small for near-optimal reduction, but also its lower counterpart A_{RT}, or the corresponding part B_T of B.

The following example gives the comparison of the balanced model reduction in Moore, Schur, and Hessenberg forms.

Example 3, from [21]. The system given by the following triple:

$$A = \begin{bmatrix} -.21053 & -.10526 & -.0007378 & 0 & .0706 & 0 \\ 1 & -.03537 & -.000118 & 0 & .0004 & 0 \\ 0 & 0 & 0 & 1 & 0 & 0 \\ 0 & 0 & -605.16 & -4.92 & 0 & 0 \\ 0 & 0 & 0 & 0 & 0 & 1 \\ 0 & 0 & 0 & 0 & -3906.25 & -12.5 \end{bmatrix},$$

$$B = \begin{bmatrix} -7.211 \\ -.05232 \\ 0 \\ 794.7 \\ 0 \\ -448.5 \end{bmatrix}, \quad C = \begin{bmatrix} 1 & 0 & 0 & .000334 & 0 & -.007728 \\ 0 & 1 & 0 & 0 & 0 & 0 \end{bmatrix},$$

is reduced from 6 to 4 state variables. The results are compared in Table 2.

Table 2

Form	δ	ε
Moore	$0.3679\ 10^{-6}$	$0.4054\ 10^{-8}$
Schur	0.006119	0.007797
Hessn.	0.04033	0.4649

The Table shows that the reduction in Moore and Schur forms is closer to the optimal solution than in Hessenberg form. For Schur representation, one obtains $\varepsilon_c = 0.1083$ and $\varepsilon_b = 0.10103$. The balanced Hessenberg form is

$$A_{bh} = \left[\begin{array}{cccc|cc} -0.0821 & -0.2977 & -0.0159 & -0.0585 & 0.0001 & 0.0144 \\ 0.3419 & -1.4002 & -16.091 & 7.3310 & -1.0583 & -3.3351 \\ 0.0000 & 17.892 & -0.6782 & 62.269 & 0.8062 & 2.7055 \\ 0.0000 & 0.0000 & 56.950 & -11.118 & 9.1326 & 9.1189 \\ \hline 0.0000 & 0.0000 & 0.0000 & -6.1402 & -0.4334 & -25.670 \\ 0.0000 & 0.0000 & 0.0000 & 0.0000 & 23.446 & -3.9540 \end{array}\right],$$

$$B_{bh} = \left[\begin{array}{c} 3.2471 \\ -3.7422 \\ 0.2124 \\ 0.8531 \\ \hline -0.0383 \\ -0.2029 \end{array}\right]$$

$$C_{bh} = \left[\begin{array}{cccc|cc} -0.8505 & 0.7360 & 0.4188 & 2.1342 & -0.2511 & -0.4779 \\ -3.1337 & -2.9258 & -0.2029 & -0.8254 & 0.0358 & 0.1996 \end{array}\right].$$

The 5,4-th entry of A_{bh}, and the norm of its counterpart are small in comparison to the other lower diagonal entries of A_{bh}. Defining $\varepsilon_h = (\|A_{RT}\| \, \|A_{TR}\|)/(\|A_R\| \, \|A_T\|)$ one finds $\varepsilon_h = 0.0270$, or if $\varepsilon_h = \|B_T\|/\|B\|$, one obtains $\varepsilon_h = 0.04190$.

B. NEAR-OPTIMAL REDUCTION IN MODAL COORDINATES

In modal coordinates, $A_{RT} = 0$, $A_{TR} = 0$ for the representation (A, B, C). Since $y = y_R + C_T x_T$, the error $y - y_R$ between the original and the reduced model is solely determined from the parameters of the truncated system, $y - y_R = C_T x_T$, and therefore its norm is

$$J^2 = \|y - y_R\|^2 = tr\left(C_T^* C_T W_{cT}\right), \tag{32a}$$

Similarly when the observability grammian is considered, the norm is

$$J^2 = tr\left(B_T B_T^* W_{oT}\right). \tag{32b}$$

or, (see [13])

$$J^2 = -tr\left(\left(A_T + A_T^*\right) W_{cT} W_{oT}\right) = -tr\left(\left(A_T + A_T^*\right) W_{oT} W_{cT}\right). \tag{32c}$$

The normalized reduction error δ, defined previously as $\delta = J/\|y\|$, can be approximately predicted from the formula

$$\delta_p^2 = tr\left(\Sigma_T^2\right)/tr\left(\Sigma^2\right),$$

The reduction in the modal coordinates does not shift the poles. The necessary and sufficient conditions are discussed in the following.

Condition 5. One of the following

$$\left.\begin{aligned} &\|C_T\| \ll \|C_R\| \text{ and } \|W_{oT}\| \ll \|W_{oR}\|, && (33a) \\ &\|B_T\| \ll \|B_R\| \text{ and } \|W_{cT}\| \ll \|W_{cR}\|, && (33b) \\ &\|W_{oT}W_{cT}\| \ll \|W_{oR}W_{cR}\| && (33c) \end{aligned}\right\}$$

is necessary and sufficient for the near-optimal reduction.

Denote $W = W_oW_c$, or $W = W_cW_o$, and let W be partitioned such that

$$W = \left[\begin{array}{c|c} W_R & W_{RT} \\ --- & --- \\ W_{TR} & W_T \end{array}\right].$$

Then the following necessary and sufficient conditions can be derived.

Condition 6. The conditions

$$\|W_{RT}\|^2 \ll \|W_R\|\ \|W_T\|, \quad \|W_{TR}\|^2 \ll \|W_R\|\ \|W_T\|, \tag{34a}$$

and

$$\|W_T\| \ll \|W_R\|, \tag{34b}$$

are necessary and sufficient for the near-optimal reduction in modal coordinates. The error is approximately determined from

$$J^2 \approx -2tr(Z_TW_T), \quad Z_T = 0.5\left(A_T + A_T^*\right). \tag{35}$$

Consider now a symmetric, or an orthogonally symmetric system [22 – 24]; its cross-grammian exists, and $W_{co}^2 = W_cW_o$. Let the cross-grammian be partitioned into the retained and truncated parts respectively

$$W_{co} = \left[\begin{array}{c|c} W_{coR} & W_{coRT} \\ ---- & ---- \\ W_{coTR} & W_{coT} \end{array}\right].$$

Then the following condition is valid.

Condition 7. The conditions

$$\|W_{coRT}\|^2 \ll \|W_{coR}\| \; \|W_{coT}\|, \; \|W_{coTR}\|^2 \ll \|W_{coR}\| \; \|W_{coT}\|, \qquad (36a)$$

and

$$\|W_{coT}\| \ll \|W_{coR}\|, \qquad (36b)$$

are necessary and sufficient for the near-optimal reduction of the symmetric systems in modal coordinates. The error is approximately determined from

$$J^2 \approx -2tr\left(Z_T W_{coT}^2\right), \; Z_T = 0.5\left(A_T + A_T^*\right). \qquad (37)$$

The broader class of near-optimal solutions can be obtained from the next condition. Denote

$$\Sigma_o^2 = -2ZW = -2ZW_cW_o, \text{ or } \Sigma_o^2 = C^*CW_c, \text{ or } \Sigma_o^2 = BB^*W_o, \qquad (38a)$$

and in modal coordinates $Z = Re(\Lambda) = \text{diag}(\zeta_i), i = 1, \ldots, n$, and $\zeta_i = Re(\lambda_i)$. Let σ_i be the i-th diagonal element of Σ_o, and

$$\Sigma = \text{diag}(\sigma_i), \; \sigma_i = 2\zeta_i w_i, \; i = 1, \ldots, n. \qquad (38b)$$

where ζ_i, w_i are diagonal entries of Z and W respectively. Reorder Σ and (A, B, C) such that $\sigma_i \geq \sigma_{i+1}$. Let $\Sigma = \text{diag}(\Sigma_R, \Sigma_T)$. Then the following condition is valid.

Condition 8. The conditions

$$\|W_{RT}\|^2 \ll \|W_R\| \; \|W_T\|, \; \|W_{TR}\|^2 \ll \|W_R\| \; \|W_T\|, \qquad (39a)$$

and

$$\|\Sigma_T\| \ll \|\Sigma_R\|, \qquad (39b)$$

are necessary and sufficient for the near-optimal reduction in modal coordinates.

Condition 8 is a weaker version of Conditions 6 and 7 and shows that small modal cost Σ_T and almost-diagonal grammians are necessary and sufficient to obtain near-optimal solution. This condition is analogous to Condition 2; however in modal coordinates, the weak correlation between state variables is demanded (in Moore balanced coordinates the state variables are uncorrelated by definition). The single condition $\|\Sigma_T\| \ll \|\Sigma_R\|$ is used by Skelton [27] and Skelton and Yousuff [3].

The procedure for choosing $n-k$ modes for reduction follows from Condition 8. Let Σ be defined by (38). Rearrange Σ such that its diagonal entries are in decreasing order, $\sigma_i \geq \sigma_{i+1}$, and accordingly rearrange the triple (A, B, C) in modal coordinates. Delete the last $n-k$ rows of A and B, and the last $n-k$ columns of A and C. For complex poles, the procedure does not need any modification, as in the balanced representation, since for a pair of complex-conjugate poles $\lambda_i = \lambda_j^*$ we have $\zeta_i = \zeta_j$, and $\gamma_i = \gamma_j$, so that $\sigma_i = \sigma_j$.

Now assume the system is partitioned into several subsystems. The question arises if the index J can be evaluated from the indices of each individual subsystem J_i, and if the obtained reduction is near-optimal. In general, the answer is negative, since the index J contains not only J_i, but also the cross-correlations of the subsystems. Let the truncated part of A be in block-diagonal form $A_T = \mathrm{diag}(A_{T1}, \ldots, A_{T\alpha})$, and B_T, C_T and grammians are partitioned accordingly $C_T = [C_{T1}, \ldots, C_{T\alpha}]$, $B_T^T = \left[B_{T1}^T, \ldots, B_{T\alpha}^T\right]$, and

$$W_{cT} = \begin{bmatrix} W_{cT11} & \cdots & W_{cT1\alpha} \\ \cdots & \cdots & \cdots \\ \cdots & \cdots & \cdots \\ W_{cT\alpha 1} & \cdots & W_{cT\alpha\alpha} \end{bmatrix}, W_{oT} = \begin{bmatrix} W_{oT11} & \cdots & W_{oT1\alpha} \\ \cdots & \cdots & \cdots \\ \cdots & \cdots & \cdots \\ W_{oT\alpha 1} & \cdots & W_{oT\alpha\alpha} \end{bmatrix},$$

$$W_T = \begin{bmatrix} W_{T11} & \cdots & W_{T1\alpha} \\ \cdots & \cdots & \cdots \\ \cdots & \cdots & \cdots \\ W_{T\alpha 1} & \cdots & W_{T\alpha\alpha} \end{bmatrix}$$

The index J is determined from (32)

$$\begin{aligned} J^2 &= \sum_{i,j=1}^{\alpha} tr\left(C_{Tj}^* C_{Ti} W_{cTij}\right) = \sum_{i,j=1}^{\alpha} tr\left(B_{Tj} B_{Ti}^* W_{oTij}\right) \\ &= -2 \sum_{i,j=1}^{\alpha} tr(z_{Ti} W_{cTij} W_{oTij}), \end{aligned} \tag{40}$$

where α is the number of subsystems deleted. If W is diagonally dominant, the index can be determined as a sum of indices of each truncated subsystem

$$J^2 = \sum_{i=1}^{\alpha} J_i^2, \tag{41}$$

where

$$\begin{aligned} J_i^2 &= tr\left(C_{Ti}^* C_{Ti} W_{cTii}\right) = tr\left(B_{Ti} B_{Ti}^* W_{oTii}\right) \\ &= -2tr\left(Z_{Ti} W_{cTii} W_{oTii}\right) = -2tr(Z_{Ti} W_{Tii}). \end{aligned} \tag{42}$$

The reduction error is close to minimal, if the subsystems with the smallest indices are truncated.

If A is simple and in modal form, so that $A = \Lambda = \text{diag}(\lambda_i)$, $\lambda_i = \zeta_i + j\omega_i$, then each subsystem A_i is the system pole λ_i itself. In this case the value of each mode is evaluated from (42) yielding, to

$$J_i^2 = -\left(|b_i|^2 |c_i|^2\right) \Big/ 2\zeta_i, \tag{43}$$

which is the modal cost obtained by Skelton and Yousuff [3]. Choosing modes with smallest J_i to be deleted, one obtains the reduced model close to the optimal one.

Example 4. The reduction in modal coordinates for the system from Example 1 is presented. Condition 8 is used for determination of the near-optimal reduction. The reduced model has four state variables and is shown as follows

$$A_R = \begin{bmatrix} -0.01909 & 0.8738 & 0 & 0 \\ -0.8738 & -0.01909 & 0 & 0 \\ 0 & 0 & -0.06562 & 5.1230 \\ 0 & 0 & -5.1230 & -0.06562 \end{bmatrix},$$

$$B_R = \begin{bmatrix} -0.8613\ 10^{-5} \\ 0.1081 \\ -0.00477 \\ 0.3085 \end{bmatrix},$$

$$C_R = [-2.2182 \ \ -0.1768\ 10^{-3} \ \ -1.0427 \ \ -0.01613],$$

with the following optimality indices $\varepsilon = 0.1911\ 10^{-3}$, the normalized error $\delta = 0.1538$, and its prediction $\delta_p = 0.1539$. Comparing the this example with Example 1, it is seen that in both cases the solutions are close to the optimal one, and that the modal reduced model is closer to the optimal model than the balanced one.

Example 5. The reduction in modal coordinates for the system from Example 4 is considered, and the reduced model is supposed to have four state variables. The reduced model is

$$A_R = \begin{bmatrix} -0.01485 & 2.4375 & 0 & 0 \\ -2.4375 & -0.01485 & 0 & 0 \\ 0 & 0 & -0.06562 & 5.1230 \\ 0 & 0 & -5.1230 & -0.06562 \end{bmatrix},$$

$$B_R = \begin{bmatrix} -0.00142 \\ 0.3975 \\ -0.00477 \\ 0.30852 \end{bmatrix},$$

$$C_R = [-0.00383 \; 0.3985 \; -0.02397 \; 0.8476],$$

with the following optimality indices $\varepsilon = 0.3683 \; 10^{-4}$, the normalized error $\delta = 0.008998$, and its prediction $\delta_p = 0.008986$. Comparing this example with Example 4 one can see that both solutions are very close to the optimal one. Note that in this case the balanced representation gave the results slightly closer the optimal ones than the modal representation.

Finally, we compare the near-optimal results with the optimal results of Wilson [10] and Hyland-Bernstein [11]. In Table 3 we compare the results from Example 6.2 of Wilson [10] with the near-optimal balanced results. The Examples 6.1 and 6.2 of Hyland and Bernstein [11] are compared here, and the results are shown in Tables 3 and 4. Both examples show that the reduced model obtained in modal or balanced coordinates is close to the optimal one obtained by Hyland and Bernstein. Note, however, that the optimal solution in Example 6.2 from Hyland and Bernstein [11] needs 14 iterations, while the near-optimal results are obtained non-iteratively.

Table 3

Order k	δ [11]	δ [9]	δ bal	ε [9]	ε bal
1	0.4268	–	0.4321	–	0.01588
2	0.03929	0.04096	0.03938	0.003163	0.000641
3	0.001306	–	0.001311	–	0.000120

Table 4

	δ	ε
Hyland [11]	0.0975329	–
Balanced	0.0975333	0.00200
Modal	0.0975333	0.00196

IV. REDUCTION OF FLEXIBLE STRUCTURE MODELS

A flexible structure is a dynamic system having complex poles with real part of the poles small in comparison to their imaginary part. The near-optimal conditions presented previously can be applied and specified to flexible structures. Later in this section, the model reduction is considered using Hankel singular values or component cost determined from test data rather than from the analytical model. If the test data include actuator and sensor dynamics along with structural dynamics, the Hankel singular values or component costs of the structure are determined from the combined dynamics. Finally, a reduction technique which is robust to the structural damping variations is presented.

A. NEAR-OPTIMALITY CONDITIONS

Consider an n-mode model for a flexible structure, modally damped, non-gyroscopic, non-circulatory with p actuators and q sensors,

$$\ddot{\eta} + \mathrm{diag}(2\zeta_i\omega_i)\dot{\eta} + \mathrm{diag}\left(\omega_i^2\right)\eta = \hat{B}u, \tag{44a}$$

$$y = \hat{C}_r\dot{\eta} + \hat{C}_d\eta, \tag{44b}$$

where η is the vector of modal coordinates, and ω_i and ζ_i are the natural frequency and damping ratio of the i-th mode, respectively. For a typical flexible structure, $\{\zeta_i\}$ are quite low (e.g. 0.005), and the $\{\omega_i\}$ occur in clusters of repeated, or nearly-repeated, frequencies. Note that all ω_i and ζ_i are positive. Defining the state vector $x = (\dot{\eta}_1, \omega_1\eta_1, \ldots, \dot{\eta}_n, \omega_n\eta_n)^T$ yields the state space representation (A, B, C) where $A = \mathrm{diag}(A_i)$, $B =$

$\left(B_1^T, \ldots, B_n^T\right)^T$ and $C = (C_1, \ldots, C_n)$, with

$$A_i = \begin{pmatrix} -2\zeta_i\omega_i & -\omega_i \\ \omega_i & 0 \end{pmatrix}, \; B_i = \begin{pmatrix} b_i \\ 0 \end{pmatrix}, \text{ and } C_i = (c_{ri}, \; c_{di}/\omega_i). \tag{45}$$

Here, b_i is the ith row of $\hat{B}$ and c_{ri} and c_{di} are the ith columns of $\hat{C}_r$ and $\hat{C}_d$, respectively.

The problem is to obtain a reduced-order model $\dot{x}_r = A_r x_r + B_r u$, $y_r = C_r x_r$ for this structure for which the output error (11) is as small as possible for the specified model order n_r. In order to specify the conditions for a near-optimal reduction, one takes advantages of the close form grammians derived in [25,26,15], which provides

$$W_{cij} = \begin{pmatrix} 2\omega_i\omega_j(\zeta_j\omega_i + \zeta_i\omega_j) & \omega_j\left(\omega_j^2 - \omega_i^2\right) \\ -\omega_i\left(\omega_j^2 - \omega_i^2\right) & 2\omega_j\omega_j\left(\zeta_i\omega_i + \zeta_j\omega_j\right) \end{pmatrix} \times \beta_{ij}/d_{ij}, \tag{46}$$

where $\beta_{ij} = b_i b_j^T$ and $d_{ij} = 4\omega_i\omega_j(\zeta_i\omega_i + \zeta_j\omega_j)(\zeta_j\omega_i + \zeta_i\omega_j) + \left(\omega_j^2 - \omega_i^2\right)^2$. The quantity d_{ij}^{-1} is essentially a measure of how closely correlated modes i and j are. It can be shown to be the square of the amplification factor between the input $\exp(\zeta_j\omega_j t)\cos\left(\omega_j\sqrt{\left(1 - \zeta_j^2\right)}t\right)$, which excites mode j to infinity, and mode i.

The general expression for W_{cij} can be simplified considerably for exactly repeated frequencies. In that case, one obtains

$$W_{cij} = I_2 \times \beta_{ij}/2(\zeta_i + \zeta_j)\omega_i \; ; \tag{47}$$

in particular, the diagonal blocks are just $W_{cii} = I_2 \times \beta_{ii}/4\zeta_i\omega_i$. Similar simplifications occur for widely separated frequencies and $\zeta_i, \; \zeta_j \to 0$. In this case,

$$W_{cij} \to \begin{pmatrix} 0 & \omega_j \\ -\omega_i & 0 \end{pmatrix} \times \beta_{ij}/\left(\omega_j^2 - \omega_i^2\right). \tag{48}$$

Thus, W_c for a structure with light damping and all frequencies widely separated has i^{th} diagonal block proportional to $1/\zeta_i$ and off-diagonal blocks independent of $\{\zeta_i\}$, and so tends to diagonal form for $\{\zeta_i\} \to 0$. Although this asymptotic result is well-known [5,6,27], what is important to realize is that it does not apply for the case of repeated (Eq. (47)) or nearly-repeated natural frequencies.

The observability grammian W_o for a system with *rate measurements* only can be obtained in a similar fashion, or more simply by noting that $A^T = PAP$, where $P = \text{diag}(1, -1, \ldots, 1, -1)$. Pre- and post-multiplying Eq. (4) by P therefore gives

$$A(PW_oP) + (PW_oP)A^T + C^TC = 0, \tag{49}$$

where the equality $CP = C$ is used. Thus, W_o is essentially as given by (46), the only alterations being that the signs of the off-diagonal entries are changed and β_{ij} is replaced by $\gamma_{rij} = c_{ri}^T c_{rj}$.

If displacement measurements are also allowed the situation is much less simple. In fact, the analytical expressions that then result for W_o are really too complicated to be useful. The only exception is the expression for the ith diagonal block of W_o for a lightly-damped structure ($\zeta_i \ll 1$), where we have the approximation

$$W_{oii} \approx I_2 \times \left(\omega_i^2 \gamma_{rii} + \gamma_{dii}\right) / 4\zeta_i\omega_i^3, \tag{50}$$

with $\gamma_{dii} = c_{di}^T c_{di}$. Although no general analytical expressions for W_o are viable, it is possible to derive a "semi-closed-form" method to evaluate this matrix which exploits the special form of (A, B, C) in (45). Let

$$W_{oij} = \begin{pmatrix} p & q \\ r & s \end{pmatrix} \tag{51}$$

be the $(i,j)^{th}$ block of W_o. Then its Lyapunov equation $A_i^T W_{oij} + W_{oij}A_j + C_i^T C_j = 0$ can be expanded and rewritten as the system of linear equations

$$\begin{pmatrix} -2(\zeta_i\omega_i + \zeta_j\omega_j) & \omega_j & \omega_i & 0 \\ -\omega_j & -2\zeta_i\omega_i & 0 & \omega_i \\ -\omega_i & 0 & -2\zeta_j\omega_j & \omega_j \\ 0 & -\omega_i & -\omega_j & 0 \end{pmatrix} \begin{pmatrix} p \\ q \\ r \\ s \end{pmatrix} = - \begin{pmatrix} c_{ri}c_{rj} \\ c_{ri}^T c_{dj}/\omega_j \\ c_{di}^T c_{rj}/\omega_i \\ c_{di}^T c_{dj}/\omega_i\omega_j \end{pmatrix}. \tag{52}$$

It is interesting to note that the determinant of the matrix in (52) is just d_{ij}, so this quantity plays a similar role in both the controllability and observability grammians.

Just as is true for W_c, the observability grammian of a lightly-damped structure with widely separated natural frequencies can be shown to be block diagonally dominant. Thus, the Hankel singular values $\{\gamma_i\}$ for such a structure are approximately given as

$$\gamma_i^4 \approx \|W_{cii}\|_2 \, \|W_{oii}\|_2 = \beta_{ii}\left(\omega_i^2\gamma_{rii} + \gamma_{dii}\right) / \left(4\zeta_i\omega_i^2\right)^2, \tag{53}$$

a result given by Gregory [5]. This expression can be used to obtain $\{\sigma_i^2\}$ for the structure directly, which are given as the diagonal elements of $-2AW_cW_o \approx -2A\text{diag}\left(\gamma_i^4\right)$, So from (45) and (53) we have the two values

$$\sigma_{i1}^2 = 4\zeta_i\omega_i\gamma_i^4 = \beta_{ii}\left(\omega_i^2\gamma_{rii} + \gamma_{dii}\right) / \left(4\zeta_i\omega_i^3\right), \; \sigma_{i2}^2 = 0 \tag{54}$$

for the i-th balanced mode. Note that σ_{i1}^2 is precisely the i-th modal cost of Skelton [1]. In practice, the idealization (54) will not hold exactly. Rather than σ_{i2} being precisely zero we will have one large and one small value for each mode, with their mean $\hat{\sigma}_i$ constituting a good measure of overall modal cost. Again, it is important to remember that the approximations (53) and (54) are based on an implicit assumption of widely separated natural frequencies. For structures with clustered modes, it is necessary to evaluate the off-diagonal blocks of W_c and W_o as well as the diagonal blocks considered above. This will be the subject of the further considerations.

Finally, if $p \geq m$ and there exists a matrix U with orthonormal columns satisfying $C = UB^TP$, then the system governed by (45) is said to be ***orthogonally symmetric*** [23,24] and its cross-grammian W_{co} is defined as the solution of the Lyapunov equation

$$AW_{co} + W_{co}A + BU^TC = 0. \tag{55}$$

Note that any flexible structure with *compatible*, i.e., physically collocated and coaxial, actuators and rate sensors is necessarily an orthogonally symmetric system. The usefulness of W_{co} for balancing and model reduction applications lies in the fact that it satisfies the relation $W_{co}^2 = W_cW_o$, so the $\{\gamma_i^2\}$ are just the absolute values of the eigenvalues of W_{co}. In fact, as $C^TC = PBU^TUB^TP = BB^T$ and $BU^TC = BU^TUB^TP = BB^T$, Eqs. (49) and (55) can be seen to reduce to the expressions $W_{co} = W_cP = PW_o$ [23]. Thus, all three grammians of an orthogonally symmetric system are given directly from (46) with suitable changes of sign, noting of course that $\gamma_{rij} = \beta_{ij}$ and $\gamma_{dij} = 0$ for such systems.

Conditions for near-optimal reduction of flexible structures are considered in both modal and balanced coordinates. Consider first modal truncation applied to an orthogonally symmetric structural model. From Condition 7, it follows that $\delta \ll 1$ if each (i,j) block W_{ij} of the cross-grammian of this modal representation, corresponding to a retained mode (i) and deleted mode (j), satisfies

$$\rho_{ij}^2 = \|W_{ij}\|^2/(\|W_{ii}\| \; \|W_{jj}\|) \ll 1. \tag{56}$$

Now, from (46), $W_{ij} = \beta_{ij}\tilde{W}_{ij}$ for $\tilde{W}_{ij}$ purely a function of $\omega_i, \omega_j, \zeta_i$, and ζ_j, so the correlation coefficient ρ_{ij} can be written as a product

$$\rho_{ij} = \kappa_{ij}\tilde{\rho}_{ij}, \tag{57}$$

where, from the Schwartz inequality, $\kappa_{ij}^2 = \beta_{ij}^2/(\beta_{ii}\beta_{jj}) \leq 1$, and $\tilde{\rho}_{ij}^2 = \|\tilde{W}_{ij}\|^2/\left(\|\tilde{W}_{ii}\| \; \|\tilde{W}_{jj}\|\right)$. The condition $\rho_{ij} \ll 1$ is now clearly

$$\kappa_{ij}\tilde{\rho}_{ij} \ll 1, \; i \neq j. \tag{58}$$

Note that $\tilde{\rho}_{ij}$ depends purely on the structural properties ζ_i, ζ_j and ω_i, ω_j rather than sensor/actuator locations. It reflects the relationship between structural parameters and the optimality of reduction. Since $0 \leq \kappa_{ij} \leq 1$, the condition $\tilde{\rho}_{ij} \ll 1$ certainly implies $\rho_{ij} \ll 1$, but not *vice versa.* By contrast, the coefficient κ_{ij} depends only on sensor/actuator locations. It can be seen that a near-optimal reduction will occur if the ith and j^{th} rows of B (columns of C) are nearly orthogonal for all retained modes i and deleted modes j. Thus, Eq. (58) decouples the contributions toward the optimality of reduction of the structural properties $(\tilde{\rho}_{ij})$ and sensor/actuator locations (κ_{ij}) in a very simple way.

We now study how the structural parameters ζ_i and ω_i influence the optimality index. Introducing dimensionless variables $\alpha = \omega_i/\omega_j, v = \zeta_i/\zeta_j$, denoting for simplicity $\omega = \omega_j$, $\zeta = \zeta_j$, and introducing $\tilde{W}_{ij}, \tilde{W}_{ii}, \tilde{W}_{jj}$ from (57) and (58) one obtains

$$\tilde{\rho}_{ij}^2 = n_{ij}/m_{ij}^2, \tag{59a}$$

where

$$m_{ij} = 4\zeta^2\alpha(\alpha + v)(1 + \alpha v) + (1 - \alpha^2)^2 \tag{59b}$$

and

$$n_{ij} = 8\zeta^2\alpha v\{4\zeta^2\alpha^2[(\alpha + v)^2 + (1 + \alpha v)^2] + (1 - \alpha^2)^2(1 + \alpha^2)\}. \tag{59c}$$

Plots of $\tilde{\rho}_{ij}$ are shown in Fig. 1 for the special case $\zeta_i = \zeta_j = \zeta$, i.e. $v = 1$. The curves show that the condition of Eq. (58) is satisfied for small ζ and widely-separated natural frequencies ($\alpha \neq 1$). In fact, the smaller the damping and the more separated the natural frequencies of the retained and deleted modes, the closer the reduction is to the optimal one. This confirms the well-known result of [5] and [6] that modal coordinates are approximately balanced, and so uncorrelated [4], for such a structure. The importance of these graphs is that they apply for any structure. Once its $\{\zeta_i\}$ and $\{\omega_i\}$ are known, the $\{\tilde{\rho}_{ij}\}$ can simply be read off, giving important modal coupling information directly.

Consider now a general, rather than orthogonally symmetric, structural model. First form the product of the controllability and observability grammians, $W = W_c W_o$. Now, block-diagonal dominance of W is a necessary and sufficient condition for near-optimal reduction using modal truncation, see Condition 6. However, since it is difficult to determine the product of grammians in closed form, we use for simplicity a sufficient condition for near-optimality, namely the block-diagonal dominance of both W_c and W_o. Defining the controllability and observability correlation coefficients (Eq. (56)),

$$\rho_{cij}^2 = \|W_{cij}\|^2 / \|W_{cii}\| \; \|W_{cjj}\| \tag{60a}$$

and

$$\rho_{oij}^2 = \|W_{oij}\|^2 / \|W_{oii}\| \; \|W_{ojj}\|, \tag{60b}$$

we can write for near-optimal reduction that $\rho_{cij} \ll 1$ and $\rho_{oij} \ll 1$, $i \neq j$. Thus we have (Eq. (58)),

$$\rho_{cij} = \kappa_{cij}\tilde{\rho}_{cij} \ll 1, \;\; \rho_{oij} = \kappa_{oij}\tilde{\rho}_{oij} \ll 1, \tag{61}$$

where $\kappa_{cij} = \kappa_{ij}$ and $\kappa_{oij} = \|c_i^T c_j\| / (\|c_i\| \; \|c_j\|)$; $\tilde{\rho}_{cij}^2 = \|\tilde{W}_{cij}\| / (\|\tilde{W}_{cii}\| \; \|\tilde{W}_{cjj}\|)$ and $\tilde{\rho}_{oij}^2 = \|\tilde{W}_{oij}\|^2 / (\|\tilde{W}_{oii}\| \; \|\tilde{W}_{ojj}\|)$, with $\tilde{W}_{cij} = W_{cij}/\beta_{ij}$ and $\tilde{W}_{oij} = W_{oij}/\|c_i^T c_j\|$.

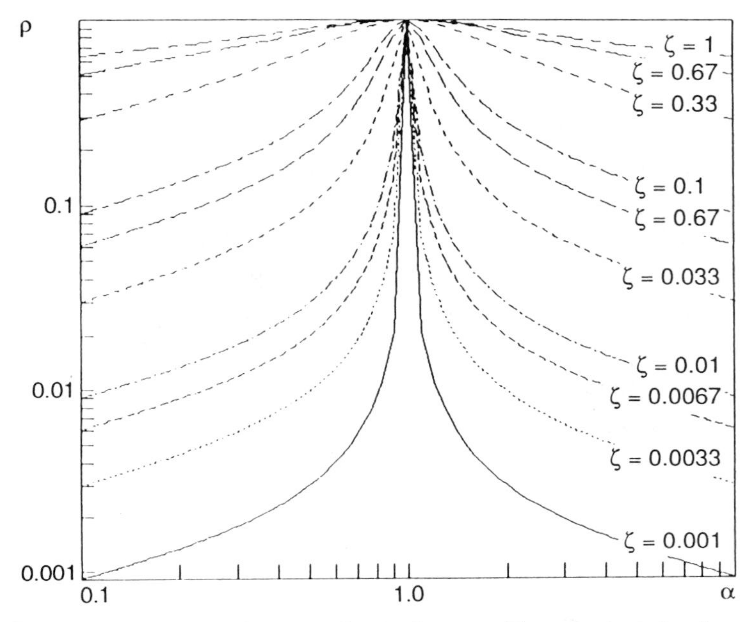

Figure 1. Modal correlations for orthogonally symmetric structure.

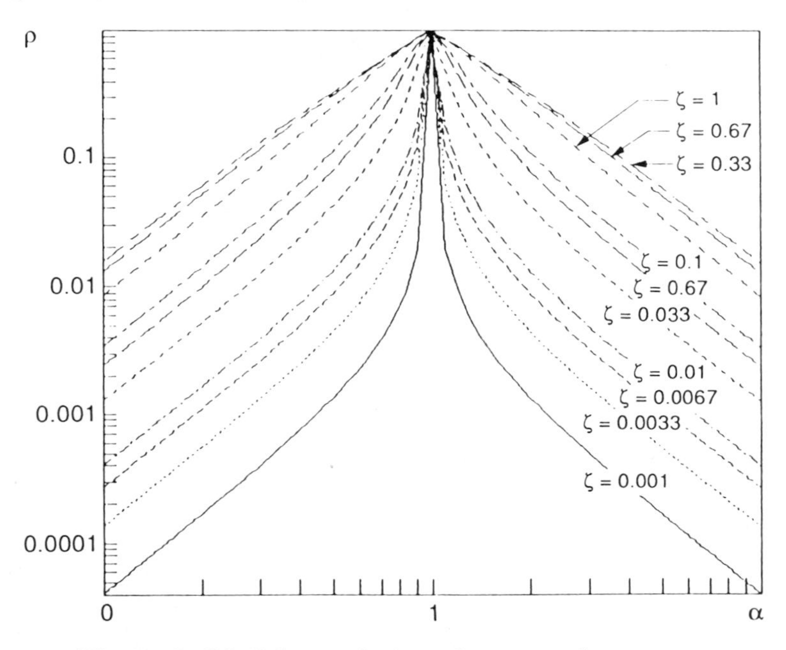

Figure 2. Modal correlations for general structure, displacement outputs.

As was shown before the controllability grammian is essentially the same as the cross-grammian in this case, up to the signs of its off-diagonal terms. Similar comments apply to the observability grammian if only rate measurements are used. Thus, the near-optimality conditions for the rate measurement case are the same as those obtained for orthogonally symmetric structures using the cross-grammian. If displacement measurements are taken though, the added complexity of the observability grammian (Eq. (52)) results in somewhat different values for the observability correlation coefficients. However, even now the same basic result applies that closely-spaced poles result in correlated modes, especially for heavy damping. This is illustrated by the plot of $\tilde{\rho}_{oij}$ in Fig. 2 for the case of displacement measurements alone, and the combined rate and displacement output example $\left(\text{for } c_{di}^T c_{dj} = c_{ri}^T c_{rj} = c_{di}^T c_{rj} = 1 \text{ and } \omega_i = 1\right)$ plotted in Fig. 3. Of course, (61) implies that near-optimal reduction can still be obtained, even for non-negligible $\tilde{\rho}_{cij}$ and $\tilde{\rho}_{oij}$, so long as κ_{cij} and κ_{oij} are small enough.

These correlation results, confirming the well-known results of Jonckherre [6] and Gregory [5], show that model reduction using either balancing or modal truncation should yield approximately the same results for structures with light damping and sufficiently separated poles. This is confirmed by the following two examples.

Example 6. The flexible system from Fig. 4 is investigated. The system has 3 degrees of freedom (d.o.f.), 2 inputs and 2 outputs, the parameters $m_1 = m_2 = m_3 = 10$, $k_1 = k_4 = 1000$, $k_2 = 500$, $k_3 = 100$, $d_i = dk_i$, $i = 1, \dots, 4$, and

$$B^T = \begin{pmatrix} 0 & 0 & 0 & 0.1 & 0.2 & -0.5 \\ 0 & 0 & 0 & -0.2 & -0.3 & 0.1 \end{pmatrix}, \; C = \begin{pmatrix} 2 & -2 & 3 & 1 & -5 & -2 \\ 1 & 3 & -1 & 2 & -2 & 1 \end{pmatrix}.$$

A near-optimal reduced system with 2 d.o.f. is desired, using modal truncation.

The system poles for $d = 0.005$, $\lambda_{1,2} = -0.01823 \pm j6.0492$, $\lambda_{3,4} = -0.05544 \pm j10.529$ and $\lambda_{5,6} = -0.08627 \pm j13.135$, are widely separated. The system grammians are diagonally dominant, as expected, and the matrix $\Sigma = \operatorname{diag}(\sigma_1, \dots, \sigma_6)$ is $\Sigma = \operatorname{diag}(5.5026,\ 3.8682,\ 4.8734,\ 0.1287,\ 0.0007,\ 0.1453)$. Deleting the last two rows of A and B, and the last columns of A and C, the cost of which is the smallest, we obtain the near-optimal reduced model. The

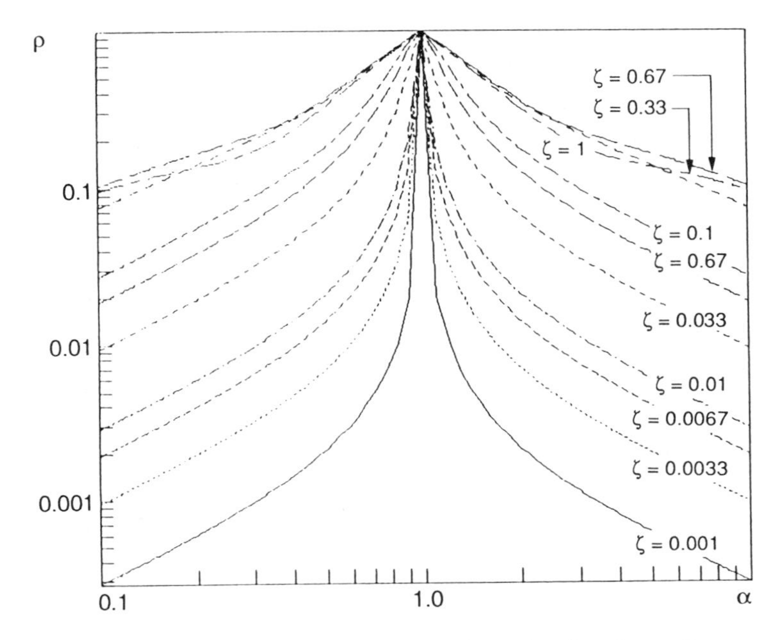

Figure 3. Modal correlations for mixed rate and displacement outputs.

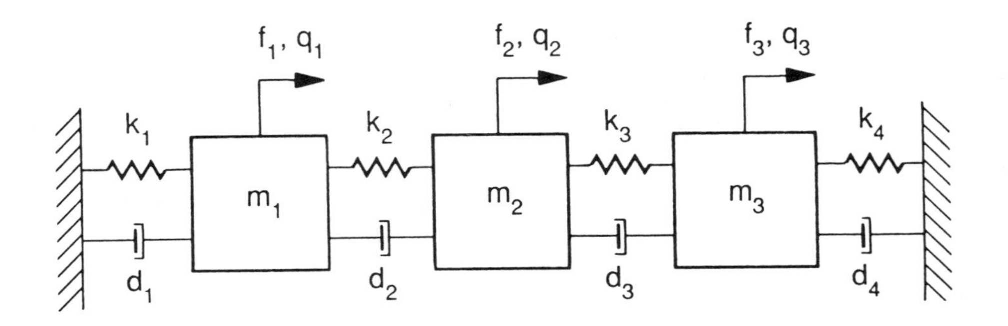

Figure 4 Simple three-spring-dashpot system.

optimality index of the obtained model is indeed small, $\varepsilon = 0.001129$, and the reduction error is $\delta = 0.05767$. The solid line in Fig. 5 shows the dependence of the optimality index ε and the reduction error δ on the damping parameter d. As expected, the smaller the damping, the closer the reduced model is to optimal.

Example 7. The same system is now reduced in balanced coordinates. The matrices Γ and Σ are $\Gamma = \text{diag}(4.80970,\ 4.80948,\ 2.17812,\ 2.17647,\ 0.80842,\ 0.80817)$ and $\Sigma = \text{diag}(5.28370,\ 3.36072,\ 0.16177, 2.22386,\ 0.38322,\ 0.02391)$. A near-optimal model in balanced coordinates is obtained by deleting the last two columns of the transformed A and C, and the last two rows of A and B, corresponding to the smallest cost. The poles of the reduced balanced model are $\lambda_{r1,2} = -0.01829 \pm j6.0492$, $\lambda_{r3,4} = -0.0554 \pm j10.529$. The obtained model is near-optimal, since its optimality index is $\varepsilon = 0.67577\ 10^{-5}$, and the reduction error is $\delta = 0.057661$. Fig. 5, dotted lines, shows the dependence of ε and δ on the damping parameter d. From the plot of ε it can be seen that, the smaller the damping is, the closer the solution is to the optimal one. For high damping the output error δ, by contrast, falls off, indicating that one of the balanced modes becomes nearly unobservable. Note that the results obtained for low levels of damping are very close to those obtained in the last Example using modal coordinates. By contrast, for high damping the modal correlations increase to the point where balancing should be expected to give significantly better results than modal, which is confirmed by Fig. 5.

Balanced reduction should also give a reduced-order model that is closer to optimal than that produced by modal truncation, even for light damping, if the structure considered has closely-spaced poles. The following pair of examples confirms that this is in fact true.

Example 8. The system from Fig. 4 is again considered, but now with the following parameters: $m_1 = m_3 = 1$, $m_2 = 2$, $k_1 = k_4 = 50$, $k_2 = k_3 = 5$, $d_i = 0.01k_i$, $i = 1, \ldots, 4$, and

$$B^T = [0\ 0\ 0\ 1\ 0\ 0], \quad C = [1\ 2\ 2\ 0\ 0\ 0].$$

This system is reduced from 3 to 2 d.o.f. The system poles are $\lambda_{1,2} = -0.02252 \pm j2.1224$, $\lambda_{3,4} = -0.27500 \pm j7.4111$, and $\lambda_{5,6} = -0.27748 \pm j7.4443$. The second and the third pair of poles are clearly closely spaced. The reduced

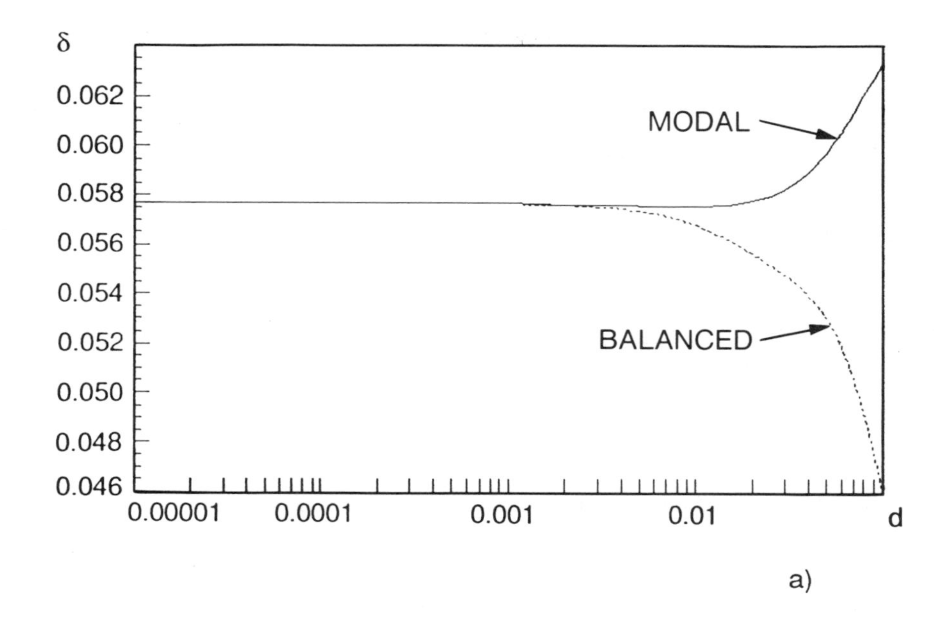

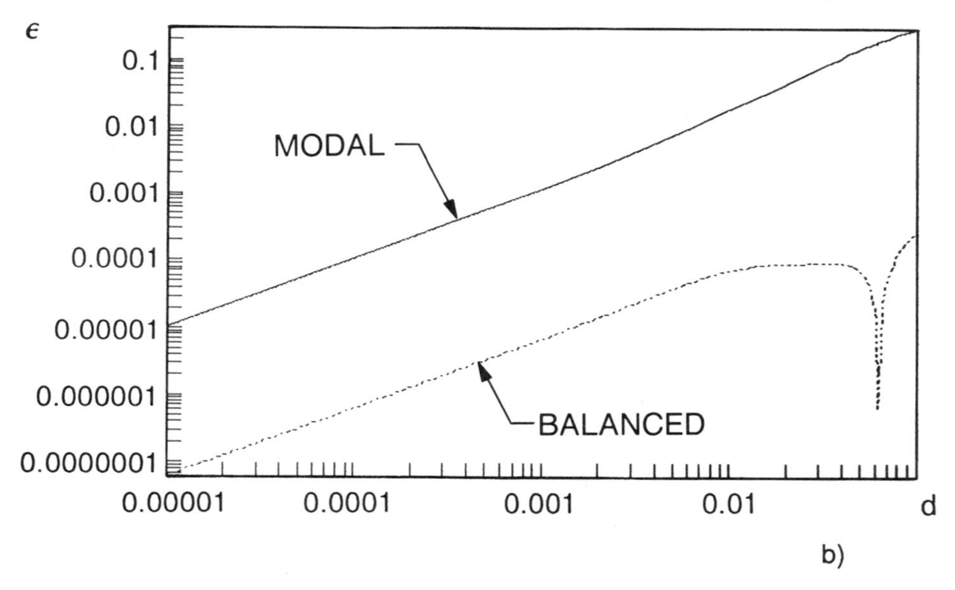

Figure 5. Effect of damping on model reduction accuracy.

model using modal truncation are:

$$A_r = \begin{pmatrix} -0.0225 & 2.1224 & 0 & 0 \\ -2.1224 & -0.0225 & 0 & 0 \\ 0 & 0 & -0.2774 & 7.4443 \\ 0 & 0 & -7.4443 & -0.2774 \end{pmatrix}, B_r = \begin{pmatrix} -0.0002 \\ 0.0495 \\ -0.1580 \\ 0.6808 \end{pmatrix}$$

and $C_r = [1.0719\ 0.0041\ 0.2598\ 0.0603]$, with $\varepsilon = 0.7050$ and $\delta = 0.30634$. In this case the result is not near-optimal, but is the best one can obtain in modal coordinates. Deleting any other mode would lead to an even larger reduction error. The source of the difficulty is, of course, the close second and third pair of poles. The correlation coefficients for these two modes, obtained from (56) are $\rho_{c23} = \rho_{o23} = 0.9964$, showing that the two modes are very highly correlated. The poor performance of this reduced-order model is further illustrated by the plots of the impulse response, h, and its power spectral density function, S, given in Fig. 6. The responses of the full (solid lines) and reduced (dotted lines) systems can be seen to be quite different.

Example 9. The same structure is now re-examined in balanced coordinates. The poles of the reduced system are $\lambda_{r1,2} = -0.02253 \pm j2.1224$ and $\lambda_{r3,4} = -0.27529 \pm j7.4630$, i.e., the two close poles have now been separated during the reduction. The optimality indices obtained are $\varepsilon = 0.5423\ 10^{-5}$ and $\delta = 0.001717$, so the reduction is near-optimal. Indeed, the plots of impulse response and power spectral density function for this reduced system overlap those of the full system in Fig. 6. This example, taken together with the preceding one, illustrates the expected result that the high correlation of close modes prevents near-optimal model reduction using modal truncation, but presents no such problems in balanced coordinates.

Finally, reduction of a flexible truss model and the Control of Flexible Structures (COFS-1) mast with sensor/actuator dynamics are presented. The truss is shown in Fig. 7. Every truss node has 2 degrees of freedom, as is numbered in Fig. 7. The first number denotes the horizontal displacement, and the second the vertical displacement. Let f_i be the force applied to the ith degree of freedom, and q_i and v_i be the displacement and velocity of this point, respectively. The system has one independent input u applied at four locations, as follows: $f_{11} = f_{12} = f_{21} = -f_{14} = 1$. Similarly, one measured output is constructed from the displacements at four points: $y = q_{15} + q_{16} + q_{17} + q_{20}$.

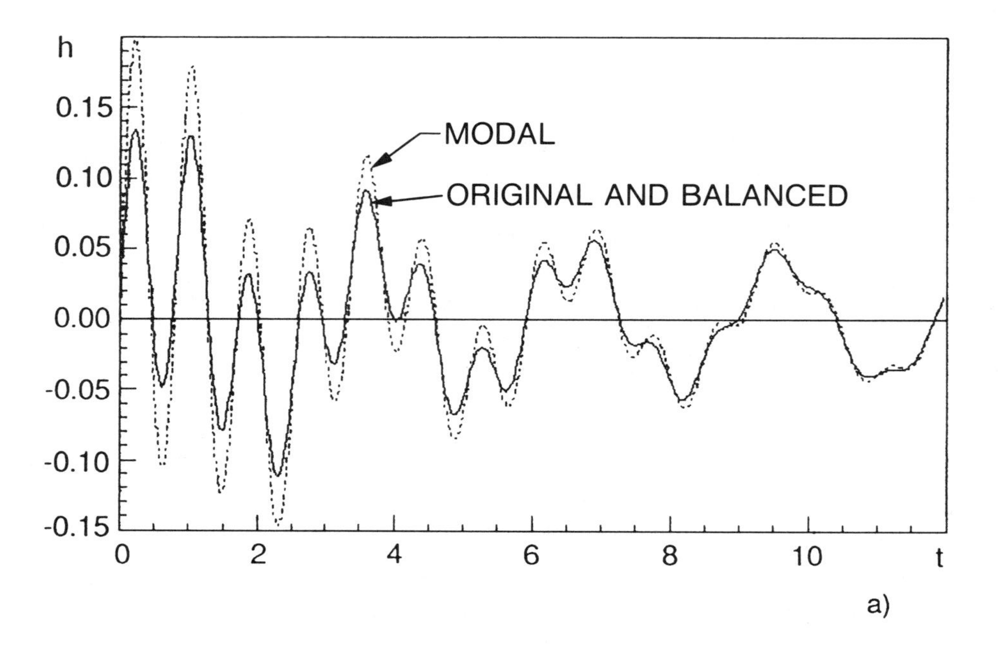

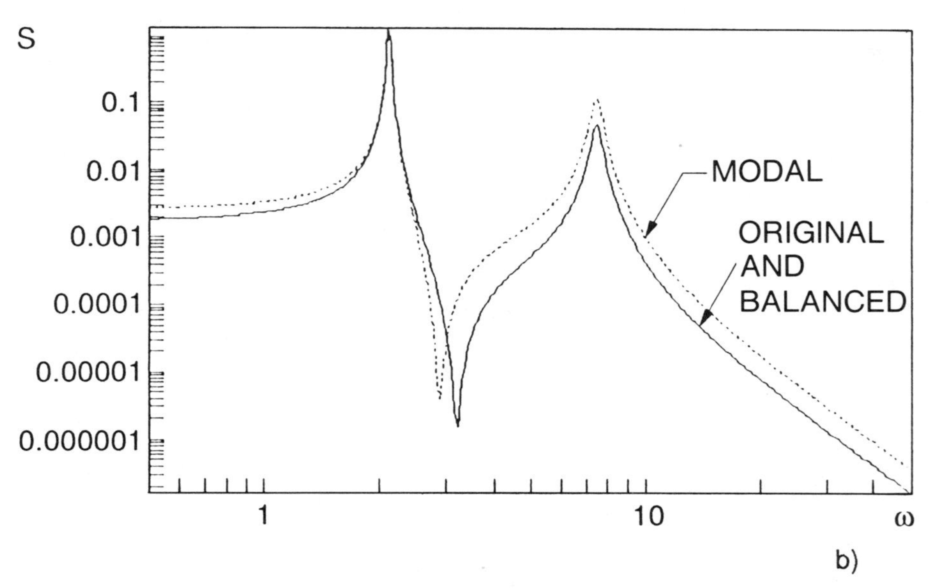

Figure 6. Responses of full and reduced-order models.

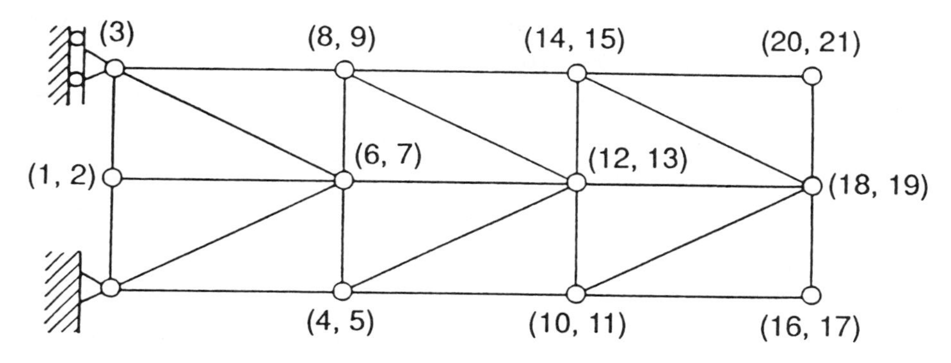

Figure 7. Flexible truss structure.

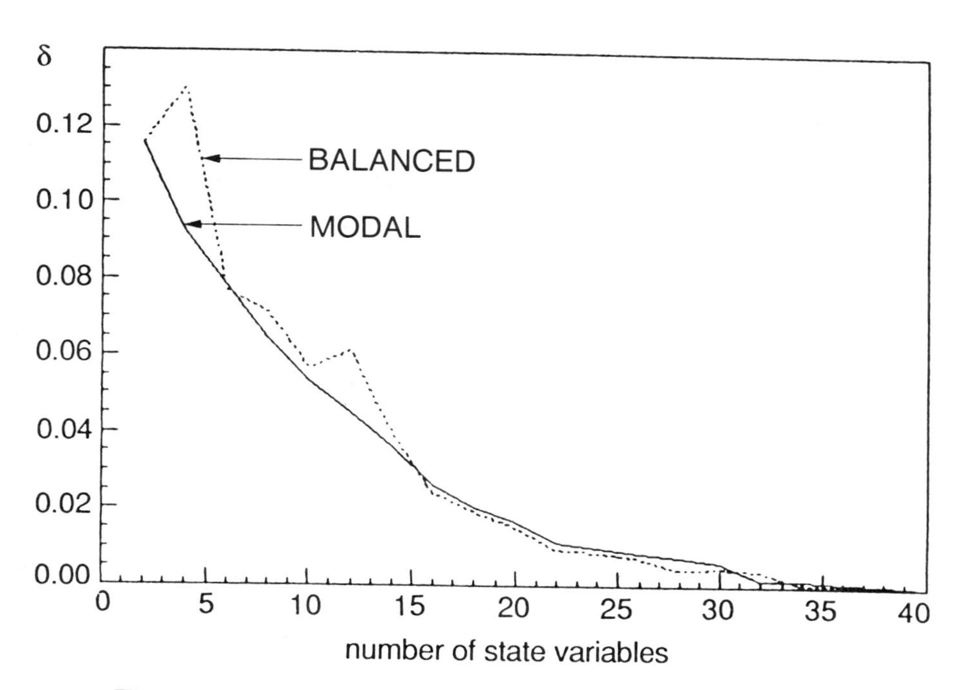

Figure 8. Effect of model order on truss reduction error.

The truss model was reduced in balanced and modal coordinates. The results are presented in Fig. 8 (balanced reduction in dotted line, modal reduction in solid line). It shows that the reduction from 42 to 4 states causes reduction error δ less than 10%, both in balanced and modal coordinates. The plots of the impulse response h, and power spectrum S are shown in Fig. 9.

The model of the COFS-1 structure, Fig. 10, consists of its structural properties as well as its actuator and sensor dynamics. It has 104 state variables, 10 inputs and 20 outputs. The reduction results, in balanced as well as modal coordinates, are presented in Fig. 11 (balanced reduction in dotted line, modal reduction in solid line). They show that reduction from 104 to 38 states causes an error about 10%, and to 60 states an error about 2%, in both balanced and modal coordinates. Plots in Figs. 12a and 12b show the impulse responses and power spectra of the full and reduced (60 state) model, with input No. 8 and outputs Nos. 17 and 18.

B. REDUCTION INDICES FROM TEST DATA

Two indices are used in this section for model reduction: Hankel singular values, and component costs. They can be obtained from the transfer function of a flexible structure. The transfer function $H = C(j\omega I - A)^{-1}B$ for the flexible structure from (45) is

$$H = \sum_{i=1}^{n} H_i, \quad H_i = C_i(j\omega I - A_i)^{-1}B_i. \tag{62}$$

For $\omega = \omega_i$ one has $\|H_k\| \ll \|H_i\|$, for $i \neq k$, therefore $H(\omega_i) \approx H_i(\omega_i)$. From (45), after some algebra, one finds that

$$H_i = a_i c_i b_i \tag{63}$$

where $a_i = 1/\left(2\zeta_i\omega_i^2\right)$, $c_i = \omega_i c_{ri} - jc_{di}$. Define $\cos\psi = \|c_i b_i\|/(\|c_i\| \; \|b_i\|)$ which is a measure of collinearity of c_i and b_i. The norm of H_i, denoted by Y_i, is the norm of the output at frequency $\omega = \omega_i$

$$Y_i = \|H_i(\omega_i)\| = a_i\|b_i\|\cos\psi\sqrt{(\|c_{di}\|^2 + \|c_{ri}\|^2\omega_i)} = a_i\|b_i\|\|c_i\|\cos\psi_i \tag{64}$$

For orthogonal c_i and b_i one obtains $H_i(\omega_i) = 0$, whereas for a single-input and a single-output or collocated sensors and actuators $\cos\psi_i = 1$.

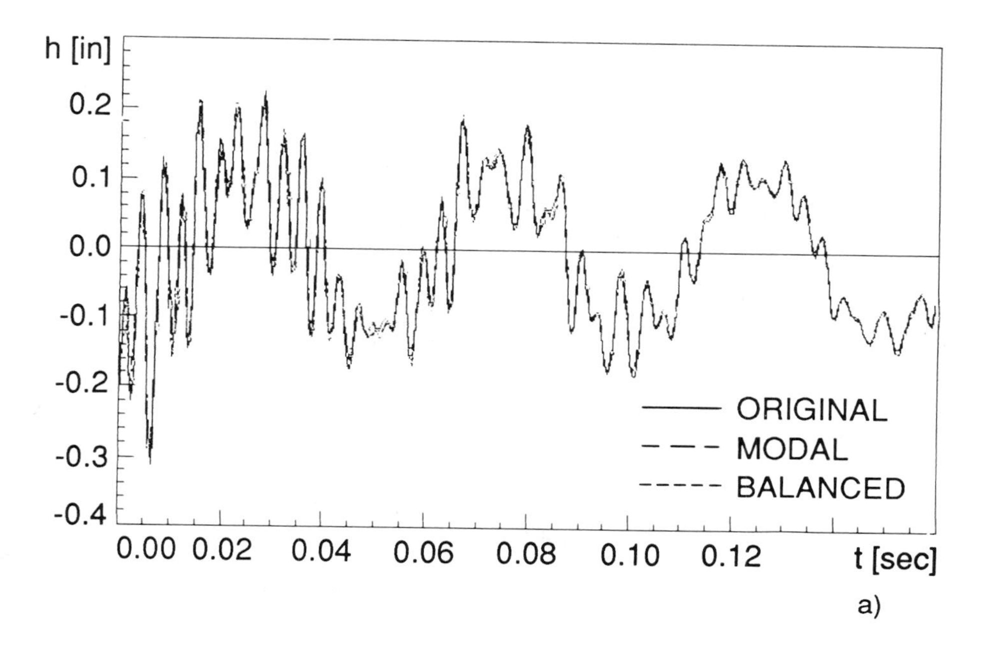

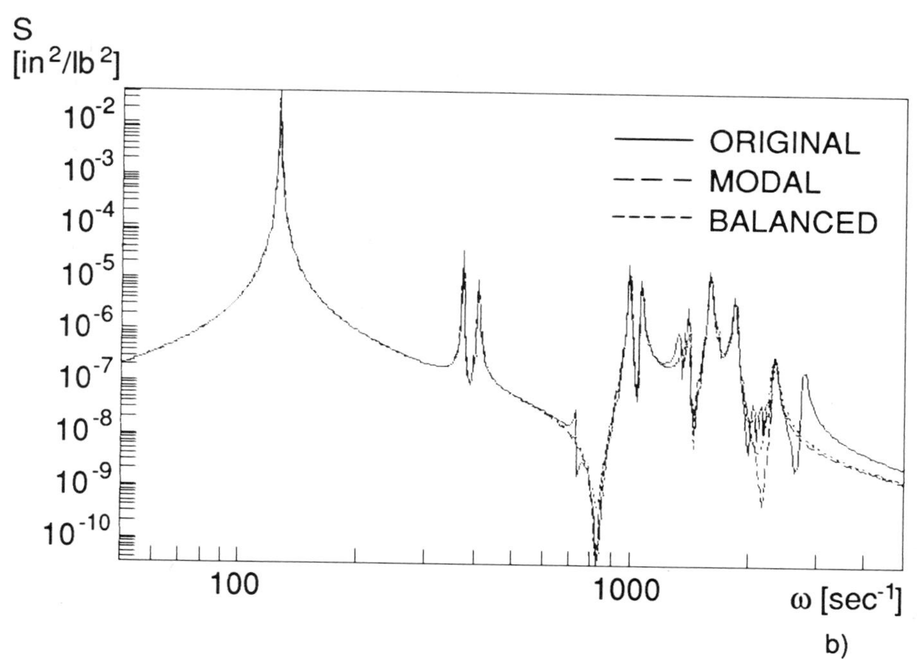

Figure 9. Responses of full and reduced truss models.

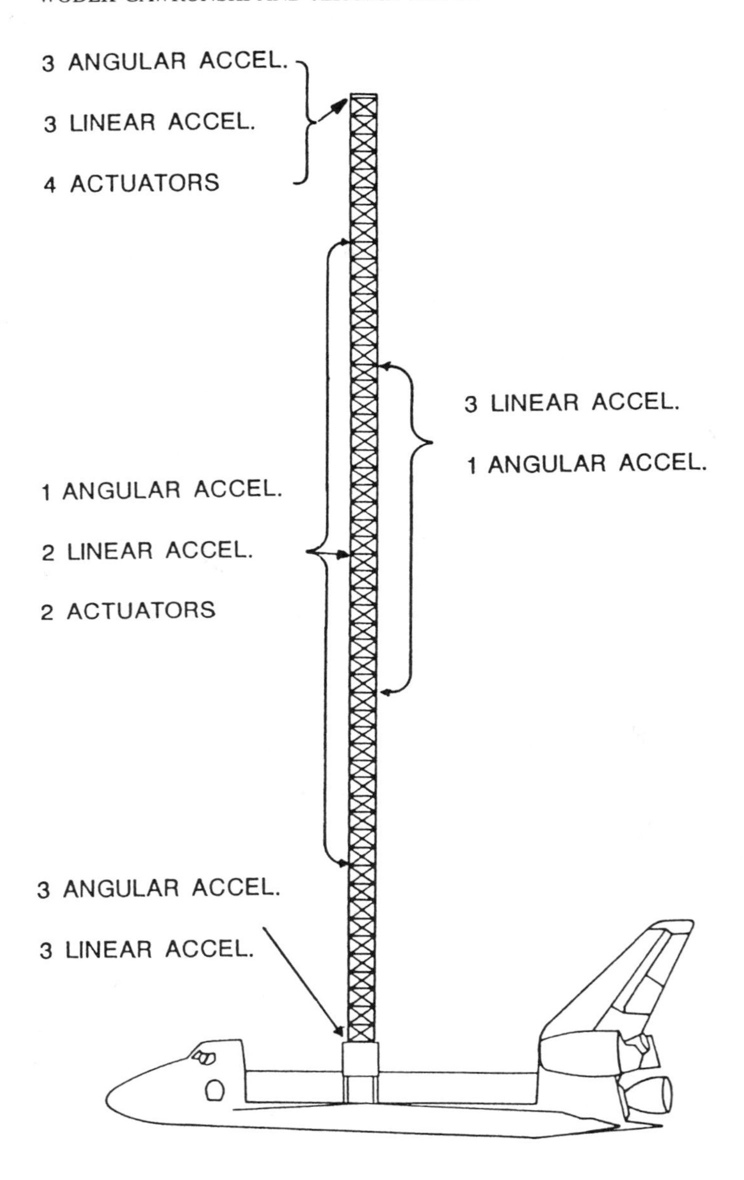

Fig. 10 Proposed COFS-1 flight experiment.

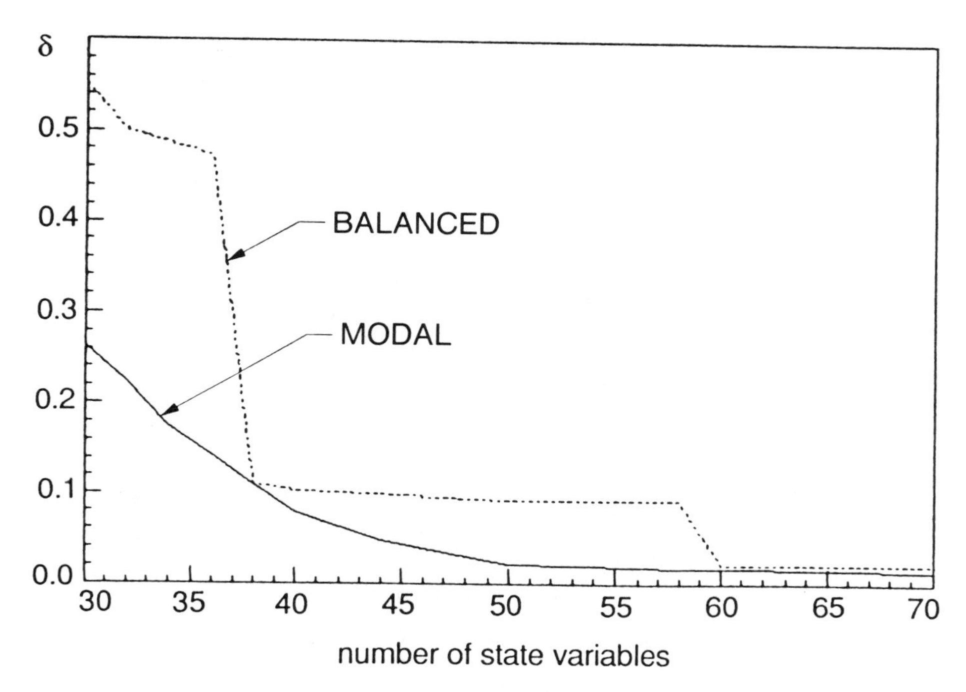

Figure 11. Effect of model order on COFS-1 reduction error.

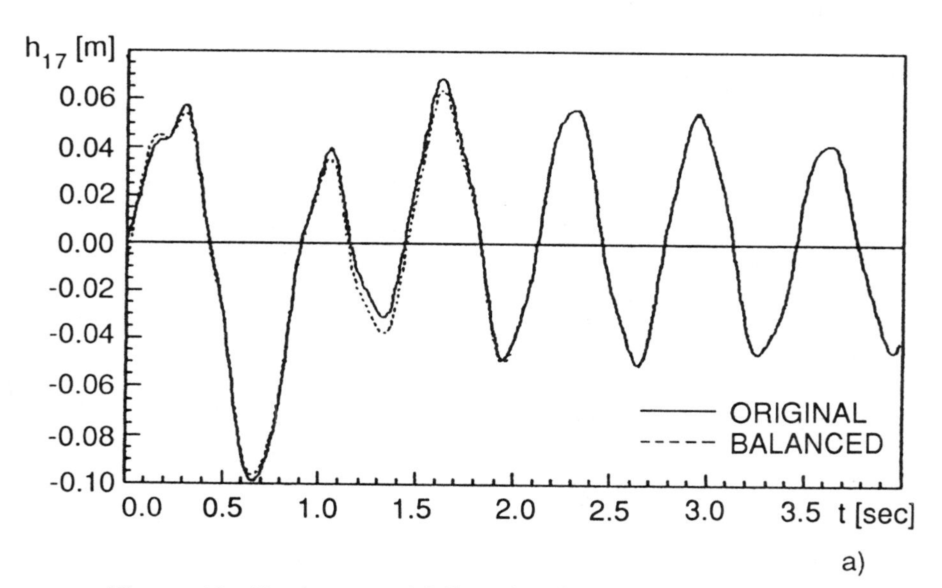

Figure 12. Responses of full and reduced COFS-1 models.

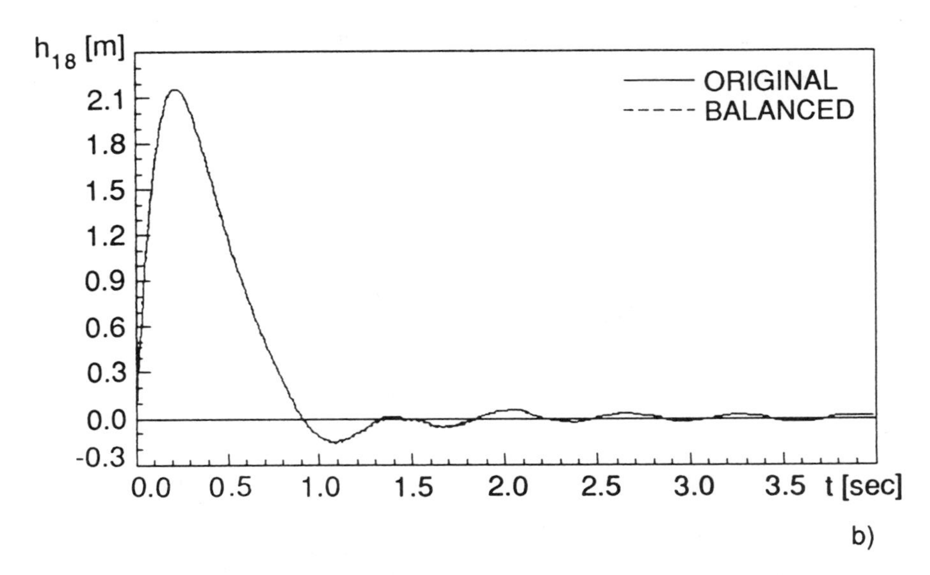

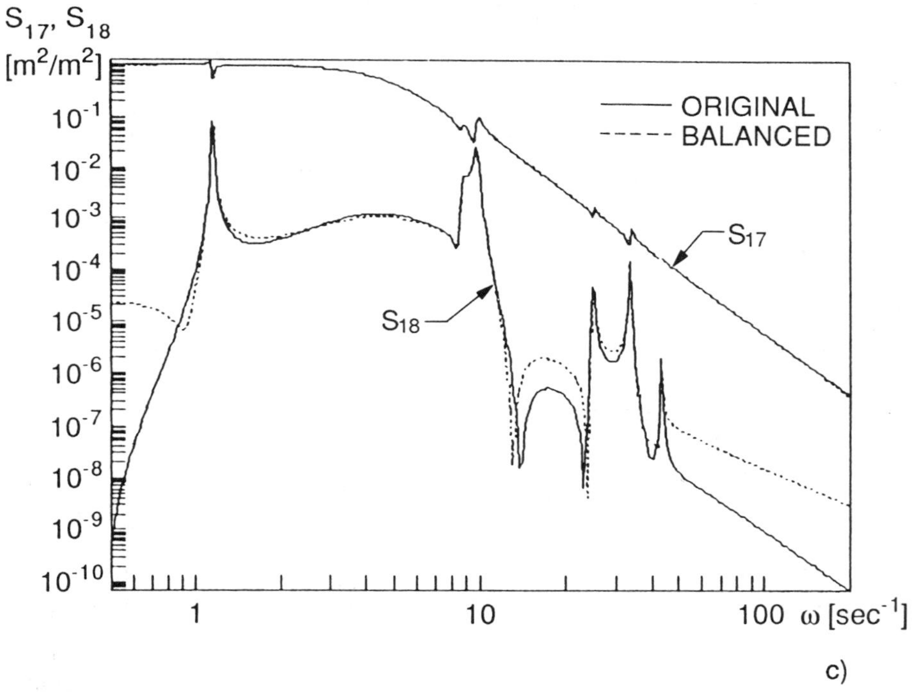

Figure 12. Responses of full and reduced COFS-1 models.

As mentioned before, flexible structures with small damping and separated poles have controllability and observability grammians diagonally dominant, $W_c = \text{diag}(W_{ci})$, $W_o = \text{diag}(W_{oi})$ where, from (48) and (50),

$$W_{ci} = I_2 \left(\|b_i\|^2 / 4\zeta_i\omega_i \right), \quad W_{oi} = I_2 \left(Y_i^2 \zeta_i\omega_i / \|b_i\|^2 \right) \tag{65}$$

and Y_i is the norm of the output at the i-th natural frequency. For these structures, the balanced grammians are computed simply as $W_{cb} = W_{ob} = (W_c W_o)^{1/2} = \text{diag}(W_{bi})$, where from (65) $W_{bi} = I_2 Y_i/2 = I_2 \gamma^2$. This shows that for flexible structures the square of the i-th Hankel singular value equals to one half of the norm of output at the i-th resonant frequency,

$$\gamma_i^2 = 0.5Y_i. \tag{66}$$

Figure 13 shows the determination of γ_2 from the test data for a single-input and single-output system, $\gamma_2^2 = 0.5Y_2 = 0.5\|H(\omega_2)\|$.

The cost of the i-th modal coordinate is determined from (21)

$$\sigma_i^2 = -tr \left(A_i \gamma_i^4 \right) = 0.5\zeta_i\omega_i Y_i^2 \tag{67}$$

where A_i is given by (45). Denoting the half power frequency [28] $\Delta_i = 2\zeta_i\omega_i$, (see Fig. 13), one may rewrite (67) as

$$\sigma_i = 0.5Y_i\sqrt{\Delta_i}. \tag{68}$$

Unlike Hankel singular values, the cost consists of a product of the resonance amplitude and the resonance width.

The results (66) and (68) show that with small damping and separated natural frequencies, the reduction technique in modal or balanced coordinates involves truncating the variables with the smallest resonance peaks (γ-truncation), or with the smallest resonance peaks and resonance width (σ-truncation). These techniques have been commonly used before Hankel singular values or component costs were introduced. However, both balanced and modal techniques develop a theoretical basis for the already-in-use techniques. More importantly, they can be used for model reduction for structures with heavy damping or closely spaced natural frequencies.

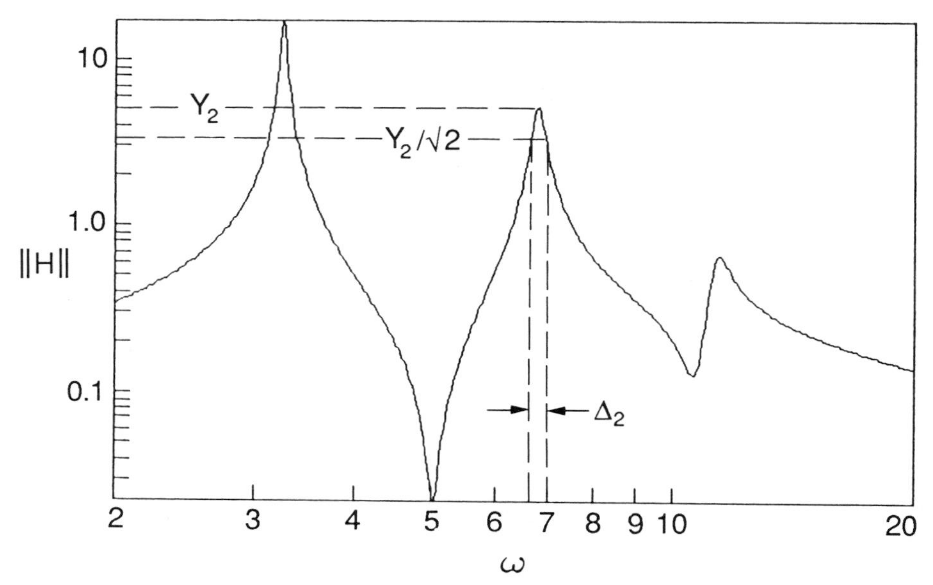

Figure 13 Hankel singular value and component cost from transfer function.

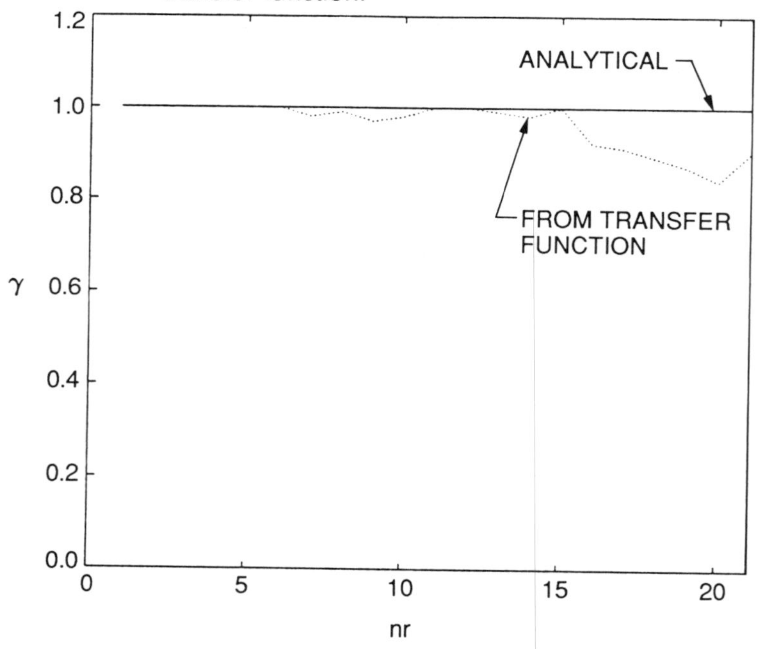

Figure 14 Comparison of Hankel singular values–analytical and from transfer function.

Example 10. A flexible system from Fig. 4 is investigated. The system parameters are as follows:
$m_1 = m_2 = m_3 = 2$, $k_1 = 20$, $k_2 = k_4 = 50$, $k_3 = 100$, $c_i = 0.001k_i$, $i = 1, 2, 3, 4$, with single input applied to the first mass, $u = f_1$, and two outputs, displacements of the second mass ($y_1 = q_2$) and velocity of the first mass ($y_2 = v_1$). Table 5 gives the system Hankel singular values, γ_i, obtained analytically, as well as the Hankel singular values, γ_{ei}, obtained from the frequency response function. It shows that the transfer function gives good approximation of Hankel singular values. The costs, σ_i, obtained analytically and the cost, σ_{ei}, from the transfer function are also given in Table 5, showing that $\sigma_i \approx \sigma_{ei}$, for $i = 1, \ldots, 6$.

Table 5

i	γ_i	γ_{ei}	σ_i	σ_{ei}
1	3.1230	3.1228	12.4475	12.4487
2	3.1226	3.1228	12.4493	12.4487
3	1.7591	1.7592	5.7215	5.7072
4	1.7591	1.7592	5.6922	5.7072
5	0.2745	0.2842	0.1800	0.1928
6	0.2744	0.2842	0.1796	0.1928

Next, a flexible structure in Fig. 7 with 42 state variables, with a single input and single output, is considered. The sensor and actuator are placed such that all Hankel singular values are equal, $\gamma_i = 1$, $i = 1, \ldots, 42$, i.e. all states are equally controlled and observed. The plots of Hankel singular values are given in Fig. 14. The analytical Hankel singular values (solid line), and those obtained from the transfer function (dotted line) are nearly the same, especially for $nr < 16$.

C. EFFECTS OF ACTUATOR DYNAMICS ON REDUCTION INDICES

Test data, besides system dynamics, also include actuator and sensor dynamics. The reduced model obtained from these data may be far from the optimal because of actuator/sensor dynamics. Without loosing generality, only actuator dynamics is considered here. In this section, reconstruction of the flexible structure indices from the joint actuator-structure indices is discussed and illustrated. The system considered is a cascade connection of actuators and a structure. Its transfer function including actuator dynamics is $H = H_s H_a$, where H_s and H_a are transfer functions of the structure and actuators respectively. Define

$$\cos\phi = \|H\|/(\|H_s\| \; \|H_a\|) \tag{69}$$

where the Froebenius norm is used, then

$$\|H_s\| = \|H\|/(\|H_a\| \cos\phi) \tag{70}$$

The value of $\cos\phi$ determines the collinearity of H_a and H_s. If the actuator transfer function is almost collinear with the structure transfer function, then $\cos\phi \approx 1$, whereas orthogonal H_a and H_s result in $H = 0$. Since $\|H_s(\omega_i)\| = Y_{si} = 2\gamma_{si}^2$ and $\|H(\omega_i)\| = Y_i = 2\gamma_i^2$, from (70) it follows that

$$\gamma_{si} = \gamma_i/\sqrt{\alpha_i}, \quad \alpha_i = \|H_{ai}\| \cos\phi_i, \tag{71}$$

and $H_{ai} = H_a(\omega_i)$, $\cos\phi_i = \cos\phi(\omega_i)$ with ω_i being the i-th resonance frequency. The Hankel singular values of the structure are obtained from the Hankel singular values of the structure with actuators by scaling the latter by the amplitude of actuator outputs at the i-th resonance frequency, and collinearity measure of H_a and H_s.

Assuming that $\|H_a(\omega)\|$ has no abrupt changes in the neighborhood of the natural frequencies, we have $\Delta_i \approx \Delta_{si}$. Therefore the component cost is similarly scaled as the Hankel singular values

$$\sigma_{si} = \sigma_i/\alpha_i. \tag{72}$$

Both (71) and (72) show that every Hankel singular value and component cost for the structure alone is obtained by individually scaling each related Hankel singular values and component cost for the combined structure-actuator system. If $\cos\phi_i$ is constant for $i = 1, \ldots, n$, then the squares of Hankel singular values and the component costs are scaled by the inverse of the actuator gain.

The determination of $\cos\phi$ from (69) using test data is difficult, if not impossible, since H_s is not known explicitly during testing. Introducing (63) into (69) yields

$$\cos\phi_i = \|c_{si}b_{si}H_{ai}\|/(\|c_{si}b_{si}\| \ \|H_{ai}\|) \tag{73}$$

Here, only the actuator transfer function and actuator-sensor location are involved. For a single-input system $\cos\phi_i = \|c_{si}H_{si}\|/(\|c_{si}\| \ \|H_{si}\|)$, and for a single-input and single-output system $\cos\phi_i = 1$.

Example 11. The system from Fig. 4, with a single output ($y = q_2$) and an actuator attached, is examined. The actuator transfer function is $H_a = 1/(1+s)$. The transfer function of the combined structure and actuator is plotted in Fig. 15 (solid line), including the structure transfer function (dotted line), and the actuator transfer function (dashed line). The Hankel singular values γ_{si} and costs σ_{si} of the structure are given in Table 5. For this system $\cos\phi_i = 1$, $i = 1,2,3$ and $\alpha_1 = 0.2938$, $\alpha_2 = 0.1459$, $\alpha_3 = 0.0879$. According to (71) and (72), the reconstructed structure Hankel singular values γ_{sri}, and reconstructed structure costs σ_{sri} are determined from the combined structure and actuator transfer function, see Table 6. One can see that $\gamma_{si} \approx \gamma_{sri}$, $\sigma_{si} \approx \sigma_{sri}$, $i = 1,\ldots,6$.

Table 6

i	γ_{si}	γ_{sri}	σ_{si}	σ_{sri}
1	2.8078	2.8059	0.2889	0.2887
2	2.7987	2.8032	0.2880	0.2884
3	0.2003	0.1997	0.0429	0.0428
4	0.1990	0.1997	0.0427	0.0428
5	0.0234	0.0236	0.0085	0.0085
6	0.0235	0.0235	0.0084	0.0084

Finally, the truss from the previous section is considered. An actuator with the transfer function $H_a = 10/(s+2000)$ is applied. The transfer function

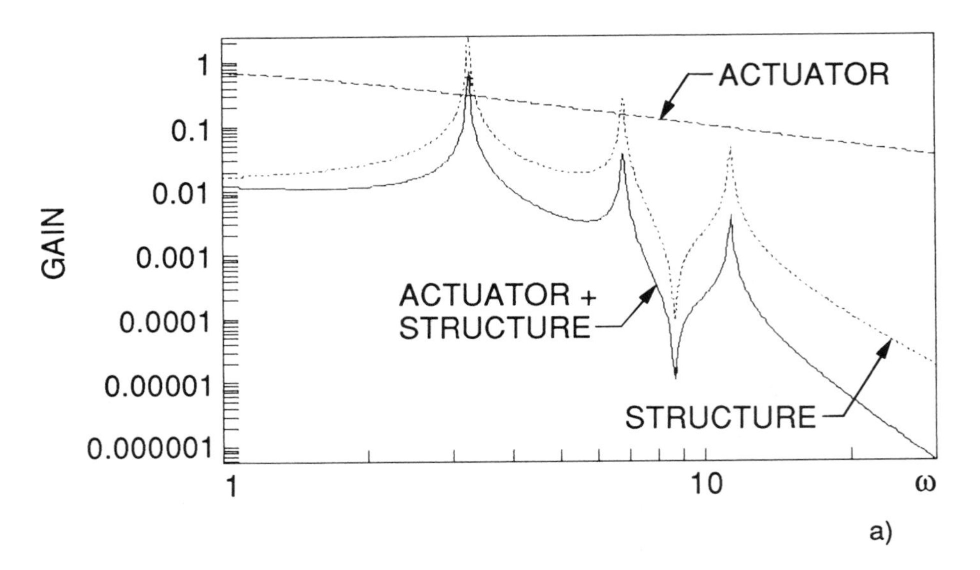

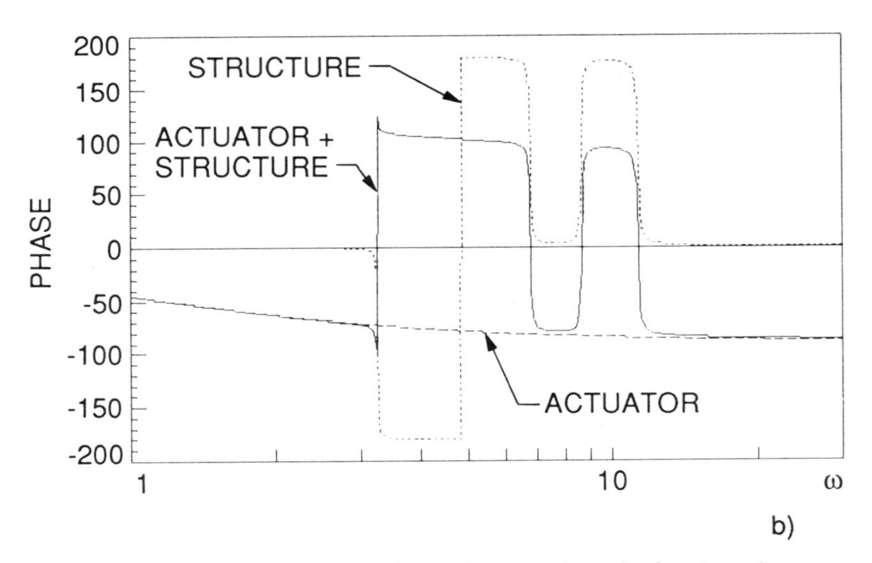

Figure 15 Transfer functions of the three-spring-dashpot system (dotted line), the actuator (dashed line), and their combined system (solid line).

of the truss and actuator connected in series is plotted in Fig. 16 in solid line, including the truss transfer function in dotted line and and the actuator transfer function in dashed line. The Hankel singular values of the truss are all equal to 1 (see solid line in Fig. 17), while Hankel singular values of the truss and actuator combination are plotted in dotted line. The reconstructed truss Hankel singular values are represented in dashed line (it overlaps most of the solid line), indicating good reconstruction results.

D. ROBUST REDUCTION

For flexible structures damping is the least accurately known variable. Its value can vary in quite a wide range, even more than 100% of its nominal value. The component costs and Hankel singular values of a structure may be highly sensitive to changes in its damping, so reduction can be quite unreliable. As a result, if a part of the system is deleted on the basis of the reduction costs obtained for the nominal damping level, there is no guarantee that the resulting reduction error will be small for the true structural damping. In this subsection we present a reduction technique which is insensitive to damping variations in the structure.

Denote the nominal modal damping by ζ_{in}, and its lower and upper variations by $\Delta\zeta_i^-(\geq 0)$ and $\Delta\zeta_i^+(\geq 0)$, respectively, so that

$$\mu^+ = \max_i \left(\Delta\zeta_i^+\omega_i\right), \quad \mu^- = \max_i \left(\Delta\zeta_i^-\omega_i\right); \tag{74}$$

are the upper and lower shifts of the real part of the system poles. For the shifted poles, the grammians can be determined by shifting all system poles by $\mu = \mu^-$ and $\mu = -\mu^+$. such that

$$(A+\mu I)W_{\mu c}+W_{\mu c}(A^*+\mu I)+BB^* = 0, \;\; (A^*+\mu I)W_{\mu o}+W_{\mu o}(A+\mu I)+C^*C = 0. \tag{75}$$

In terms of flexible structures, the shift to the right, for $\mu = \mu^-$, means subtracting structural damping, while the shift to the left, for $\mu = -\mu^+$, means adding structural damping.

The following robust reduction procedure can then be applied:

1. Determine μ^- and μ^+ from (74) for given lower $\left(\Delta\zeta_i^-\right)$ and upper $\left(\Delta\zeta_i^+\right)$ damping estimates, and from (75) find the grammians W_c^+, W_o^+, and W_c^-, W_o^- respectively.

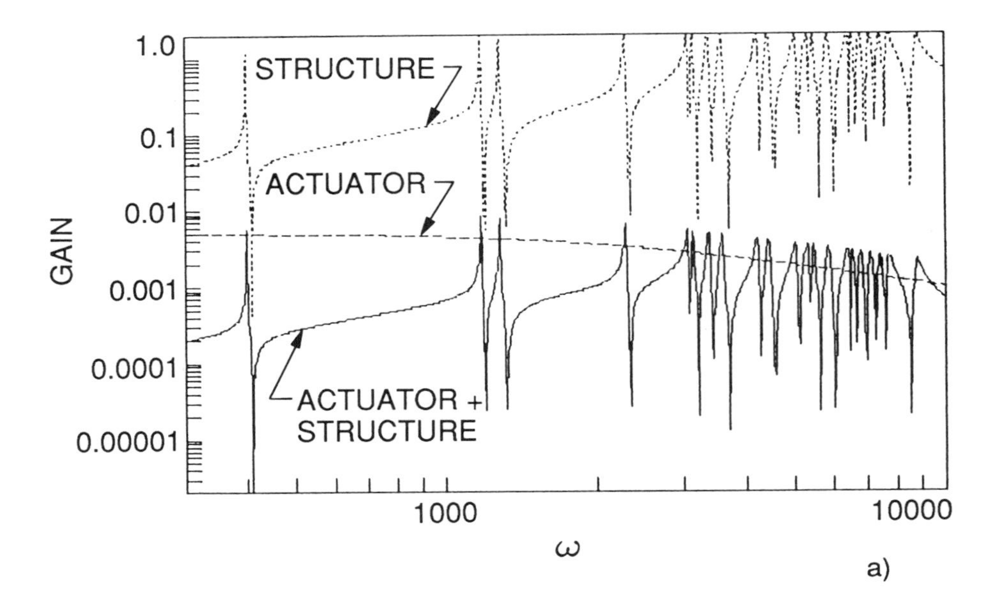

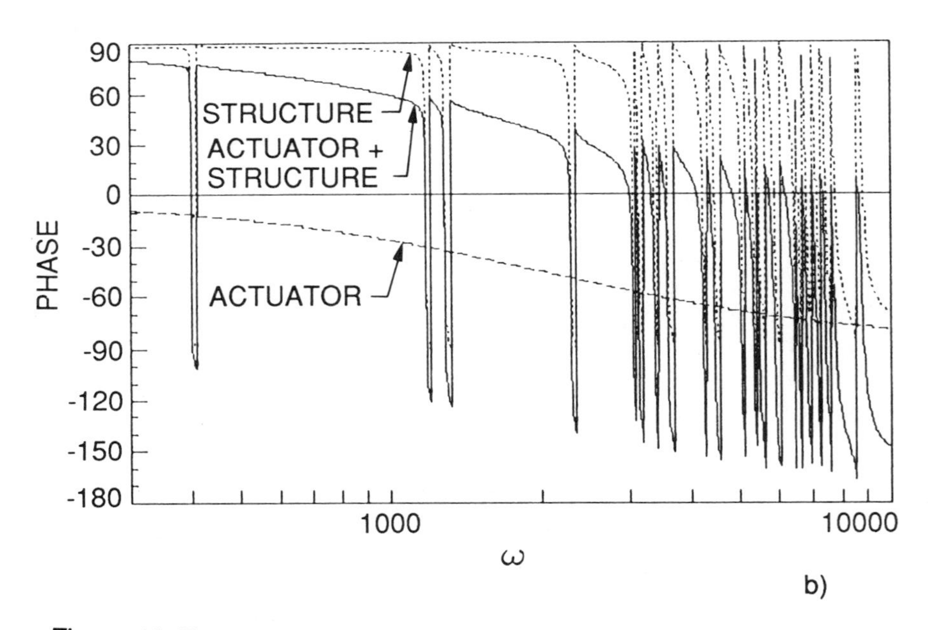

Figure 16 Transfer functions of the truss (dotted line), the actuator (dashed line), and their combined system (solid line).

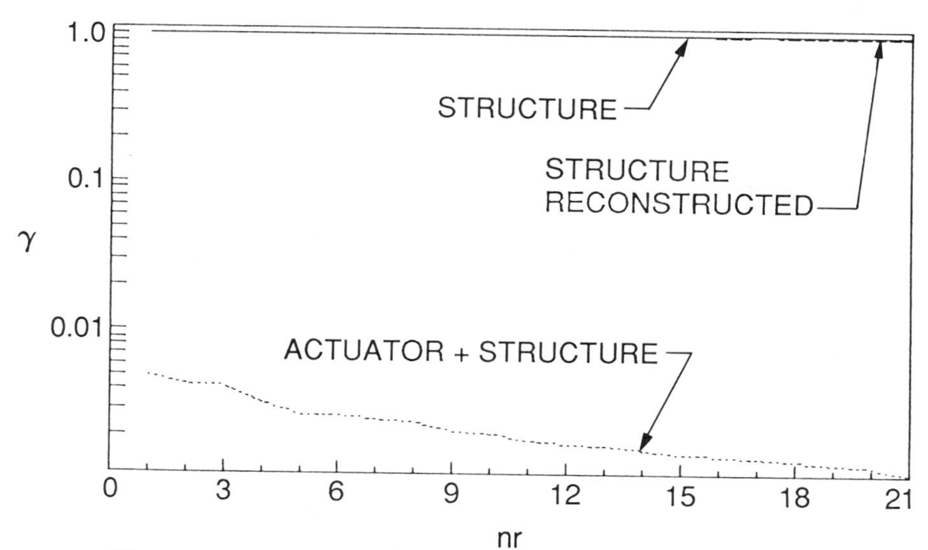

Figure 17 Hankel singular values of the truss (solid line), the truss and actuator combined (dotted line), and the reconstructed Hankel singular values of the truss (dashed line).

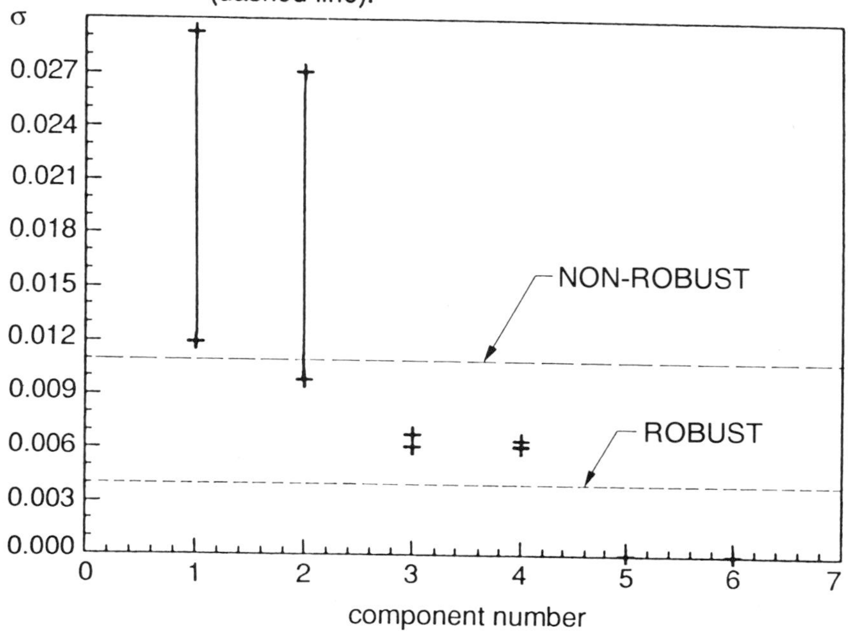

Figure 18 Modal cost uncertainties for a three-spring-dashpot system.

2. Determine the upper and lower costs $\left(\sigma_i^+ \text{ and } \sigma_i^-\right)$, and the cost interval $\Delta\sigma_i = \left[\sigma_i^-, \sigma_i^+\right]$, for upper $\left(W_c^+, W_o^+\right)$, and lower $\left(W_c^-, W_o^-\right)$ grammians
3. Plot the cost interval of each component (see Fig. 18 for a typical case). If the cost intervals of the truncated parts of the system do not overlap the cost intervals of the retained parts of the system (dotted line, Fig. 18), then the reduction is robust with respect to the damping variations. Any damping value taken from the intervals $Z_i = \left[\zeta_{in} - \Delta\zeta_i^-, \ \zeta_{in} - \Delta\zeta_{in}^+\right]$ gives the same reduced model. If the cost intervals of the truncated and retained subsystems overlap (dashed line, Fig. 18), the reduction is not robust. Its results depend on the particular damping values from the intervals Z_i.

Example 12. The system from Fig. 4 with the parameters as in Example 8 is re-examined in balanced coordinates. Its damping is not exactly known but the modal damping variations are known to be $\mu^+ = \mu^- = 0.01$. The cost bounds obtained for this case are displayed in Fig. 18, showing that the reduction from 6 to 4 (or 2) states is robust.

Example 13. The system from Fig. 4 is again considered in balanced coordinates but now with the parameters: $m_1 = 4$, $m_2 = 2$, $m_3 = 7$, $k_1 = 10$, $k_2 = k_3 = 15$, $k_4 = 20$, damping $d_i = 0.005k_i$, $i = 1, \ldots, 4$, and

$$B^T = (0\ 0\ 0\ 1\ 2\ -1), \quad C = \begin{pmatrix} 10 & 2 & 2 & 0 & 3 & 0 \\ 0 & 1 & 1 & 1 & 1 & 1 \end{pmatrix}.$$

The system poles are $\lambda_{1,2} = -0.0056 \pm j1.4913$, $\lambda_{3,4} = -0.0138 \pm j2.3529$ and $\lambda_{5,6} = -0.0462 \pm j4.2997$. The lower bound for modal damping perturbation is $\mu^- = 0.005$, and the upper bound is $\mu^+ = 0.003$. The bounds of the reduction cost are shown in Fig. 19. The reduction from 6 to 4 states, based on the nominal damping levels, retains states 1,2,5, and 6, while deleting states 3 and 4. One can see from Fig. 19 that this reduction is not robust.

Finally, the robustness of the truss reduction was tested (the truss is shown in Fig. 7). The lower variation of damping is $\mu^- = 0.01$, and the upper variation is $\mu^+ = 3$. The cost bounds are plotted in Fig. 20, showing that the reduction in most cases is not robust.

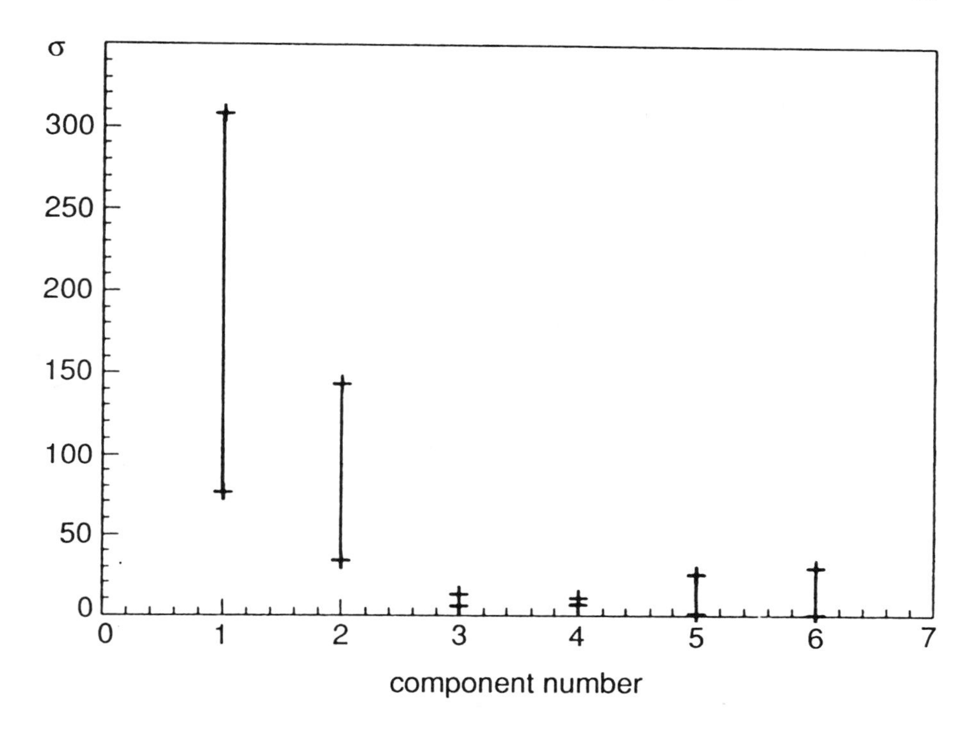

Figure 19 Modal cost uncertainties for a three-spring-dashpot system.

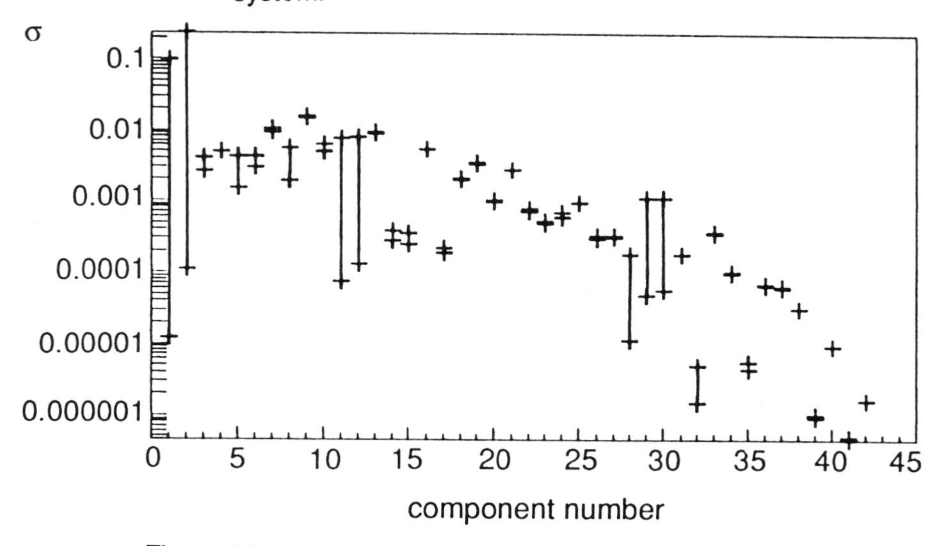

Figure 20 Modal cost uncertainties for a flexible truss.

V. REDUCTION IN LIMITED TIME AND FREQUENCY INTERVALS

In this section, the concepts of grammians defined over a finite time or/and frequency interval are presented including computational procedures. Applying the model reduction technique to these grammians leads to near-optimal reduced models which closely reproduce the full system output in the time and/or frequency interval of interest. This approach has clear connections with a filter design, and can be applied to unstable systems without encountering the divergence problems inevitable with an infinite interval.

A. MODEL REDUCTION IN LIMITED TIME INTERVAL

Assume that a system is excited, and the response is measured within the time interval $T = [t_1, t_2]$, $t_2 > t_1 \geq 0$. The grammians over the time interval $T = [t_1, t_2]$ are defined as

$$W_c(T) = \int_{t_1}^{t_2} \exp(A\tau)BB^* \exp(A^*\tau)d\tau, \quad W_o(T) = \int_{t_1}^{t_2} \exp(A^*\tau)C^*C \exp(A\tau)d\tau. \tag{76}$$

Note that $W_c(T) \geq 0$, $W_o(T) \geq 0$, if $t_2 \geq t_1$. The grammians are determined from the formulas (see [14])

$$W_c(T) = W_c(t_1) - W_c(t_2), \quad W_o(T) = W_o(t_1) - W_o(t_2), \tag{77}$$

$$W_c(t) = S(t)W_cS^*(t), \quad W_o(t) = S^*(t)W_oS(t), \tag{78}$$

$$S(t) = -\exp(At), \tag{79}$$

where W_c, W_o are the solutions of (4). Computations of the limited-time grammians from (77)–(79) may not be effective for large-scale systems. In this case the modal version of (77)–(79) is used, as given in Appendix A.

Stability conditions for the reduced balanced model are studied in a limited time interval, assuming the original system stable. Denote

$$Q_c(t) = S(t)BB^*S^*(t), \quad Q_o(t) = S^*(t)C^*CS(t), \tag{80}$$

$$Q_c(T) = Q_c(t_1) - Q_c(t_2), \quad Q_o(T) = Q_o(t_1) - Q_o(t_2), \tag{81}$$

then the following condition is valid:

Theorem 1. For a stable A and positive semidefinite $Q_c(T)$, $Q_o(T)$, the reduced balanced model is stable.

Proof. For positive semidefinite $Q_c(T)$, $Q_o(T)$ one can find $B(T)$, $C(T)$ such that $Q_c(T) = B(T)B^*(T)$, $Q_o(T) = C^*(T)C(T)$, and that grammians $W_c(T)$, $W_o(T)$ satisfy the Lyapunov equations

$$AW_c(T)+W_c(T)A^*+B(T)B^*(T) = 0, \quad A^*W_o(T)+W_o(T)A+C^*(T)C(T) = 0, \tag{82}$$

According to Appendix B, the reduced balanced model obtained for the triple $(A, B(T), C(T))$ is stable, which completes the proof.

Denote $\Delta t = t_2 - t_1$, and $Q_c(\Delta t) = BB^* - S(\Delta t)\ BB^*S^*(\Delta t)$, $Q_o(\Delta t) = C^*C - S^*(\Delta t)C^*CS(\Delta t)$. In this case one obtains $Q_c(T) = S(t_1)Q_c(\Delta t)S^*(t_1)$, $Q_o(T) = S^*(t_1)\ Q_o(\Delta t)S(t_1)$, from which $Q_c(T)$, $Q_o(T)$ are positive semidefinite if $Q_c(\Delta t)$ and $Q_o(\Delta t)$ are positive semidefinite. For stable systems $\lim_{\Delta t\to 0} S(\Delta t) = \lim_{\Delta t\to 0} e^{A\Delta t} = 0$. Consequently, from the definitions of $Q_c(\Delta t)$, $Q_o(\Delta t)$, the reduced balanced model in the finite time interval is stable, if the time interval Δt is long enough. If the time interval is too short to obtain stable balanced reduction and a stable model is required, then either the interval should be enlarged, or another reduction (modal or Schur balanced) applied. The last two reductions give always a stable model while reducing a stable system.

The computational procedure is summarized as follows.

1. Determine grammians W_c, W_o from (4) with given (A, B, C).
2. Compute $S(t_i)$, $W_c(t_i)$, $W_o(t_i), i = 1, 2$, from (78) and (79).
3. Determine $W_c(T), W_o(T)$ from (77).
4. Apply the reduction procedure to obtain (A_R, B_R, C_R) for $W_c(T)$, $W_o(T)$.
5. Compute J, δ, ε, for the reduced model (A_R, B_R, C_R) from (12), (13), (24).

The following examples illustrate the application of the grammians in model reduction. Since the reduction is performed in balanced, as well as in modal coordinates, the variables in modal coordinates are denoted with the ***subscript*** "m," while those in balanced coordinates are indicated by the ***subscript*** "b". The resulting plots present the full system response as well as the reduced model responses. The full system response is plotted in solid line, the reduced balanced response in dotted line, and the reduced modal response in dashed line.

Example 14. Figure 4 shows a system with the masses $m_1 = 11$, $m_2 = 5$, $m_3 = 10$, stiffnesses $k_1 = k_4 = 10$, $k_2 = 50$, $k_3 = 55$, and dampings $d_i = 0.01k_i$, $i = 1,2,3,4$. The single input u is applied giving $f_1 = u$, $f_2 = 2u$, $f_3 = -5u$, the output is $y = 2q_1 - 2q_2 + 3q_3$, where q_i is the displacement of the i-th mass, and f_i is the force applied to that mass. The system triple (A, B, C) is then shown as follows

$$A = \begin{bmatrix} 0 & 0 & 0 & 1 & 0 & 0 \\ 0 & 0 & 0 & 0 & 1 & 0 \\ 0 & 0 & 0 & 0 & 0 & 1 \\ -5.4545 & 4.5455 & 0 & -0.0545 & 0.0455 & 0 \\ 10 & -21 & 11 & 0.1000 & -0.2100 & 0.1100 \\ 0 & 5.5000 & -6.5000 & 0 & 0.0550 & -0.0650 \end{bmatrix},$$

$B^T = [0\ 0\ 0\ 0.0909\ 0.4\ -0.5]$, $C = [2\ -2\ 3\ 0\ 0\ 0]$, and its poles are $\lambda_{1,2} = -0.0038 \pm j0.8738$, $\lambda_{3,4} = -0.0297 \pm j2.4374$, $\lambda_{5,6} = -0.1313 \pm j5.1217$. Two cases are considered for the model reduction from 6 to 4 state variables: Case 1, the reduction over the interval $T_1 = [0, 8]$, and Case 2, the reduction over the interval $T_2 = [10, 18]$. In Case 1, the first and the third pair of poles are preserved in the reduced modal model, and $\lambda_{b1,2} = -0.0009487 \pm j0.8712$, $\lambda_{b3,4} = -0.1395 \pm j5.0890$ are the poles of the reduced balanced model. The reduction error is $\delta_b = 0.2954$, or $\delta_m = 0.2911$, and the optimality index is $\varepsilon_b = 0.0005683$, or $\varepsilon_m = 0.01291$. The optimality indices are much less than 1, and thus the reduced models are near-optimal. The impulse responses for the balanced and modal reductions are compared in Fig. 21a. Case 2 is obtained from Case 1 by shifting the interval T_1 by 10. In this case, the first two pair of poles are preserved in modal reduction, and $\lambda_{b1,2} = -0.003865 \pm j0.87424,)$ $\lambda_{b3,4} = -0.03047 \pm j2.4439$ are the poles of the reduced balanced model. The reduction error in this case is $\delta_b = 0.2919$, or $\delta_m = 0.3012$, and the reduced model is near-optimal, since $\varepsilon_b = 0.06253$, $\varepsilon_m = 0.04900$. The impulse responses of the reduced and original systems are presented in Fig. 21b for Case 2. Comparison of Figs. 21a and 21b shows that the third mode is less visible for $t > 10$.

Example 15. Example 14 is re-examined, with $d_i = -0.006k_i$, $i = 1,2,3,4$,

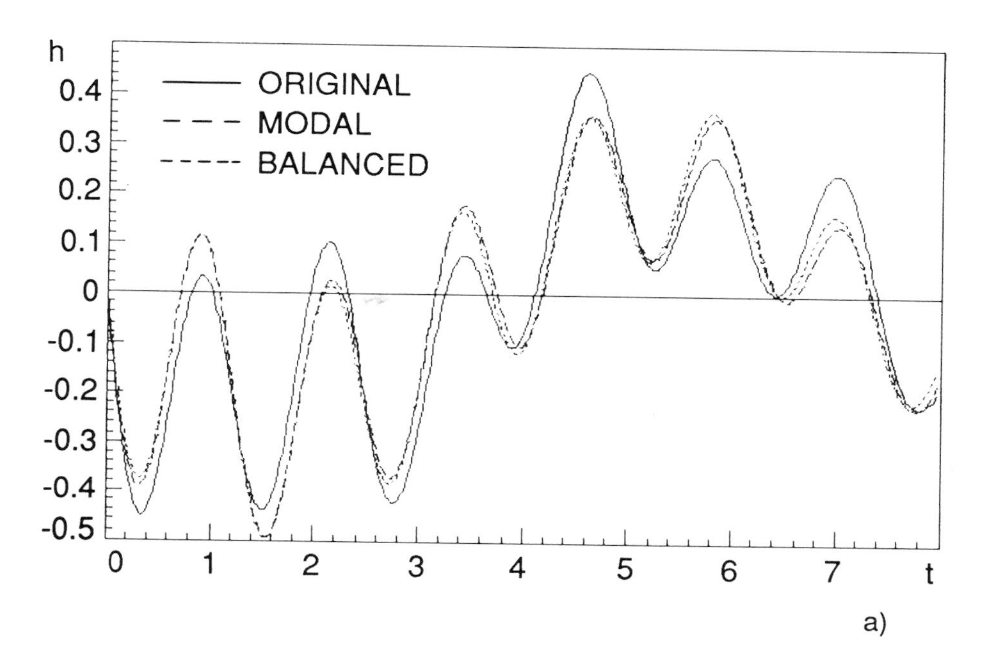

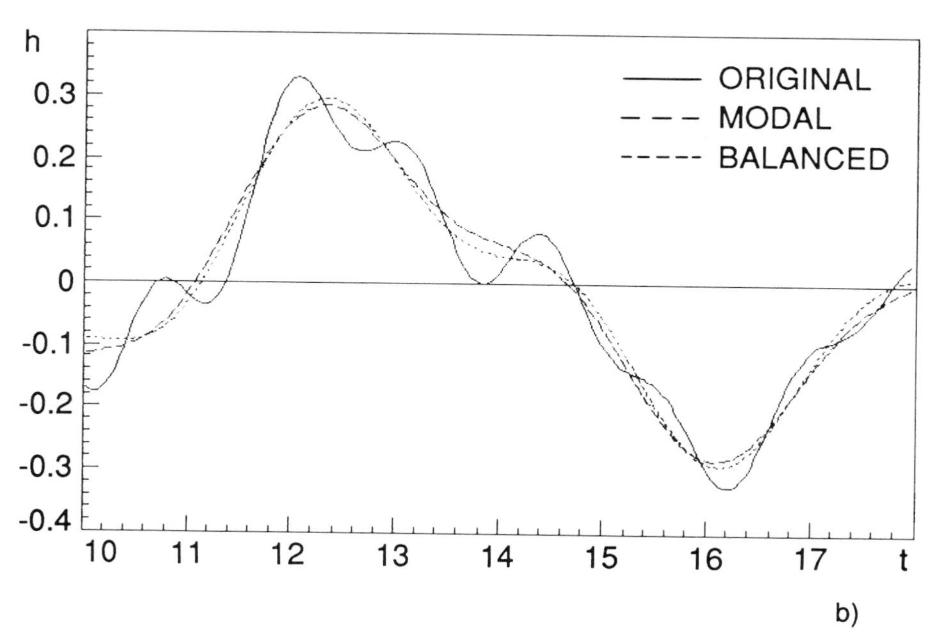

Figure 21 Impulse responses of a three-spring-dashpot system and its reduced models, for a) $T = [0, 8]$, b) $T = [10, 18]$.

which makes the system unstable with the poles $\lambda_{1,2} = 0.0023 \pm j0.8738$, $\lambda_{3,4} = 0.0178 \pm j2.4375$, $\lambda_{5,6} = 0.0787 \pm j5.1228$. The system grammians are determined over the interval [0, 8]. The reduced balanced model of order 4 has the following poles $\lambda_{b1,2} = 0.00594 \pm j0.8716$, $\lambda_{b3,4} = 0.0748 \pm j5.1238$, whereas the first and the third pairs of poles are preserved in the reduced modal model. The reduction error is $\delta_b = 0.2117$, or $\delta_m = 0.2122$, and optimality index is $\varepsilon_b = 0.03201$, or $\varepsilon_m = 0.0200$. The reduced models are obviously near-optimal. The impulse responses of the original and reduced models, within the interval [0,8], are shown in Fig. 22.

The reduced modal model, obtained for a stable system, is stable, since it preserves the system poles. However, the balanced reduction within limited time interval does not guarantee the stability of the reduced model. This can be heuristically explained by the fact that within the finite time interval, especially within a short interval, the response of an unstable system can approximate the response of a stable system with an acceptable error. The following example illustrates the case.

Example 16. Example 14 is re-examined, with damping $d = 0.006k$, so that the system is stable. The obtained reduced balanced model has the following poles $\lambda_{1,2} = 0.0008736 \pm j0.8713$, $\lambda_{3,4} = -0.08569 \pm j5.1127$, i.e., it is unstable. The reduced modal model is stable, with poles $\lambda_{m1,2} = -0.002291 \pm j0.8738$, $\lambda_{m3,4} = -0.07875 \pm j5.1228$. The reduction error is $\delta_b = 0.2807$, or $\delta_m = 0.2810$, and the optimality index $\varepsilon_b = 0.01192$, or $\varepsilon_m = 0.008634$. The impulse responses of the original and reduced models are shown in Fig. 23. The matrices $Q_c(T)$, $Q_o(T)$ for the reduced balanced model are not positive semidefinite, and their eigenvalues are $\lambda_{Q_cT} = \{0.4168,\ -0.1999,\ 0,\ 0,\ 0,\ 0\}$, $\lambda_{Q_cT} = \{16.6741,\ -4.7071,\ 0,\ 0,\ 0,\ 0\}$, which indicate an unstable reduced balanced model.

B. MODEL REDUCTION IN LIMITED FREQUENCY INTERVAL

In this section, an approach is developed for determining the reduced model for which its output fits the full model output within a limited frequency interval $\Omega = [\omega_1, \omega_2]$, $\omega_2 \geq \omega_1 \geq 0$. A limited frequency output is obtained,

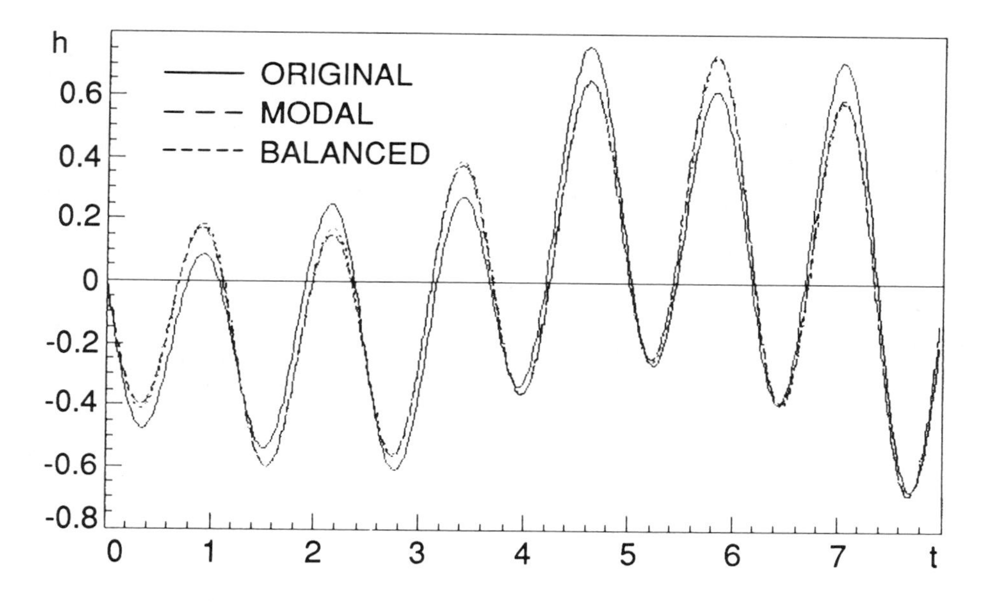

Figure 22 Impulse responses of an unstable system, and its reduced models.

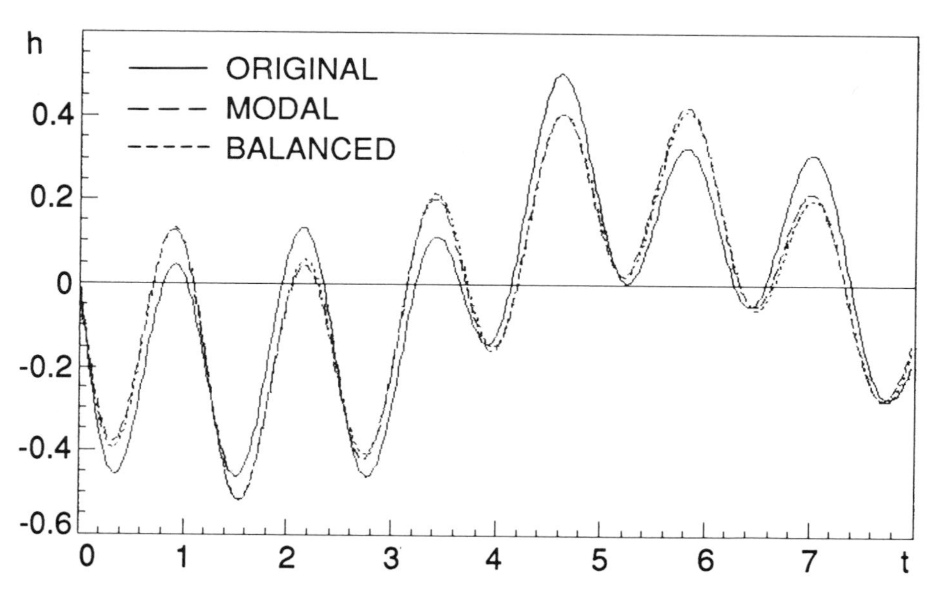

Figure 23 Impulse responses of a three-spring-dashpot system and its reduced models.

for example, for a system excited by a limited frequency input. According to the Parseval theorem, the integrals (76) over the time interval $[0, \infty]$ can be determined in frequency domain as follows

$$W_c = \int_{-\infty}^{\infty} H(v)BB^*H^*(v)dv/2\pi, \quad W_o = \int_{-\infty}^{\infty} H^*(v)C^*CH(v)dv/2\pi, \tag{83}$$

Using (4) and noting that

$$AW_c + W_cA^* = H(v)^{-1}W_c + W_cH^{*-1}(v),$$

where

$$H(v) = (jvI - A)^{-1}, \tag{84}$$

is a Fourier transforms of $\exp(At)$. In order to obtain grammians in the limited frequency band Ω, let the impulse responses pass through the frequency window $[\omega_1, \omega_2]$, so that from (83) one obtains

$$W_c(\Omega) = W_c(\omega_2) - W_c(\omega_1), \quad W_o(\Omega) = W_o(\omega_2) - W_o(\omega_1), \tag{85}$$

where

$$W_c(\omega) = \int_{-\omega}^{\omega} H(v)BB^*H^*(v)dv/2\pi, \quad W_o(\omega) = \int_{-\omega}^{\omega} H^*(v)C^*CH(v)dv/2\pi. \tag{86}$$

Note that $W_c(\omega)$, $W_o(\omega)$ are positive definite, and that they act as low-pass filters for signals. By using (4) and noting that $-AW_c - W_cA^* = H^{-1}(v)W_c + W_cH^{*-1}(v)$, (86) becomes

$$W_c(\omega) = \int_{-\omega}^{\omega} (H(v)W_c + W_cH^*(v))\, dv/2\pi, \tag{87}$$

which produces

$$W_c(\omega) = W_cS^*(\omega) + S(\omega)W_c, \tag{88}$$

where

$$S(\omega) = \int_{-\omega}^{\omega} H(v)dv/2\pi = (j/2\pi)ln((j\omega I + A)(-j\omega I + A)^{-1}), \tag{89}$$

and W_c is the controllability grammian obtained from (4). In (88) the controllability grammian in a finite frequency interval is obtained from the stationary controllability grammian by a frequency-dependent weighting. Similarly, the observability grammian in the limited frequency interval is determined

$$W_o(\omega) = S^*(\omega)W_o + W_oS(\omega), \tag{90}$$

where W_o is the observability grammian obtained from (4). For large scale systems, calculating a matrix logarithm (89) may be ineffective, unless in modal coordinates. Modal version of limited frequency grammians is given in Appendix A.

Finally, it is easy to see, that for the frequency interval $\Omega = [\omega_1, \omega_2]$, $\omega_2 > \omega_1$, the grammians are alternatively obtained from (85), (89), (90), in the form

$$W_c(\Omega) = W_cS^*(\Omega) + S(\Omega)W_c, \quad W_o(\Omega) = S^*(\Omega)W_o + W_oS(\Omega), \tag{91}$$

where $S(\Omega) = S(\omega_2) - S(\omega_1)$. Note that $H(v_1)$, $H(v_2)$ commute for any v_1 and v_2, $H(v_1)H(v_2) = H(v_2)H(v_1)$. From this property, and definitions (84), (89) one obtains

$$\begin{aligned} S(\omega)W_c &= \int_{-\omega}^{\omega}\int_{-\infty}^{\infty} H(v)H(\mu)BB^*H^*(\mu)d\mu\ dv/4\pi^2 \\ &= \int_{-\infty}^{\infty} H(\mu)S(\omega)BB^*H^*(\mu)d\mu/2\pi \end{aligned} \tag{92}$$

and from (88)

$$W_c(\omega) = S(\omega)W_c + W_cS^*(\omega) = \int_{-\infty}^{\infty} H(\mu)Q_c(\omega)H^*(\mu)d\mu/2\pi, \tag{93}$$

where

$$Q_c(\omega) = S(\omega)BB^* + BB^*S^*(\omega). \tag{94}$$

For a stable matrix A, the integral (93) can be obtained by solving the following Lyapunov equation

$$AW_c(\omega) + W_c(\omega)A^* + Q_c(\omega) = 0, \tag{95}$$

and consequently from (85), the grammian $W_c(\Omega)$ is computed from

$$AW_c(\Omega) + W_c(\Omega)A^* + Q_c(\Omega) = 0, \tag{96}$$

where

$$Q_c(\Omega) = Q_c(\omega_2) - Q_c(\omega_1). \tag{97}$$

Similarly, the observability grammian $W_o(\Omega)$ is obtained by solving the following Lyapunov equation

$$A^* W_o(\Omega) + W_o(\Omega)A + Q_o(\Omega) = 0, \tag{98}$$

where

$$Q_o(\Omega) = Q_o(\omega_2) - Q_o(\omega_1), \; Q_o(\omega) = S^*(\omega)C^*C + C^*CS(\omega). \tag{99}$$

With these establishments, the stability condition for the balanced reduction in a limited frequency interval is as follows.

Theorem 2. For a stable A and positive semidefinite $Q_c(\Omega)$, $Q_o(\Omega)$, the reduced balanced model is stable.

Proof. For positive semidefinite $Q_c(\Omega)$, $Q_o(\Omega)$ one can find $B(\Omega)$, $C(\Omega)$ such that $Q_c(\Omega) = B(\Omega)B^*(\Omega)$, $Q_o(\Omega) = C^*(\Omega)C(\Omega)$. From the Lyapunov equations (96), (98), according to the Theorem from Appendix B, the reduced balanced model, for the triple $(A, B(\Omega), C(\Omega))$, is stable, which completes the proof.

Note that $S(\omega) = 0$ for $\omega = 0$, and $S(\omega) = 0.5I$ for $\omega \to \infty$. Hence, for the frequency interval $\Omega = [0, \omega]$, one can find large enough ω such that the obtained reduced balanced model is stable.

The computational procedure is summarized as follows.

1. Determine grammians W_c, W_o, from (4) for given (A, B, C).
2. Determine $W_c(\omega_1)$, $W_c(\omega_2)$ from (88), $W_o(\omega_1)$, $W_o(\omega_2)$ from (90).
3. Determine $W_c(\Omega), W_o(\Omega)$ from (85).
4. Apply the reduction procedure to obtain (A_R, B_R, C_R) for $W_c(\Omega)$, $W_o(\Omega)$.
5. Determine J, δ, ε from (12), (13), (25) for the reduced model (A_R, B_R, C_R).

Example 17. Example 14 is considered in two cases: Case 1, in the frequency interval $\Omega_1 = [0.7, 3.2]$, and Case 2, in the frequency interval

$\Omega_2 = [1.5, 3.2]$. In the first case the reduced model with 4 state variables is determined, whereas, in the second case the reduced model with 2 state variables is obtained. The results are as follows. In Case 1 the reduction error is $\delta_b = 0.03405$, or $\delta_m = 0.03531$, the optimality index $\varepsilon_b = 0.002773$, or $\varepsilon_m = 0.02340$, while in Case 2 $\delta_b = 0.1086$, or $\delta_m = 0.1331$, and $\varepsilon_b = 0.01087$, or $\varepsilon_m = 0.1145$. In either case the obtained reduced model is close to the optimal one ($\varepsilon \ll 1$). Figure 24a and 24b show a good fit of the power spectral density functions within the considered frequency interval, indicating that the reduction in the limited frequency interval can serve as a filter design method. For example, in Case 2 the output signal is filtered such that the resulting output of the reduced model fits in the best way the output of the original system within the interval $\Omega_2 = [1.5, 3.2]$.

C. MODEL REDUCTION IN LIMITED TIME AND FREQUENCY INTERVAL

Here, the grammians and model reduction in the limited time and frequency intervals are presented. The grammians are determined either first in time and then frequency domain, or *vice versa*. The results are identical in both cases, since the frequency and time domain operators commute, as it is shown below.

Consider the controllability grammian in the limited time interval, defined by (76) and determined from (77). According to the Parseval theorem, the grammian in the infinite time interval $[0, \infty]$ can also be determined in the infinite frequency domain $[-\infty, \infty]$ from (83). Assuming that signals $H_c(\omega) = H(\omega)B$ in (83) are measured only inside the limited frequency interval $[-\omega, \omega]$, W_c can be determined from (88). Introducing (88) to (78) yields

$$W_c(t, \omega) = S(t)W_c(\omega)S^*(t), \tag{100}$$

where $S(t)$ is given by (79) and $W_c(\omega)$ by (88). In the next case, determining the controllability grammian in frequency domain and then applying the Parseval theorem in the time domain, one obtains

$$W_c(t, \omega) = W_c(t)S^*(\omega) + S(\omega)W_c(t), \tag{101}$$

where $W_c(t)$ is given in (78), and $S(\omega)$ is in (89). The question arises whether $W_c(t, \omega)$ in (100) and (101) are identical. In order to show it, first note that

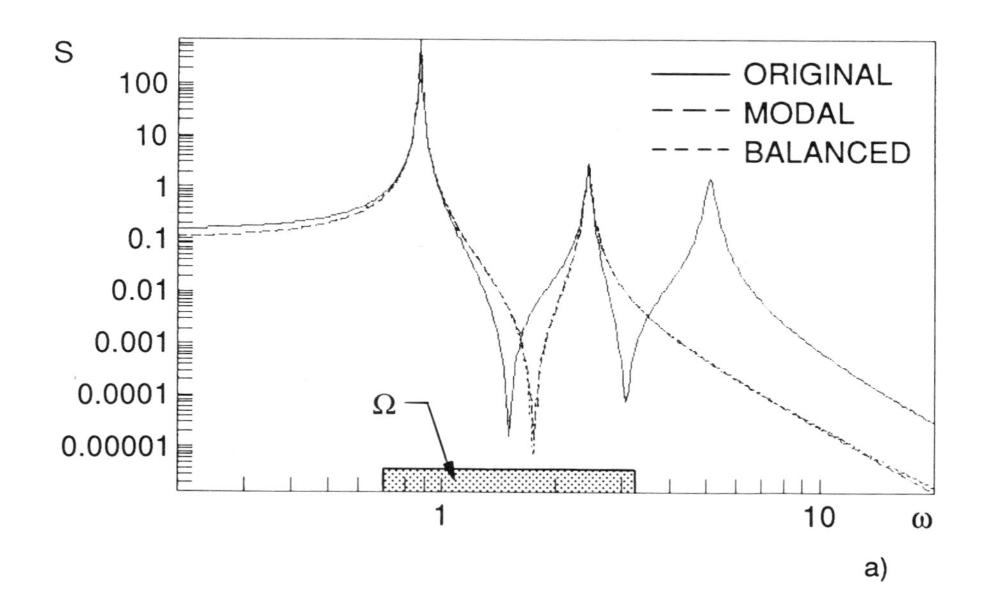

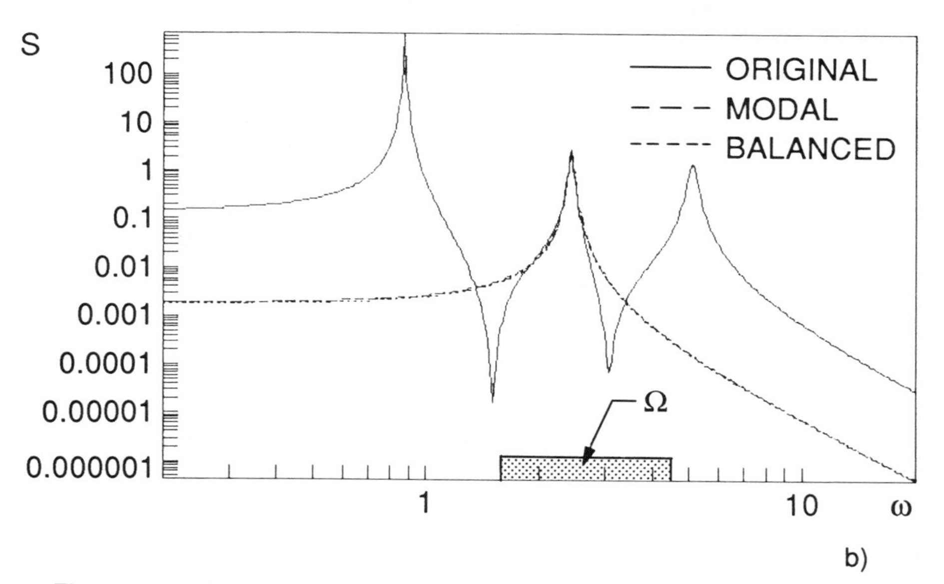

Figure 24 Power spectral density functions of a three-spring-dashpot system and its reduced models for a) $\Omega = [0.7, 3.2]$, b) $\Omega = [1.5, 3.2]$.

$S(t)$ and $S(\omega)$ commute

$$S(t)S(\omega) = S(\omega)S(t) \tag{102}$$

for any ω and t. Indeed, since

$$S(\omega) = \int_{-\omega}^{\omega} (j\nu I - A)^{-1} d\nu = \int_{-\omega}^{\omega} F(e^{A\tau}) d\nu, \tag{103}$$

where $F(.)$ denote the Fourier transformation operator, one obtains

$$S(t)S(\omega) = \int_{-\omega}^{\omega} e^{At} F(e^{A\tau}) d\nu = \int_{-\omega}^{\omega} F(e^{A\tau}) e^{AT} d\nu = S(\omega)S(t). \tag{104}$$

Next, introduction of (88) to (100), with the aid of (102), yields

$$W_c(t,\omega) = S(t)W_c S^*(t)S^*(\omega) + S(\omega)S(t)W_c S^*(t). \tag{105}$$

Recalling (78), the equality of (100) and (101) is thus proved.

As a consequence of the commuting property, the controllability grammian over the limited time interval $T = [t_1, t_2]$, $t_2 \geq t_1$, and the frequency interval $\Omega = [\omega_1, \omega_2]$, $\omega_2 \geq \omega_1$, is determined from

$$W_c(T,\Omega) = W_c(T,\omega_2) - W_c(T,\omega_1) = W_c(t_1,\Omega) - W_c(t_2,\Omega), \tag{106a}$$

where

$$W_c(T,\omega) = W_c(t_1,\omega) - W_c(t_2,\omega), \quad W_c(t,\Omega) = W_c(t,\omega_2) - W_c(t,\omega_1), \tag{106b}$$

and

$$W_c(t,\omega) = S(t)W_c(\omega)S^*(t) = W_c(t)S^*(\omega) + S(\omega)W_c(t). \tag{106c}$$

Similarly the observability grammian over the time interval T and frequency interval Ω is obtained from

$$W_o(T,\Omega) = W_o(T,\omega_2) - W_o(T,\omega_1) = W_o(t_1,\Omega) - W_o(t_2,\Omega), \tag{107a}$$

where

$$W_o(T,\omega) = W_o(t_1,\omega) - W_o(t_2,\omega), \quad W_o(t,\Omega) = W_o(t,\omega_2) - W_o(t,\omega_1), \tag{107b}$$

and

$$W_o(t,\omega) = S^*(t)W_o(\omega)S(t) = W_o(t)S(\omega) + S^*(\omega)W_o(t). \tag{107c}$$

The modal grammians in limited time and frequency intervals are given in the Appendix A.

In order to determine the stability conditions for the reduced balanced model in limited time and frequency intervals, note that $W_c(T,\Omega)$, $W_o(T,\Omega)$ satisfy the Lyapunov equations

$$\begin{aligned} W_c(T,\Omega)A^* + AW_c(T,\Omega) + Q_c(T,\Omega) &= 0, \\ W_o(T,\Omega)A + A^*W_o(T,\Omega) + Q_o(T,\Omega) &= 0, \end{aligned} \tag{108}$$

where

$$Q_c(T,\Omega) = Q_c(t_1,\Omega) - Q_c(t_2,\Omega), \quad Q_c(t,\Omega) = Q_c(t,\omega_2) - Q_c(t,\omega_1), \tag{109a}$$

$$Q_c(t,\omega) = Q_c(t)S^*(\omega) + S(\omega)Q_c(t) = S(t)Q_c(\omega)S^*(t), \tag{109b}$$

and $Q_c(t)$, $Q_c(\omega)$ are given by (80) and (94) respectively. Similarly

$$Q_o(T,\Omega) = Q_o(t_1,\Omega) - Q_o(t_2,\Omega), \quad Q_o(t,\Omega) = Q_o(t,\omega_2) - Q_o(t,\omega_1), \tag{110a}$$

$$Q_o(t,\omega) = Q_o(t)S(\omega) + S^*(\omega)Q_o(t) = S^*(t)Q_o(\omega)S(t), \tag{110b}$$

and $Q_o(t)$, $Q_o(\omega)$ are given by (80) and (99) respectively.

Theorem 3. For a stable A and positive semidefinite $Q_c(T,\Omega)$, $Q_o(T,\Omega)$, the reduced balanced model is stable.

Proof. For positive semidefinite $Q_c(T,\Omega)$, $Q_o(T,\Omega)$ one can find $B(T,\Omega)$, $C(T,\Omega)$ such that $Q_c(T,\Omega) = B(T,\Omega)B^*(T,\Omega)$, $Q_o(T,\Omega) = C^*(T,\Omega)C(T,\Omega)$. According to the Theorem from Appendix B, the reduced balanced model obtained for the triple $(A, B(T,\Omega), C(T,\Omega))$ is stable, which completes the proof.

Again, large time and frequency intervals result in a stable reduced balanced model. The computational procedure can be set up alternatively: either by first applying frequency and then time transformation of grammians, or by first applying time and then frequency transformation. Only the first procedure is further considered, since the second one is similar.

1. Determine grammians W_c, W_o from (4) with given (A, B, C).
2. Compute $W_c(\omega_1)$, $W_c(\omega_2)$, $W_o(\omega_1)$, $W_o(\omega_2)$ from (88), (90).
3. Compute $W_c(t_\alpha, \omega_\beta)$, $W_o(t_\alpha, \omega_\beta)$ from (106c), (107c), for $\alpha, \beta = 1, 2$.
4. Compute $W_c(t_\alpha, \Omega)$, $W_o(t_\alpha, \Omega)$ from (106b), (107b), for $\alpha = 1, 2$.
5. Compute $W_c(T, \Omega)$, $W_o(T, \Omega)$ from (106a), (107a).
6. Apply the reduction procedure to obtain (A_R, B_R, C_R) for $W_c(T, \Omega)$ $W_o(T, \Omega)$.
7. Compute J, δ, ε, from (12), (13), (25) for the reduced model (A_R, B_R, C_R).

Example18. Example 14 is considered in the finite time interval $T = [0, 8]$, and the finite frequency interval $\Omega = [0.7, 3.2]$. The reduction errors are $\delta_b = 0.1153$, or $\delta_m = 0.1358$, and the optimality indices $\varepsilon_b = 0.0406$, or $\varepsilon_m = 0.02595$. Figures 25a,b show the impulse responses and the power spectral density functions of the original system, the reduced balanced model, and the reduced modal model. This example is a combination of Examples 14 and 17, however, the results are different. In Example 14 for $T = [0, 8]$, the second pole was deleted. In the present example, the third pole is deleted, as a result of the additional restriction on the frequency interval.

D. APPLICATIONS

In this section model reduction of the Advanced Supersonic Transport and the flexible truss structure are analyzed. The model of the Advance Supersonic Transport, given by Colgren [29], is a linear unstable system of order eight, with four inputs and eight outputs. In order to make our presentation concise, the model presented here is restricted to one output. The system triple (A, B, C) is

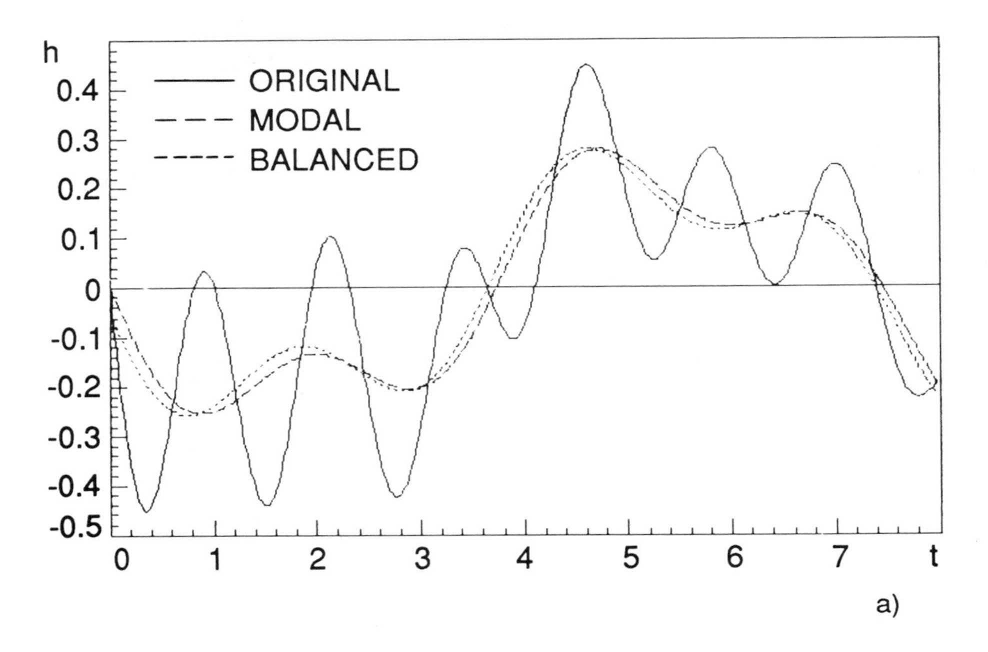

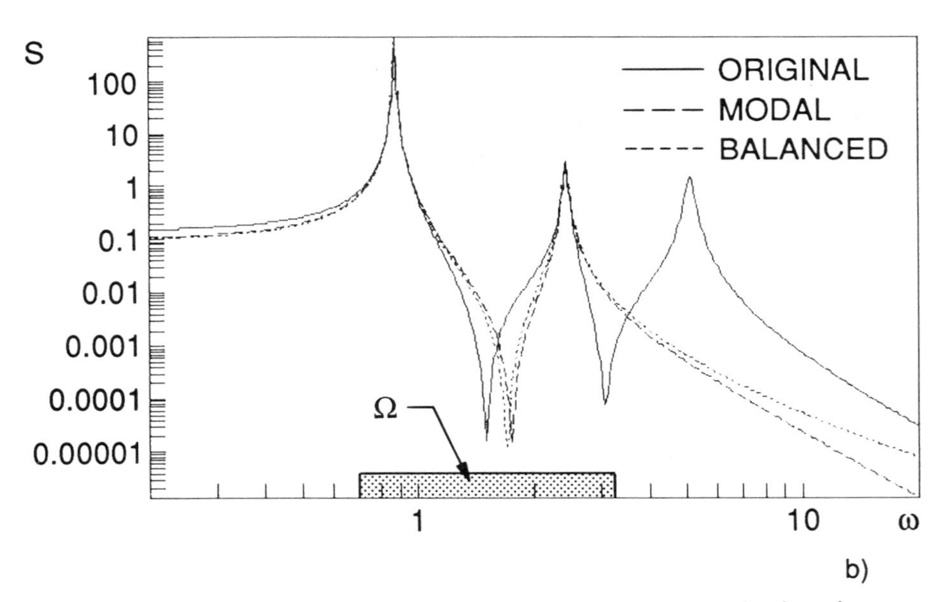

Figure 25 Impulse responses and power spectral density functions of a three-spring-dashpot system and its reduced model for $T = [0, 8]$, and $\Omega = [0.7, 3.2]$.

$$A = \begin{bmatrix} -.0127 & -.0136 & -.0360 & .0000 & .0000 & .0000 & .0000 & .0000 \\ -.0969 & -.4010 & .0000 & .9610 & 19.59 & -.1185 & -9.200 & -.1326 \\ .0000 & .0000 & .0000 & 1.000 & .0000 & .0000 & .0000 & .0000 \\ -.2290 & 1.726 & .0000 & -.7220 & -12.02 & -.3420 & 1.842 & .8810 \\ .0000 & .0000 & .0000 & .0000 & .0000 & 1.000 & .0000 & .0000 \\ .0000 & .1204 & .0000 & .0496 & -44.00 & -1.274 & -4.030 & -.5080 \\ .0000 & .0000 & .0000 & .0000 & .0000 & .0000 & .0000 & .0000 \\ .0000 & .1473 & .0000 & .3010 & -7.490 & -.1257 & -21.70 & -.8030 \end{bmatrix}$$

$$B^T = \begin{bmatrix} .0000 & -.0215 & .0000 & -1.097 & .0000 & -.6400 & .0000 & -1.882 \\ .0194 & .0000 & .0000 & .0000 & .0000 & .0000 & .0000 & .0000 \\ .0000 & -.0040 & .0000 & .3660 & .0000 & .1625 & .0000 & .4720 \\ .0000 & -1.786 & .0000 & -.0569 & .0000 & -.0370 & .0000 & -.0145 \end{bmatrix}$$

$$C = [0\ 1\ 0\ 0\ 0\ 0\ 0\ 0].$$

The system poles are $\Lambda = \text{diag}(0.6687, -1.7756, -0.0150 \pm j0.0886, -0.3122 \pm j4.4485, -0.7257 \pm 6.7018)$, so there is one unstable mode. The classical approach to the model reduction with an unstable mode consists of removing the unstable pole from the model, and applying the reduction procedure in the infinite time interval [29]. Then the unstable pole is added back to the reduced model. Here, the finite time balanced reduction is applied to the aircraft model, without removal of the unstable mode. The time interval $T = [0, 1.4]$ is chosen. Model reduction from 8 to 5 state variables within the time interval T is performed. The obtained reduced model (A_R, B_R, C_R) is

$$A_R = \begin{bmatrix} .60190 & .13115 & -.56625 & .04536 & -.07113 \\ -.00496 & -.27179 & -2.8503 & -4.1174 & .29035 \\ -.17438 & 2.2471 & -2.1435 & .26343 & 4.7254 \\ .14697 & 4.2541 & .56207 & -.09845 & -.25075 \\ -.03516 & .30043 & -3.1991 & .11915 & -1.3764 \end{bmatrix}$$

$$B_R = \begin{bmatrix} -.01360 & .00150 & -.01633 & 1.1356 \\ .75034 & .00027 & -.18467 & .02078 \\ -.68156 & -.00184 & .19596 & .38055 \\ .14032 & .00138 & -.04967 & -.30202 \\ -.35104 & -.00110 & .09182 & .03674 \end{bmatrix}$$

$$C_R = [-1.1348\ -.74659\ -.94062\ .31464\ .41958]$$

with its poles $\Lambda_R = \text{diag}(0.6151, -1.4217 \pm j3.2551, -0.5260 \pm j5.2109)$. The reduction error is $\delta_b = 0.008462$, with the optimality index $\varepsilon_b = 0.01947$, so that the obtained reduced model is near-optimal. Its step responses (dotted line) are compared with the step responses of the full model (solid line) in Fig. 26.

Next, model reduction of a flexible truss structure given in Fig. 7 is investigated. Every truss node has 2 degrees of freedom, and is numbered as in Fig. 7. The first number denotes the horizontal displacement, and the second one the vertical displacement. Let f_i be the force applied to the i-th degree of freedom, q_i and v_i be the displacement and the velocity at the i-th degree of freedom respectively. The system has one input u, applied as follows $f_2 = f_3 = f_{11} = -u$, $f_{10} = f_{20} = f_{21} = u$, and two outputs $y_1 = q_{20}$, $y_2 = \alpha v_{17}$, where the weighting coefficient $\alpha = 0.001$ is used to scale y_2. The system poles are shown as follows:

$$
\begin{array}{lll}
\lambda_{1,2} = -0.0796 \pm j399.0, & \lambda_{3,4} = -0.6955 \pm j1179.4, & \lambda_{5,6} = -0.8276 \pm j1286.6., \\
\lambda_{7,8} = -2.6429 \pm j2299.1, & \lambda_{9,10} = -4.7173 \pm j3070.9 & \lambda_{11,12} = -4.9259 \pm j3138.8, \\
\lambda_{13,14} = -5.6674 \pm j3366.7, & \lambda_{15,16} = -6.4074 \pm j3579.8, & \lambda_{17,18} = -8.9431 \pm j4229.2, \\
\lambda_{19,20} = -9.9084 \pm j4451.6, & \lambda_{21,22} = -12.775 \pm j5054.6, & \lambda_{23,24} = -14.261 \pm j5340.6, \\
\lambda_{25,26} = -14.814 \pm j5443.1, & \lambda_{27,28} = -17.137 \pm j5854.4, & \lambda_{29,30} = -20.846 \pm j6456.9, \\
\lambda_{31,32} = -21.170 \pm j6506.9, & \lambda_{33,34} = -22.994 \pm j6781.4, & \lambda_{35,36} = -25.208 \pm j7100.4, \\
\lambda_{37,38} = -27.386 \pm j7400.7, & \lambda_{39,40} = -29.647 \pm j7700.1, & \lambda_{41,42} = -38.775 \pm j8806.1.
\end{array}
$$

The reduction from 42 to 22 state variables is considered in five cases: Case 1, grammians and balanced representation are obtained in the time interval $T_1 = [0,\ 0.04]$; Case 2, in the time interval $T_2 = [0,\ 0.2]$; Case 3, grammians are determined and balanced in the frequency interval $\Omega_1 = [0,\ 5500]$; Case 4, in the frequency interval $\Omega_2 = [1700,\ 7000]$; Case 5, in the finite time interval T_2, and the finite frequency interval Ω_1. In Case 1 the reduction index is $\delta_b = 0.09272$, or $\delta_m = 0.09748$, and the optimality index is $\varepsilon_b = 0.01761$, or $\varepsilon_m = 0.01595$. The outputs of the original as well as the reduced models are shown in Fig. 27. In Case 2 the following indices $\delta_b = 0.08539$, $\delta_m = 0.08915$, $\varepsilon_b = 0.001797$, and $\varepsilon_m = 0.003063$ are obtained. In Case 3 the results are $\delta_b = 0.02193$, $\delta_m = 0.02242$, $\varepsilon_b = 0.001372$, and $\varepsilon_m = 0.008987$, whereas in Case 4 $\delta_b = 0.04327$, $\delta_m = 0.08702$, $\varepsilon_b = 0.005830$, and $\varepsilon_m = 0.02107$. The power spectrum of both outputs for Case 3 are shown in Fig. 28, and for Case 4

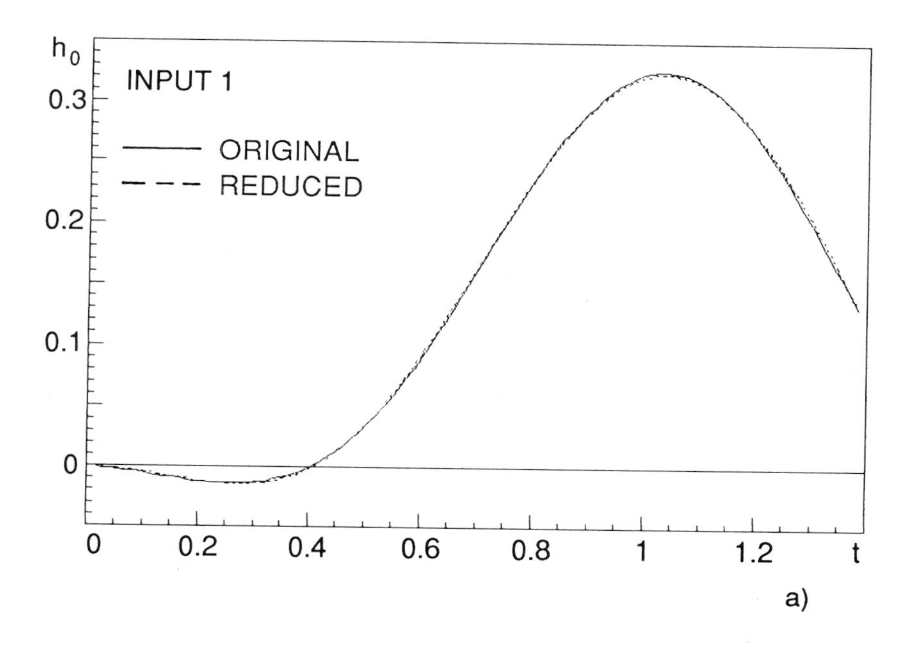

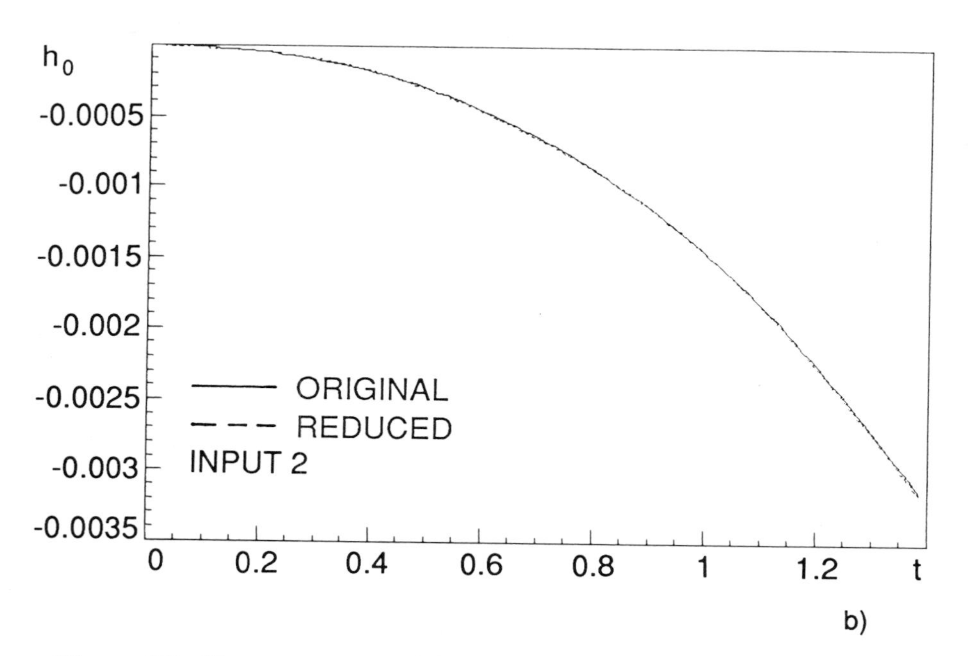

Figure 26. Step responses of the Advanced Supersonic Transport and its reduced model for T = [0, 1.4].

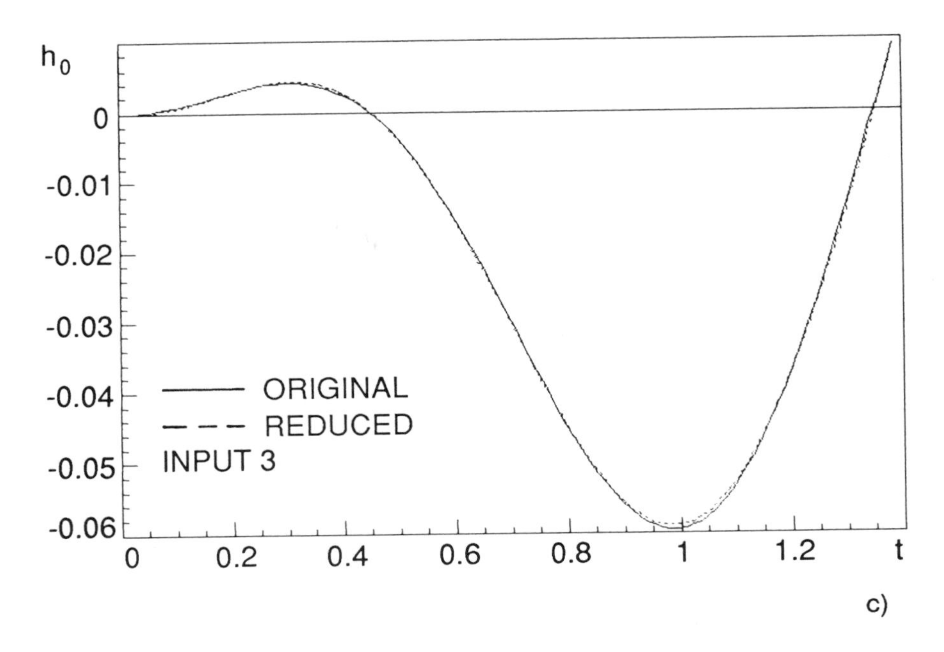

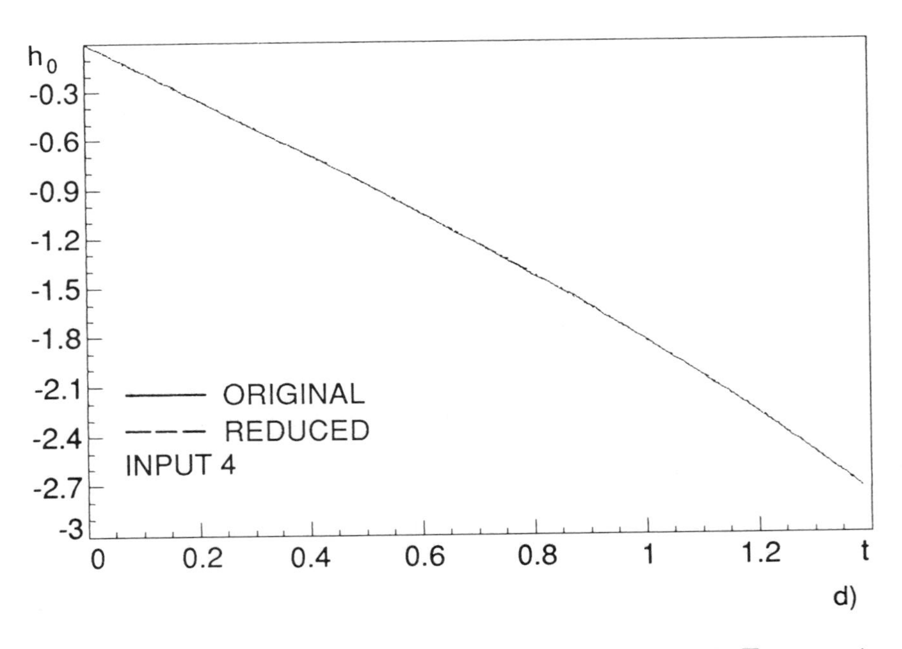

Figure 26. **Step responses of the Advanced Supersonic Transport and its reduced model for T = [0, 1.4].**

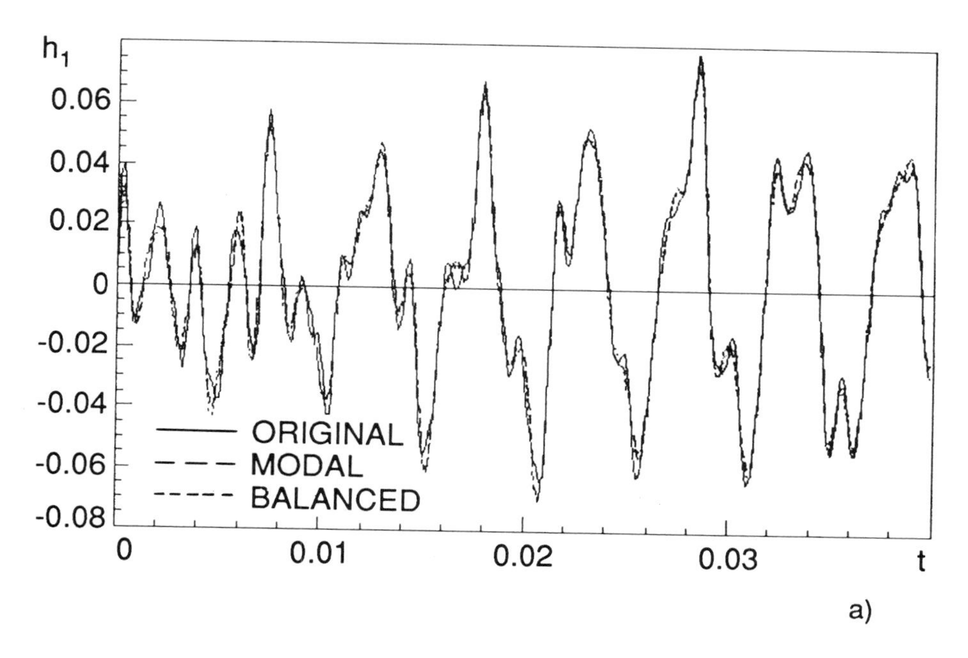

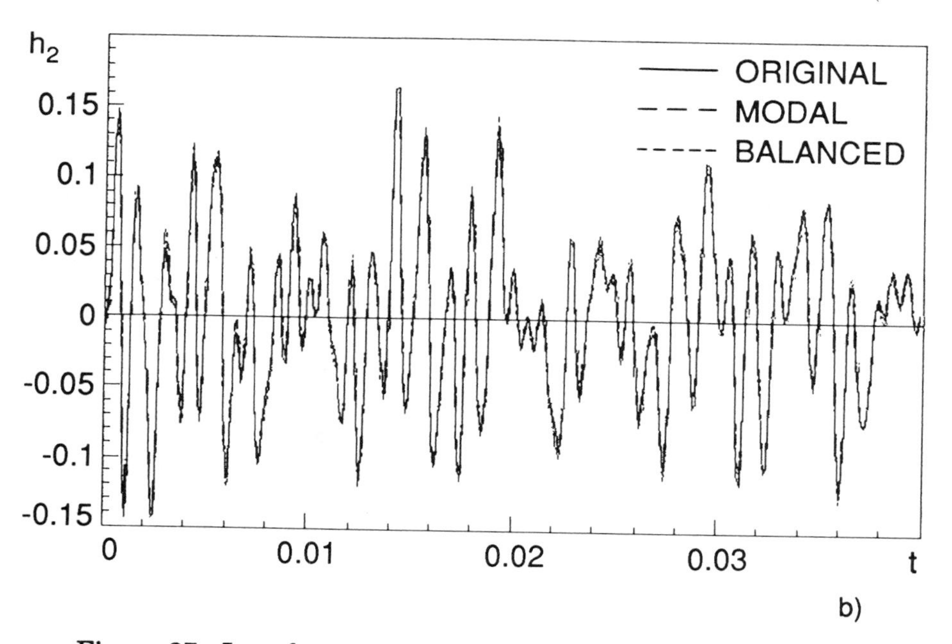

Figure 27. Impulse responses of the truss structure, and its reduced models for T = [0, 0.04].

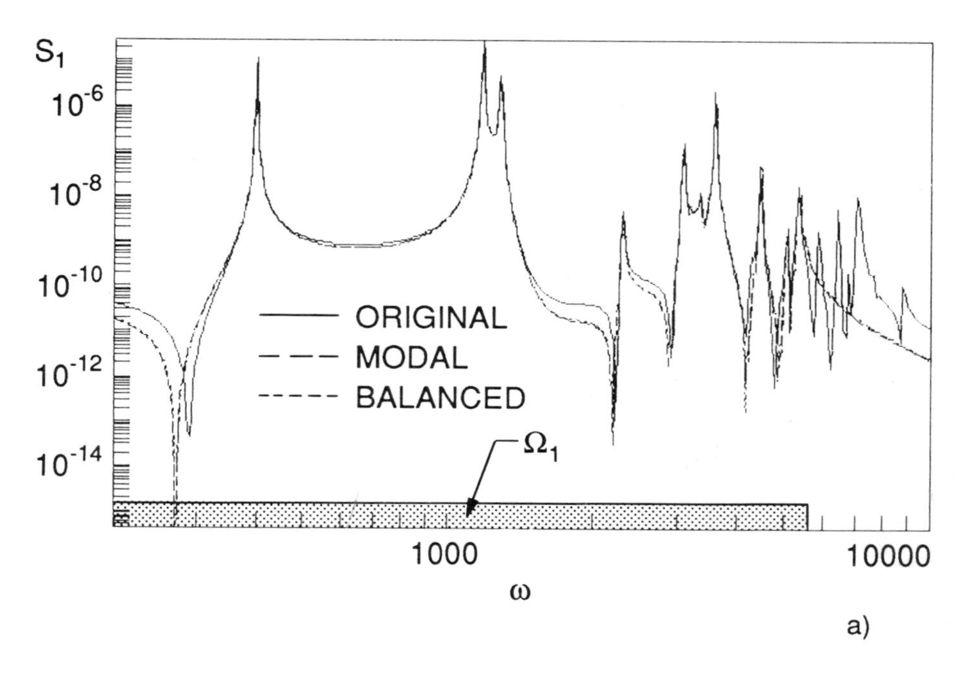

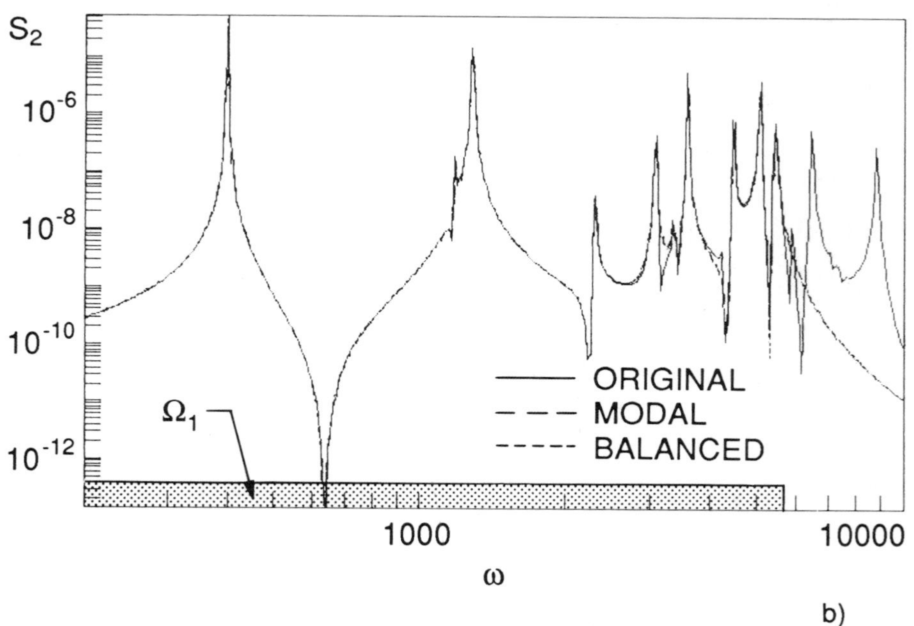

Figure 28. Power spectral density functions of the truss structure and its reduced models for $\Omega = [0, 5500]$.

in Fig. 29. In Case 5, which is a combination of Case 2 and 3, the results are $\delta_b = 0.04061$, or $\delta_m = 0.03710$, and $\varepsilon_b = 0.006334$, or $\varepsilon_m = 0.01444$, and the impulse responses and power spectra are shown in Fig. 30. The reduced models, for all presented cases are near-optimal, due to the small value of ε. The reduction error is also reasonably small. The model reduction from 42 to 22 state variables has less than 10% error in each case.

VI. CONCLUSIONS AND SUMMARY

It was shown in this chapter that the reduction in the balanced or modal coordinates, under certain conditions, is near-optimal. These conditions were used by the authors to design computer codes for near-optimal reduction in the balanced or modal coordinates; the codes have appeared useful in most applications. Note again, that while the presented method does not give the optimal solution, it gives in many cases results acceptable in applications, since the obtained solutions are near the global optimum.

This study has also considered the specific problem of model reduction for flexible structures by means of either modal truncation or the technique of balancing. While it is well established that these methods are essentially the same for a lightly damped structure with widely separated natural frequencies, it is emphasized that the same is *not* true for the typical flexible space structure with many closely-spaced frequencies. These considerations are quantified by using the concept of modal correlations, which are derived from closed-form expressions for the grammians of a structure and displayed in a particularly simple graphical form. Another question addressed was that of the sensitivity of modeling error to variations in the (generally ill-defined) damping present in the structure. A robust model reduction technique is developed to cope with this problem.

The determination of reduction indices (Hankel singular values and component costs) directly from the test data of a flexible structure is presented. The effect of actuator dynamics on reduction indices is discussed. It is shown that the reduced model obtained from test data may be far from the optimal one because of actuator dynamics. An approach is given to reconstruct the indices of the structure from the indices of the combined structure and actuator dynamics.

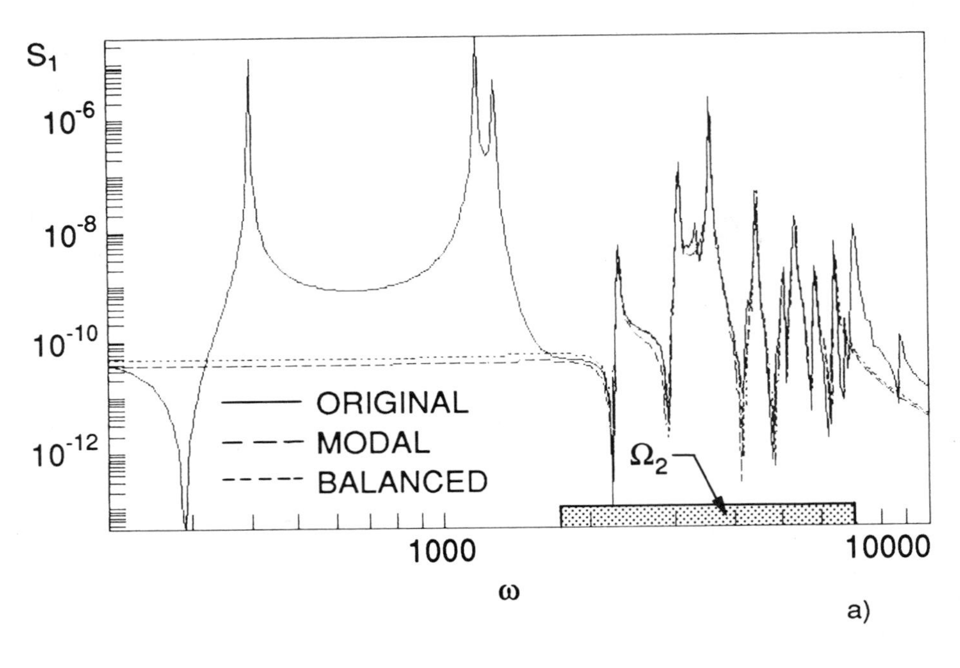

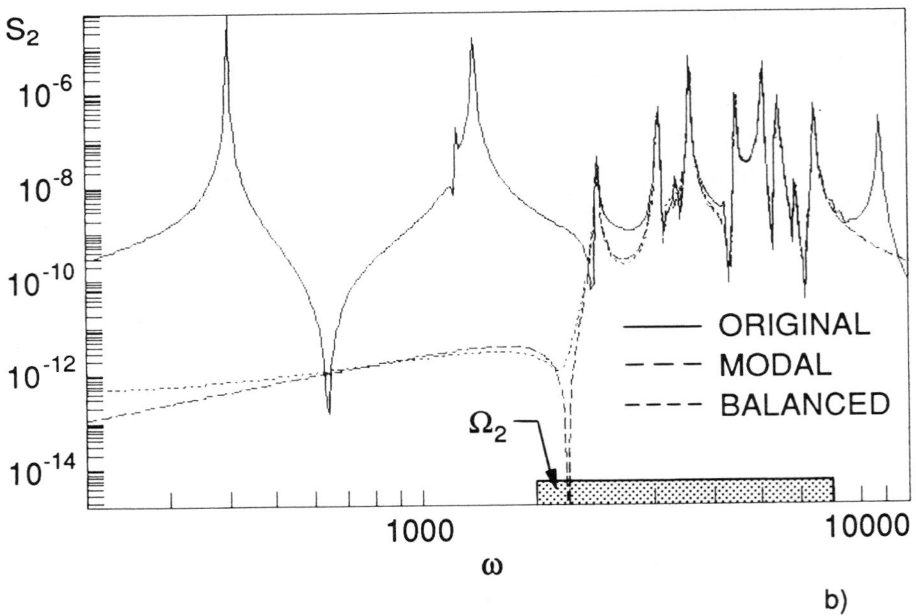

Figure 29. Power spectral density functions of the truss structure and its reduced models for $\Omega = [1700, 7000]$.

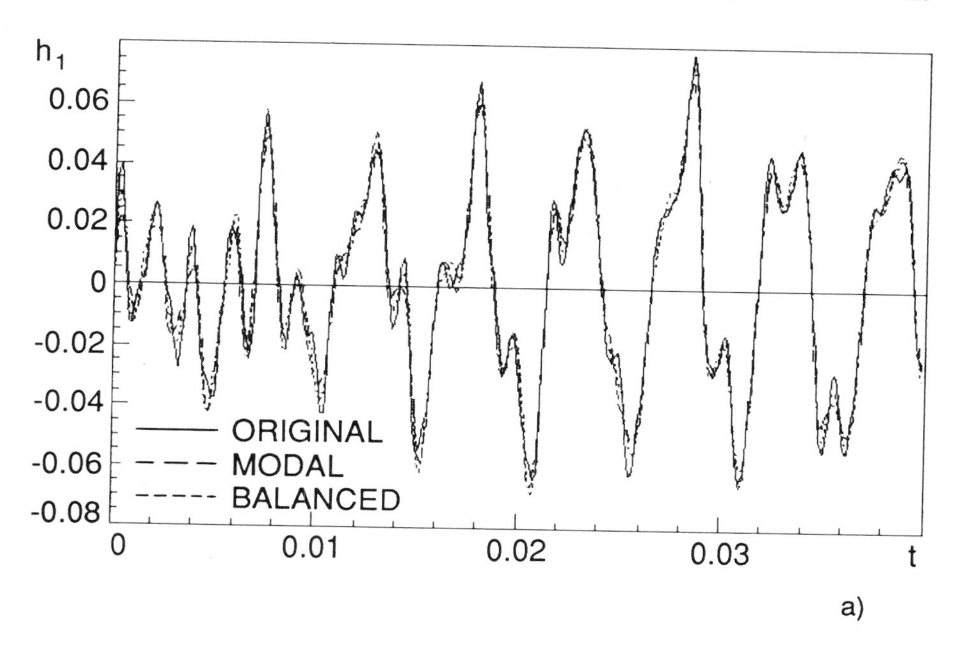

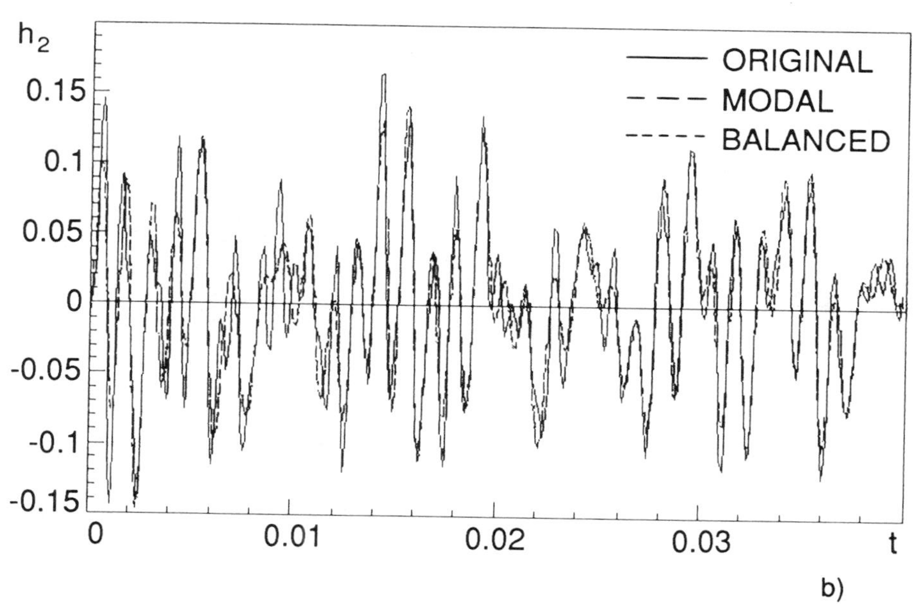

Figure 30. Impulse responses and power spectral density functions of the truss structure and its reduced models for $T = [0, 0.2]$ and $\Omega = [0, 5500]$.

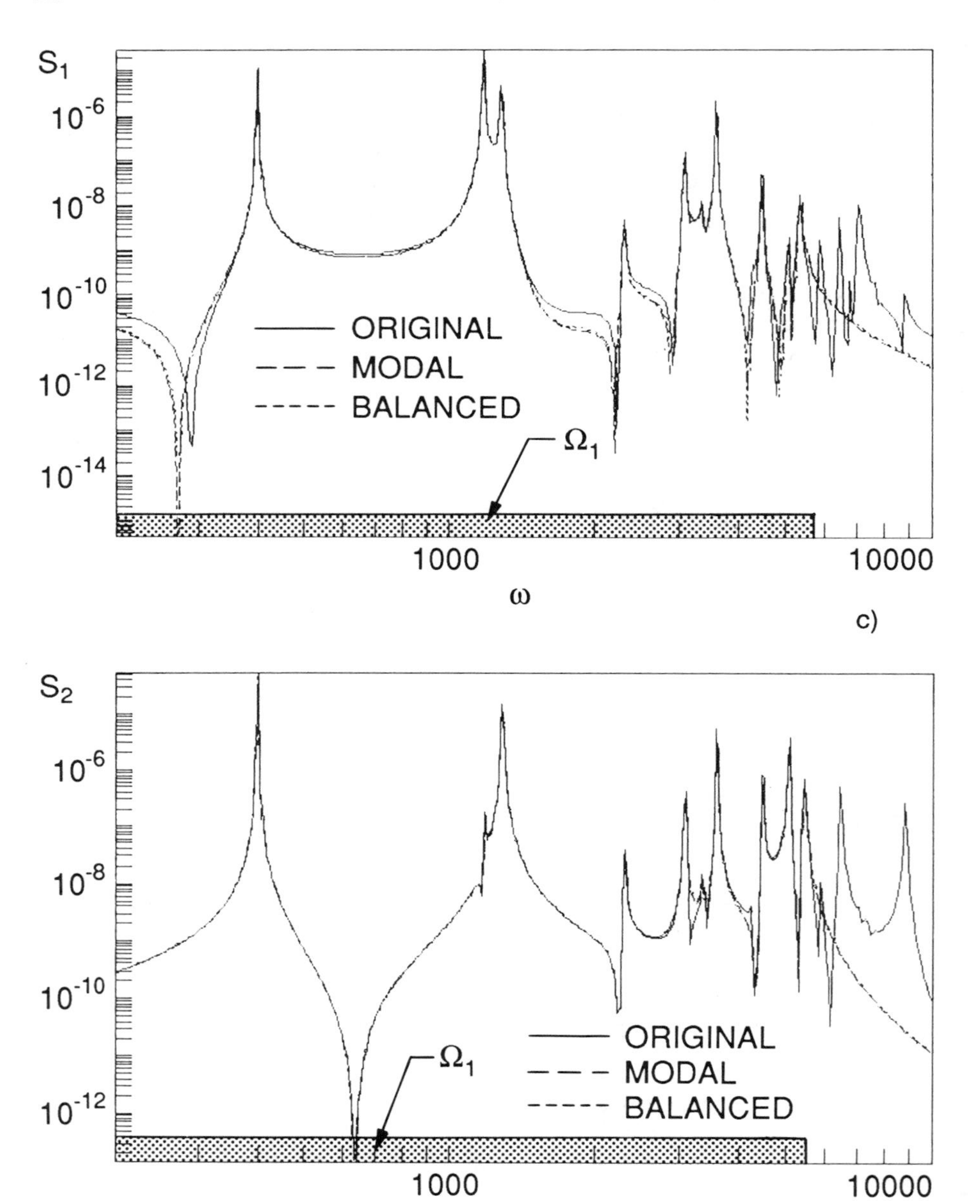

Figure 30. Impulse responses and power spectral density functions of the truss structure and its reduced models for $T = [0, 0.2]$ and $\Omega = [0, 5500]$.

The system balancing as well as the model reduction can be performed in finite time and/or frequency intervals. This fact makes the method more useful in practical cases, since data is always available in limited time and frequency intervals. In order to obtain grammians in limited time and frequency intervals, the stationary grammians are computed first from the standard Lyapunov equations, and then transformed into the finite time and/or frequency interval. The method presented in this study allows one to check if the obtained results are close to the optimal one. The reduction in the balanced or modal coordinates gives, in most cases, the reduced model close to the optimal one, and the illustrative examples support this statement.

It is well known that if the system is stable, the reduced model in the infinite time interval obtained from the balanced coordinates is also stable. In a finite time interval the stability of the balanced reduced model may not be assured. In the case where the reduction in balanced coordinates gives an unstable model, the use of modal coordinates, or Schur balanced coordinates, is recommended. On the other hand, an unstable system can be reduced using limited-time balancing, since the grammians exist for unstable systems within finite time interval.

APPENDIX

APPENDIX A. GRAMMIANS IN MODAL COORDINATES

Modal grammians are important from the point of view of computational efficiency. For large-scale systems computational effort in modal coordinates is much smaller than in other coordinates.

The $ik-th$ entry of the limited-time controllability grammian, $W_c(T)$, or the observability grammian, $W_o(T)$, in modal coordinates is determined as follows

$$w_{cik}(T) = w_{cik}(t_1) - w_{cik}(t_2), \quad w_{oik}(T) = w_{oik}(t_1) - w_{oik}(t_2), \tag{A.1}$$

where

$$w_{cik}(t) = w_{cik} s_i(t) s_k^*(t), \quad w_{oik}(t) = w_{oik} s_i^*(t) s_k(t), \tag{A.2}$$

and $w_{cik} = -b_i b_k^* / \left(\lambda_1 + \lambda_k^*\right)$, $w_{oik} = -c_i^* c_k / \left(\lambda_i^* + \lambda_k\right)$, $s_i(t) = \exp(\lambda_i t)$, b_i is the i-th row of B, and c_i is the i-th column of C.

The $ik-th$ entry of the controllability grammian, $W_c(\Omega)$, or the observability grammian, $W_o(\Omega)$, in modal coordinates in a finite frequency interval, Ω, is determined as follows

$$w_{cik}(\Omega) = w_{cik}(\omega_2) - w_{cik}(\omega_1), \quad w_{oik}(\Omega) = w_{oik}(\omega_2) - w_{oik}(\omega_1), \tag{A.3a}$$

where

$$w_{cik}(\omega) = w_{cik}\left(s_i(\omega) + s_k^*(\omega)\right), \quad w_{oik}(\omega) = w_{oik}\left(s_i^*(\omega) + s_k(\omega)\right), \tag{A.3b}$$

$$s_i(\omega) = (j/2\pi)ln((j\omega + \lambda_i)/(-j\omega + \lambda_i)), \tag{A.3c}$$

and w_{cik}, w_{oik} are the ik-th entries of the modal grammians, as above. Alternatively,

$$w_{cik}(\Omega) = w_{cik}(s_i(\Omega) + s^*k(\Omega)), \quad w_{oik}(\Omega) = w_{oik}\left(s_i^*(\Omega) + s_k(\Omega)\right), \tag{A.4}$$

where $s_i(\Omega) = s_i(\omega_2) - s_i(\omega_1)$.

The $ik-th$ entry of $W_c(T,\Omega)$, $W_o(T,\Omega)$ in modal coordinates in a finite time, T, and finite frequency interval, Ω, is determined as

$$w_{cik}(T,\Omega) = w_{cik}(T,\omega_2) - w_{cik}(T,\omega_1), \; w_{oik}(T,\Omega) = w_{oik}(T,\omega_2) - w_{oik}(T,\omega_1), \tag{A.5a}$$

or

$$w_{cik}(T,\Omega) = w_{cik}(t_1,\Omega) - w_{cik}(t_2,\Omega), \; w_{oik}(T,\Omega) = w_{oik}(t_1,\Omega) - w_{oik}(t_2,\Omega), \tag{A.6a}$$

where

$$w_{cik}(T,\omega) = w_{cik}(t_1,\omega) - w_{cik}(t_2,\omega), \quad w_{oik}(T,\omega) = w_{oik}(t_1,\omega) - w_{oik}(t_2,\omega), \tag{A.5b}$$

or

$$w_{cik}(t,\Omega) = w_{cik}(t,\omega_2) - w_{cik}(t,\omega_1), \quad w_{oik}(T,\omega) = w_{oik}(t,\omega_2) - w_{oik}(t,\omega_1), \tag{A.6b}$$

and

$$w_{cik}(t,\omega) = w_{cik}\left(s_i(\omega) + s_k^*(\omega)\right) s_i(t) s_k^*(t), \tag{A.7a}$$

$$w_{oik}(t,\omega) = w_{oik}\left(s_i^*(\omega) + s_k(\omega)\right) s_i^*(t) s_k(t). \tag{A.7b}$$

The terms $w_{cik}, w_{oik}, s_i(t), s_i(\omega)$ are defined earlier.

APPENDIX B. STABILITY THEOREM

The presented theorem is a modification of Moore, [4] Parnebo and Silverman [17]. Glover [30] stability theorem. In this theorem the requirement of positive definiteness of grammians is dropped.

Theorem. Let (A, B, C) be a triple satisfying

$$A\Gamma^2 + \Gamma^2 A^* + BB^* = 0, \quad A^*\Gamma^2 + \Gamma^2 A + C^*C = 0, \tag{B.1}$$

where $\Gamma = \text{diag}(\gamma_i), i = 1, \ldots, n$, with the entries γ_i in decreasing order, $\gamma_i \geq \gamma_{i+1} \geq 0$. Let Γ be partitioned such that $\Gamma = \text{diag}(\Gamma_R, \ \Gamma_T)$, where Γ_R is positive definite, and the triple (A, B, C) partitioned accordingly, then (A_R, B_R, C_R) is stable.

Proof. Equations (B.1) and (9) imply

$$A_R\Gamma_R^2 + \Gamma_R^2 A_R^* + B_R B_R^* = 0, \quad A_R^*\Gamma_R^2 + \Gamma_R^2 A_R + C_R^* C_R = 0.$$

Since Γ_R is positive definite, (A_R, B_R, C_R) is completely controllable and observable, and consequently, from the Theorem 3.3 of Glover [30], A_R is stable.

ACKNOWLEDGEMENTS

This work was done while the first author held a National Research Council-NASA Langley Research Center Senior Research Associateship. Participation of Dr. Trevor Williams in preparation of Section IV is greatly appreciated.

REFERENCES

1. R. E. Skelton and P. C. Hughes, "Modal Cost Analysis for Linear Matrix-Second-Order Systems," *Journal of Dynamic Systems, Measurements and Control*, **102**, pp. 151–158, 1980.
2. R. E. Skelton, "Cost Decomposition of Linear Systems with Application to Model Reduction," *International Journal of Control*, **32**, pp. 1031–1055, 1980.

3. R. E. Skelton and A. Yousuff, "Component Cost Analysis of Large Scale Systems," *International Journal of Control*, **37**, 1983, pp. 285–304.
4. B. C. Moore, "Principal Component Analysis in Linear Systems: Controllability, Observability, and Model Reduction," *IEEE Transactions on Automatic Control*, **AC-26**, pp. 17–32, 1981.
5. C. Z. Gregory, Jr., "Reduction of Large Flexible Spacecraft Models Using Internal Balancing Theory," *Journal of Guidance, Control and Dynamics*, **7**, pp. 725–732, 1984.
6. E. A. Jonckheere, "Principal Component Analysis of Flexible Systems—open loop case," *IEEE Transactions on Automatic Control*, **AC-29**, pp. 1095–1097, 1984.
7. W. Gawronski and H. G. Natke, "Balancing Linear Systems," *International Journal of Systems Science*, **17**, pp. 237–249, 1987.
8. W. Gawronski and H. G. Natke, "Balanced State Space Representation in the Identification of Dynamic Systems," *Application of System Identification in Engineering*, ed. H. G. Natke, Springer, Wien-New York, 1988.
9. D. A. Wilson, "Optimum Solution of Model-Reduction Problem, *IEE Proceedings*," **119**, pp. 1161–1165, 1970.
10. D. A. Wilson, "Model Reduction for Multivariable Systems," *International Journal of Control*, **20**, pp. 57–64, 1974.
11. D. C. Hyland and D. S. Bernstein, "The Optimal Projection Equations for Model Reduction and the Relationships among the Methods of Wilson, Skelton and Moore," *IEEE Transactions on Automatic Control*, **AC-30**, pp. 1201–1211, 1985.
12. R. E. Skelton, P. C. Hughes and H. B. Hablani, "Order Reduction for Models of Space Structures Using Modal Cost Analysis," *Journal of Guidance, Control, and Dynamics*, **5**, pp. 351–357, 1982.
13. W. Gawronski and J.-N. Juang, "Near-Optimal Model Reduction in Balanced and Modal coordinates," *Proc. 26th Annual Allerton Conference on Communication, Control and Computing*, Monticello, IL, Sept. 1988.
14. W. Gawronski and J.-N. Juang, "Grammians and Model Reduction in Limited Time and Frequency Range," *Proc. Guidance, Navigation and Control Conference*, Minneapolis, MN, pp. 275–285, Aug. 1988.
15. Gawronski, W. and Williams, T. "Model Reduction for Flexible Space

Structures," *Proceedings of 30th AIAA/ASME/ASCE/AHS/ASC 30th Structures, Structural Dynamics and Materials Conference*, Mobile, Alabama, pp. 1555–1565, 1989.

16. Gawronski, W. "Model Reduction for Flexible Structures: Test Data Approach," *Proc. AIAA Guidance, Navigation, and Control Conf.*, Boston, MA, pp. 1195–1199, August 1989.
17. Parnebo, L., and Silverman, L. M., "Model Reduction via Balanced State Space Representation," *IEEE Transactions on Automatic Control*, **AC-27**, pp. 382–387, 1982.
18. P. T. Kabamba, "Balanced Gains and their Significance for L_2 Model Reduction," *IEEE Transactions on Automatic Control*, **AC-30**, pp. 690–693, 1985.
19. R. E. Skelton and P. T. Kabamba, "Comments on Balanced Gains and their Significance for L_2 Model Reduction," *IEEE Transactions on Automatic Control,*, **AC-31**, pp. 796–797, 1985.
20. Luenberger, D. G., "**Optimization by Vector Space Methods**", New York: Wiley, 1969
21. C. P. Kwong, "Optimal Chained Aggregation for Reduced Order Modelling," *Int. J. Control*, **35**, pp. 965–982, 1982.
22. K. V. Fernando, and H. Nicholson, "On the Cross-Grammians for Symmetric MIMO Systems," *IEEE Transactions On Circuits and System*, **CAS-32**, 1985, pp. 487–489.
23. J. A. De Abreu-Garcia and F. W. Fairman, "A Note on Cross-Grammians for Orthogonally Symmetric Realizations," *IEEE Transactions on Automatic Control*, **AC-31**, pp. 866–868, 1986.
24. J. A. De Abreu-Garcia and F. W. Fairman, "Balanced Realization of Orthogonally Symmetric Transfer Function Matrices," *IEEE Transactions on Circuits and Systems*, **CAS-34**, pp. 997–1010, 1987.
25. T. Williams, "Closed-Form Grammians and Model Reduction for Flexible Space Structures," *Proc. 27th IEEE Conference on Decision and Control*, Austin, TX, pp. 1157–1158, 1988.
26. R. E. Skelton, R. Singh and J. Ramakrishnan, "Component Model Reduction by Component Cost Analysis," *Proc. Guidance, Navigation and Control Conference*, Minneapolis, pp. 264–274, 1988.

27. E. A. Jonckheere and L. M. Silverman, "Singular Value Analysis of Deformable Systems," *Proc. 20th IEEE Conference on Decision and Control*, San Diego, pp. 660–668, 1981,

28. R. W. Clough and J. Penzien, "**Dynamics of Structures**," New York: McGraw Hill, 1975.

29. Colgren, R. D., "Methods for Model Reduction," *AIAA Guidance, Navigation and Control Conference*, Minneapolis, MN, pp. 777–790, 1988.

30. Glover, K., 1984, "All Optimal Hankel-Norm Approximations of Linear Multivariable Systems and their L^{∞}-Error Bounds," *International Journal of Control*, **39**, pp. 1115–1193, 1984.

Distributed Transducers for Structural Measurement and Control

Shawn E. Burke

James E. Hubbard, Jr.

The Charles Stark Draper Laboratory, Inc.
Cambridge, Massachusetts 02139

I. INTRODUCTION

Modern structural systems are required to meet increasingly stringent performance specifications. Often these specifications necessitate structural optimization and the addition of active control to meet desired performance goals. For example, consider the class of systems represented by aerospace structures. These structures are in general lightweight, compact, and may incorporate components which are lightly damped and have low mass density. Such designs often have relatively slow structural responses and long decay times which tend to degrade performance and undermine system stability.

Historically, the control system designer did not need to explicitly address structural control. Aircraft flight control systems, for example, have had bandwidths that were usually well below the aerospace structure's first resonance frequency. It was therefore possible to model a small aircraft as a rigid body. For large flexible aircraft, the response could be controlled implicitly via attenuation using notch filters, stiffeners, and passive damping treatments. These techniques were viable because of the widely separated control and structural bandwidths. Modern performance requirements, however, often exceed the capabilities of passive methodologies, and require active control systems.

Active structural control falls under the general heading of distributed parameter control, or the control of systems described using space and (usually) time variables. Distributed parameter control has been and continues to be an area of intense research interest. While there exists a wealth of distributed parameter control theory in the open literature, one finds few applications and practical implementations of the theory by way of experiment. One of the reasons for this disparity is the difficulty of marrying distributed parameter systems and lumped parameter control theory. Current design practice often relies upon lumped parameter techniques, with distributed parameter system theory and concepts being used to study the design tradeoffs concomitant with such an approach.

The distributed nature of continuum vibrating systems makes it very difficult to apply modern lumped parameter control philosophy and methods. In practice one finds that a substantial amount of precise information is needed to describe these systems, as well as a large number of sensors and actuators to identify and control the structural response. Modal expansions are commonly employed to approximate the structure's behavior [1]. This representation is complete, as the series expansion captures the true system behavior when all terms are included. However, computational limitations generally necessitate a truncation of the modal expansion to provide a more tractable design model.

Control designs based upon reduced-order structural models lead to the phenomena of control and observation spillover, as the truncated higher modes can destabilize the control system by contaminating sensor measurements with high frequency signals. This spillover problem is further compounded by the practical need for low order compensators. Many of the popular methods used for compensator order reduction use further model trucations in order to achieve a desired compensator order [2]. With this additional approximation, the spillover problem can become even more acute. It is often difficult to determine the number of modes required to accurately model a structure, and to reconcile the location of actuators and sensors [3,4]. In addition, experimental implementations of the resulting control strategies suffer from poor stability-robustness characteristics [5]. Balas has demonstrated that for even a simple flexible beam problem, control and observation spillover can cause closed-loop instability of an otherwise open-loop stable beam [6].

It is the goal of the research efforts described in this chapter to design and implement distributed parameter control schemes which allow both distributed sensing and actuation through the use of new and innovative technology. This union of distributed parameter systems with distributed parameter transducers and concepts leads to simple, realizable control system designs. The ensuing design methods do not require plant model truncations, and offer the possibility of controlling all modes of flexible structural components using simple compensator topologies. The analytical development provides insight into the utilization of both distributed *and* discrete transducers for structural measurement and control. Finally, all of the design examples presented in this chapter have been implemented and validated experimentally; summaries of these experiments are presented throughout.

In the next section unified models of linear distributed and discrete actuators and sensors are developed using concepts from the theory of singularity functions. Examples are presented for a particular class of distributed piezoelectric film transducers. In section III, these representations are combined with novel distributed parameter control design methods to yield stabilizing compensators for the vibration control of elastic beams and plates. Experimental implementations are summarized. Modal interpretations of these compensators are developed. The section concludes with a discussion of the smart structures control concept, where distributed sensors and actuators are embedded into structural components to provide "built in" active vibration control.

II. TRANSDUCER REPRESENTATIONS

A. COMPACT REPRESENTATIONS OF DISTRIBUTED TRANSDUCERS

In this section we introduce a compact analytical representation of distributed and discrete transducers. Transducers are parameterized by their placement, type (i.e. translational versus rotary devices), and distribution, or spatial aperture. In its most general form a distributed transducer is one which has the capability to reconfigure itself over space as a function of time, vis

$$u = u(x,t). \tag{1}$$

If the distributed transducer has temporal dynamics which do not vary over its spatial aperture, then this distribution can be represented as a separable product of space and time functions, and is referred to as degenerate [7]. For the class of degenerate distributed transducers Eq. (1) takes the special form

$$u(x,t) = \Lambda(x)\,u(t). \tag{2}$$

This is the mathematical form used to describe many common sensors and actuators (i.e. piezoceramic stacks, accelerometers, shakers, strain gages, etc.).

From a distributed parameter system design standpoint one must determine both the necessary spatial and temporal characteristics of the transducer. To model transducers for structural estimation and control it is necessary to choose a notation which can adequately represent a broad class of devices, both spatially discrete and distributed. It is also desirable to have a notation that is applicable to both sensors and actuators. Historically, generalized functions have provided a particularly elegant means of modeling these distributions, providing a compact notation using the theory of distributions in the form of singularity functions [8]. These functions were originally applied to structural components by Macauley in 1919, and later by Clebsch [9,10]. A brief review is offered here for the purpose of the analysis which follows; the reader is refered to references [11,12] for a more complete exposition.

Consider the family of singularity functions represented by

$$f_n = \langle x-a\rangle^n \equiv \begin{cases} 0 & \text{if } x < a, \\ (x-a)^n & \text{if } x \geq a. \end{cases} \tag{3}$$

The function f_n is defined to have a value only when the argument is positive. This is the so-called Macauly notation for singularity functions. Singularity functions obey the integration law

$$\int_{-\infty}^{x} \langle x-a\rangle^n \, dx = \frac{\langle x-a\rangle^{n+1}}{n+1}, \quad n \geq 0. \tag{4}$$

The functions $\langle x-a\rangle^{-1}$ and $\langle x-a\rangle^{-2}$ are exceptions, equalling zero everywhere except when x equals a where they are infinite, such that (3) and (4) are true. Table I presents a list of singularity functions of different degrees and their common properties. This table is a convenient means of tabulating the integrals and derivatives of these functions and, as will be shown, provides insight into the selection, design and use of distributed transducers for structural sensing and actuation.

B. DISTRIBUTED ACTUATION USING DEGENERATE TRANSDUCERS

It is important to note at this juncture that the results and analysis presented herein are generic, and apply to a broad class of actuators and sensors. In order to illustrate the application of these results, however, we choose to employ distributed piezoelectric devices, and in particular the piezoelectric polymer film polyvinylidene flouride (PVF_2) [13].

This type of transducer is made piezoelectrically active by a polarization process applied during manufacture [14]. Once the polymer film has been polarized, a displacement field applied to the film results in the production of a charge distribution at the film's surfaces which is proportional to the ensueing strain field. Consequently,

Table I. Properties of singularity functions.

Names	Definition and Integration Property	Graphical Representation
Doublet Dipole Concentrated Moment function	$\langle x-a\rangle^{-2} = 0 \quad \text{if } x \neq a$ $\int_{-\infty}^{x} \langle x-a\rangle^{-2} dx = \langle x-a\rangle^{-1}$	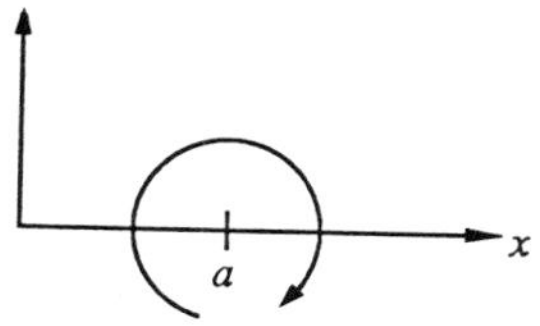 a, x
Dirac delta Delta function Impulse Concentrated force function	$\langle x-a\rangle^{-1} = 0 \quad \text{if } x \neq a$ $\int_{-\infty}^{x} \langle x-a\rangle^{-1} dx = \langle x-a\rangle^{0}$	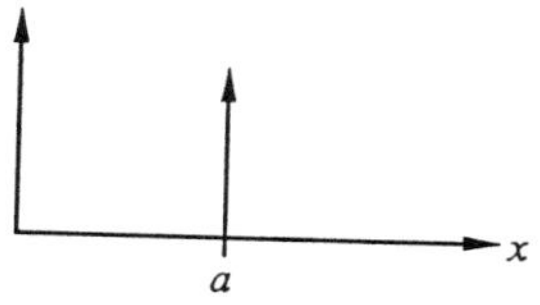 a, x
Unit step function	$\langle x-a\rangle^{0} = \begin{cases} 0 & \text{if } x < a \\ 1 & \text{if } x \leq a \end{cases}$ $\int_{-\infty}^{x} \langle x-a\rangle^{0} dx = \langle x-a\rangle^{1}$	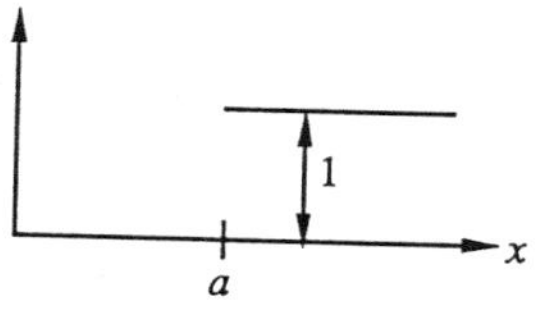 1, a, x
Ramp function	$\langle x-a\rangle^{1} = \begin{cases} 0 & \text{if } x < a \\ x-a & \text{if } x \leq a \end{cases}$ $\int_{-\infty}^{x} \langle x-a\rangle^{1} dx = \frac{\langle x-a\rangle^{2}}{2}$	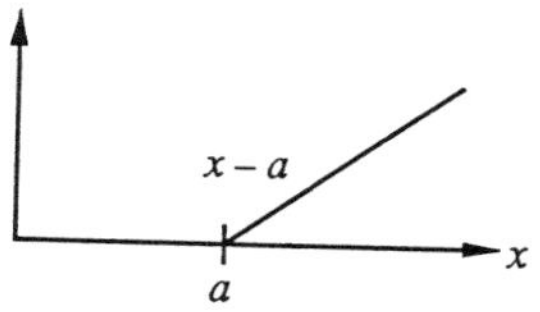 $x-a$, a, x
General Macauley notation	$\langle x-a\rangle^{n} = \begin{cases} 0 & \text{if } x < a \\ (x-a)^{n} & \text{if } x \leq a \end{cases}$ $\int_{-\infty}^{x} \langle x-a\rangle^{n} dx = \frac{\langle x-a\rangle^{n+1}}{n+1}$	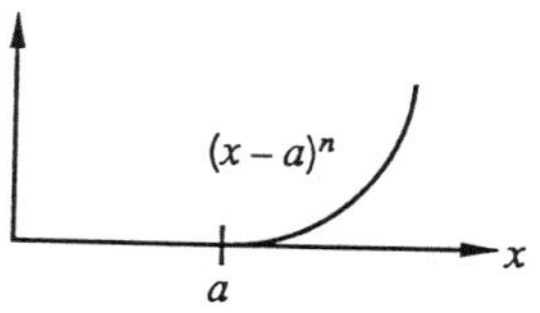 $(x-a)^n$, a, x

PVF_2 film can be used as a distributed sensor [15]. Conversely, the application of an electric field across the faces of the polarized film results in a longitudinal strain field, and hence the film may be used as a distributed actuator as well [16]. The specific transducer models developed here apply equally well to other piezoelectric devices, e.g. crystals.

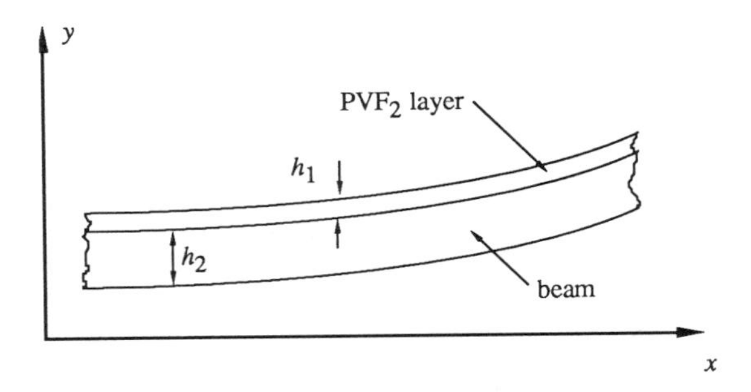

Fig. 1. Beam/film composite geometry.

As an example, consider the the bending vibrations of an elastic beam component having a distributed piezoelectric actuator bonded to one face, as shown in Fig. 1. The governing equation of the composite beam/film system is given by [16]

$$\left[EI\hat{y}_{\hat{x}\hat{x}} - m\ \hat{V}(\hat{x},\hat{t})\right]_{\hat{x}\hat{x}} + \rho A\ \hat{y}_{\hat{t}\hat{t}} = 0, \tag{5}$$

where the mass length density and composite flexural rigidity are defined in terms of beam $(\cdot)_1$ and film $(\cdot)_2$ properties by

$$\rho A \equiv \rho_1 A_1 + \rho_2 A_2, \tag{6}$$

$$EI \equiv E_1 I_1 + E_2 I_2, \tag{7}$$

and the film's piezoelectric gain coefficient m is given by

$$m \equiv \frac{-d_{31}(h_1 + h_2)(E_1 h_1 E_2 B)}{2(E_1 h_1 + E_2 h_2)}. \tag{8}$$

The parameters h_1 and h_2 are the thicknesses of the beam and film, respectively, B is the beam's width, and d_{31} is a piezoelectric constituitive constant for the film. The notation $(\cdot)_x$ represents partial differentiation with respect to the indicated parameter. The linear inhomogeneous Eq. (5) is the well known Bernoulli-Euler beam model with a control actuator distribution defined by $m\widehat{V}(\widehat{x},\hat{t})$. The actuator provides a distributed bending moment. The control moment is characterized by the constant m (for a beam of given material and geometry), which represents the applied bending moment per volt. If the material properties or the geometry of the composite beam/film structure change along its length, then the control moment is a function of x also, producing a spatially-varying distributed moment. Similarly, if the control voltage is varied over space, this too results in a spatially-varying distributed control moment. This "spatial weighting" is compactly represented using the singularity function notation.

Equation (5) can be nondimensionalized as

$$y_{xxxx} + y_{tt} = V_{xx}, \tag{9}$$

for the nondimensional variables

$$t \equiv \hat{t}\sqrt{\frac{EI}{\rho A L^4}}, \quad V \equiv \frac{\widehat{V} m L}{EI}, \\ x \equiv \frac{\widehat{x}}{L}, \ y \equiv \frac{\widehat{y}}{L}. \tag{10}$$

The parameter L is a characteristic length scale, which for a finite beam is most conveniently chosen to equal the beam's length. The nondimensional governing equation can be combined with appropriate boundary conditions to determine the complete response of the system. The primary interest here, however, is not in solving Eq. (9) for the system response, but instead using it to investigate the implications of distributed actuation.

The film control input $V(x,t)$ appears in the governing equation via its Laplacian, due to the beam's moment curvature relation. The control is assumed to be separable, i.e.

$$V(x,t) = \Lambda(x)\,u(t). \tag{11}$$

The Laplacian operates on the chosen spatial control distribution to determine its overall effect on the beam. To illustrate this, consider a spatially uniform control distribution represented by

$$\Lambda(x) \equiv \langle x - a\rangle^0, \tag{12}$$

so that

$$V(x,t) = \langle x-a \rangle^0 u(t). \tag{13}$$

The effective loading that this uniformly distributed film actuator presents to the beam is given by the Laplacian of its distribution, in accord with Eq. (9). Using Table I, the total effect on the beam of a spatially uniform distribution is given by a time-modulated *point moment* acting at $x = a$ described by

$$V_{xx} = \langle x-a \rangle^{-2} u(t). \tag{14}$$

One can use Table I to determine the effective loading of a broad class of film actuator distributions as represented by an appropriate linear combination of singularity functions.

Another interesting example is given by the choice of a linear spatial distribution or "ramp" function $\langle x-a \rangle^1$, whose Laplacian is given by

$$V_{xx}(x,t) = \langle x-a \rangle^{-1} u(t). \tag{15}$$

This effective loading is equivalent to a time modulated *point force* located at $x = a$ due to the action of the spatial derivative operator. This distribution is achieved in practice, for example, by shaping the electrode deposited on the piezoelectric film's surface.

It is now evident that distributed actuators may be spatially weighted to achieve the same effective loading that discrete actuators give. Discontinuities in the film distribution's amplitude give rise to point moments, while discontinuities in slope yield point forces. In addition, if the distribution is chosen such that $n \geq 2$ in the functional representation (3), for example

$$\Lambda(x) = \langle x-a \rangle^2, \tag{16}$$

then the Laplacian operation upon the film distribution produces an effective load which is in fact distributed rather than concentrated. For $n = 2$, this is given by

$$V_{xx}(x,t) = \langle x-a \rangle^0 u(t). \tag{17}$$

This is a *uniformly distributed load*, as shown in Table I. Such a distribution may be effectively used to target and control certain vibrational modes of a structure [17]. Consequently, singularity functions are also a convenient method of representing spatial distributions for more general degenerate *distributed* loadings. And finally, those spatial weightings that cannot be represented solely using singularity functions can instead be defined by a product of a suitably differentiable function $g(x)$ and singularity functions (over a suitable spatial aperture $x \in [a,b]$), vis

$$V(t) = g(x)\left[\langle x-a\rangle^0 - \langle x-b\rangle^0\right]. \tag{18}$$

The effective loading of a film actuator with such a distribution is evaluated using the product rules for differentiation of generalized functions [11].

C. DISTRIBUTED SENSING USING DEGENERATE TRANSDUCERS

The behavior and design of distributed sensors can also be investigated using singularity functions to represent the sensor apertures. As discussed earlier, when a spatially-varying strain field is applied to a layer of the piezoelectric polymer PVF_2, a charge distribution is developed on the electrode surface of the film. The total charge accumulated over the length L of the entire film surface is given in dimensional form by

$$Q(\hat{t}) = \int_0^L \left[e_{31}\, b(\hat{x})\, \varepsilon(\hat{x},\hat{t})\right] d\hat{x}, \tag{19}$$

where e_{31} is the film piezoelectric strain constant, ε is the applied strain field, and $b(x)$ is the spatial distribution of the charge collection enforced by appropriately shaping the sensing film's collecting electrodes. If the sensing film is bonded to a structural beam element, as shown in Fig. 1, the ability of the piezoelectric film to convert the mechanical energy of the structure to electric energy in the form of a surface charge is characterized by an electromechanical coupling factor k_{31}, and Eq. (19) may be re-written to explicitly include coupling efficiency;

$$Q(\hat{t}) = \int_0^L \left[\left(\frac{k_{31}^2}{g_{31}}\right) b(\hat{x})\, \varepsilon(\hat{x},\hat{t})\right] d\hat{x}. \tag{20}$$

The applied strain field distribution $\varepsilon(x,t)$ can be described in terms of the curvature of the sensing film/beam composite component member through the strain-curvature relation

$$\varepsilon(\hat{x},\hat{t}) = -EI\,\hat{y}_{\hat{x}\hat{x}}, \tag{21}$$

hence

$$Q(\hat{t}) = -EI\int_0^L \left[\left(\frac{k_{31}^2}{g_{31}}\right) b(\hat{x})\,\hat{y}_{\hat{x}\hat{x}}\right] d\hat{x}. \tag{22}$$

Equation (22) relates the charge produced on the electrode surface of polarized PVF_2 film to the strain distribution produced by the structural curvature of the component member to which it is attached. This accumulated charge can be represented in terms of a sensor output voltage $\hat{V}_0$ by considering the capacitive effects of the film as a dielectric material, and integrating Eq. (22) by parts to yield

$$\hat{V}_0(\hat{t}) = -\left(\frac{q_0}{C_0}\right)\left[\, b(\hat{x})\hat{y}_{\hat{x}}(\hat{x},\hat{t})\Big|_0^1 - b'(\hat{x})\hat{y}(\hat{x},\hat{t})\Big|_0^1 + \int_0^L \hat{y}(\hat{x},\hat{t})b''(\hat{x})\,d\hat{x}\right], \tag{23}$$

where

$$q_0 \equiv \frac{D\,k_{31}^2}{g_{31}}. \tag{24}$$

The notation $(\cdot)'$ denotes ordinary differentiation with respect to the argument.

Equation (23) is the dimensional governing equation for PVF_2 film as a distributed sensor when bonded to a beam. The aperture parameter $b(\hat{x})$ allows the spatial sensing distribution to be weighted according to the designer's choice. In evaluating Eq. (23) in terms of weighting functions $b(\hat{x})$ of the form (3), we assume that the spatial extent of the sensor lies wholely within the interval $[(0 + \Delta),(L - \Delta)]$. The parameter Δ is an infinitesimally small quantity that vanishes in the limit. Given these assumptions, $b(\hat{x})$ is identically zero at the system boundaries, hence the first two terms in brackets in Eq. (23) are zero. The output of the sensor is proportional to the integral in the third term, hence

$$\hat{V}_0(\hat{t}) = -\left(\frac{q_0}{C_0}\right)\int_0^L \hat{y}(\hat{x},\hat{t})b''(\hat{x})\,d\hat{x}. \tag{25}$$

Utilizing Eqs. (10), and the further definition

$$V_0 \equiv \widehat{V}\left(\frac{C_0}{q_0}\right) \tag{26}$$

permits the sensor output Eq. (25) to be expressed nondimensionally as

$$V_0(t) = -\int_0^1 y(x,t)b''(x)dx. \tag{27}$$

From the integrand of Eq. (27) it is apparent that the effective voltage output of the sensor is determined by the Laplacian of the spatial weighting function $b(x)$. Therefore, the singularity functions of Table I may be employed to give insight into the design and construction of degenerate distributed sensors using a strategy identical to that developed for actuators.

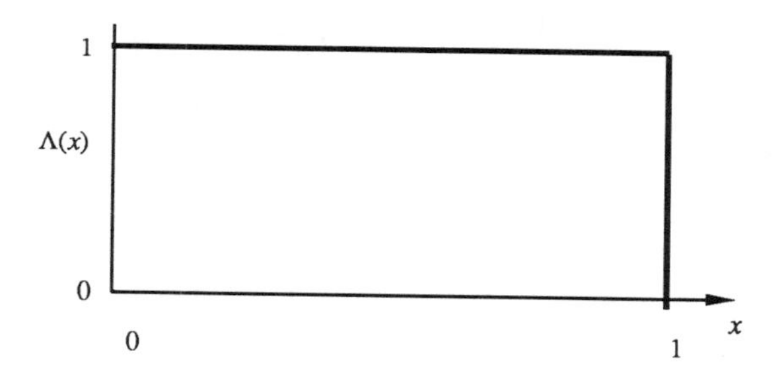

Fig. 2. Uniform transducer spatial distribution.

As an example of distributed sensor design, we consider measurements on a cantilever beam of unit length. A piezoelectric film sensor is applied along the entire length of the beam. The electrode distribution is chosen to be spatially uniform, as represented by the singularity functions

$$b(x) = \langle x \rangle^0 - \langle x - 1 \rangle^0. \tag{28}$$

This distribution is sketched in Fig. 2. Combining Eqs. (27) and (28), and performing the necessary manipulations using appropriate boundary conditions yields

$$V_0(t) = -y_x(1,t). \tag{29}$$

The voltage output of this uniform sensor distribution applied to a cantilever beam will be proportional to the beam's *angular displacement* at the free end $x = 1$.

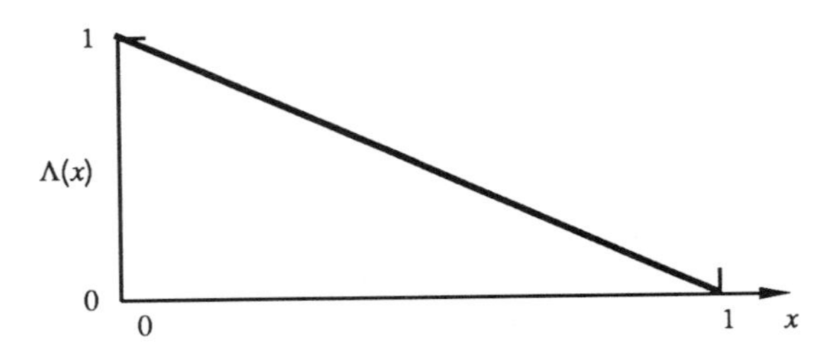

Fig. 3. Linear ("ramp") transducer spatial distribution.

Similarly, when the ramp collecting electrode spatial distribution shown in Fig. 3, expressed mathematically by

$$b(x) = \langle x\rangle^0 - \langle x\rangle^1 + \langle x-1\rangle^1 , \tag{30}$$

is applied to the same cantilever beam, this results in a sensor whose output takes the form

$$V_0(t) = y(1,t) . \tag{31}$$

The output is now proportional to the beam's *translational displacement* at the free end $x = 1$. Other aperture weightings can be synthesized to sense further parameters of interest.

An experimental implementation of the both the uniform and ramp distributions utilizing degenerate piezoelectric film sensors has been demonstrated by Miller and Hubbard [18]. Both sensors were applied to cantilever beams which contained discrete sensors at the beam tip in order to correlate the film sensor's output with an appropriately calibrated measurement. Angular and linear accelerometer measurements were used to calibrate the uniform and linearly distributed sensors, respectively. The cantilever beam structures were excited using a bandlimited continuous sinusoidal sweep near the first few structural resonance frequencies. The relevant physical parameters of the beam and PVF_2 film are presented in Table II.

Figure 4 shows the data obtained from the uniformly distributed sensor. The dashed and hash-marked lines are analytical predictions, while the solid line represents experimental data. The data is presented in the form of a spectrum of the ratio of PVF_2 film transducer output and accelerometer output versus excitation frequency. The dashed lines are analytical predictions using the nominal value of 12% for the PVF_2 electromechanical coupling factor. This represents an upper bound. The hash-marked line is the analytical prediction for an adjusted electromechanical coupling factor of 6 to

7%, which is more indicative of that achieved in the implementation tested here. There is excellent correlation between the PVF_2 sensor output and the analytical predictions.

Table II. Beam/film static properties.

Parameter	Value
Length	0.2762 m
Width	1.27×10^{-2} m
Beam Thickness	3.81×10^{-4} m
Film Thickness	6.35×10^{-5} m
Beam Modulus	2.1×10^{11} Pa
Film Modulus	1.5×10^{9} Pa
Beam Density	7.8×10^{3} kg/m^3
Film Density	1.8×10^{3} kg/m^3
Film d_{31}	1.2×10^{-11} m/V

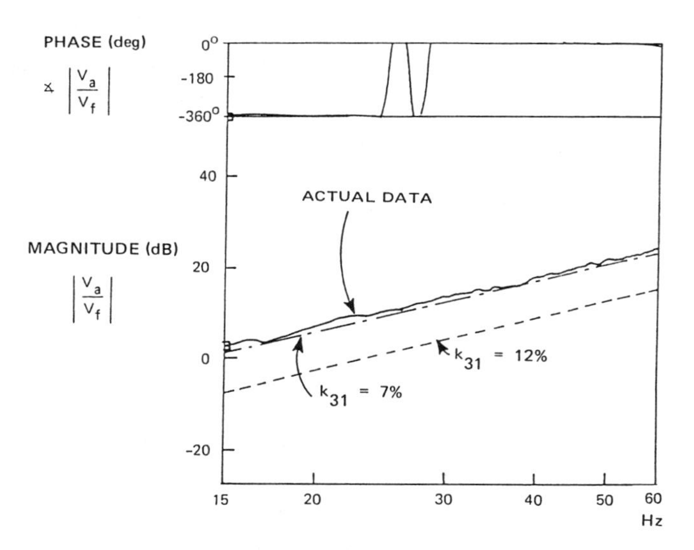

Fig. 4. Spatially uniform film sensor data.

Similar measurements were performed using the linearly distributed sensor. These results are presented in Fig. 5. This sensor distribution measures the translational displacement at the cantilever beam's free end. The analytical and experimental data correlate well at lower frequencies. Because of the boundary conditions there is less correlation between the measured and predicted outputs near the second structural mode. The higher modes of a cantilever beam are characterized by large angular tip motions and minimal translational tip displacements and accelerations. Because there is little translational tip motion at these frequencies, neither the film sensor nor the accelerometer adequately senses its respective parameter.

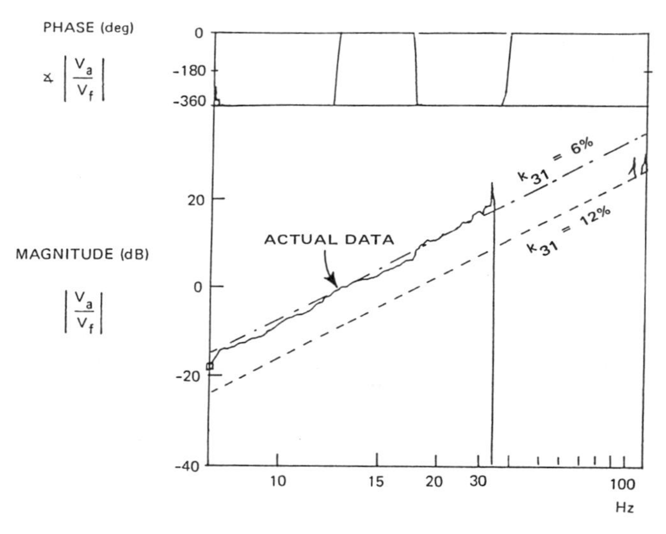

Fig. 5. Ramp film sensor data.

It is clear from the discussions above that distributed sensors and actuators may be designed, via appropriate spatial weighting functions, to provide both discrete and distributed sensing and loading parameters. For example, a strain energy sensor utilizing a parabolic film distribution has been designed for aircraft wing structural health monitoring [19]. Results illustrating the application and performance of distributed transducers for structural control are presented in the next section.

III. CONTROL SYSTEM SYNTHESIS

A.1. CONTROL DESIGN BASED UPON LYAPUNOV'S DIRECT METHOD

The second or direct method of Lyapunov provides a convenient means for designing stabilizing compensators for structural control [20,21]. The Lyapunov method is based upon global energy and power flow considerations. The design goal is to synthesize a transducer distribution, using the representations developed in the previous section, and a feedback control law that removes energy from the system, thus damping its resonant response. The method is very appealing in that it produces very low order compensators that can be readily implemented. Further, it does not require a modal plant representation, hence no model truncations are required.

The technique will now be derived for the control of thin beams with arbitrary linear boundary conditions. The Bernoulli-Euler beam governing equation is expressed in nondimensional form as

$$y_{xxxx} + y_{tt} = u(x,t), \quad 0 < x < 1. \tag{32}$$

The associated nondimensional boundary conditions are given by

$$y(\xi,t) = y_x(\xi,t) = 0; \qquad \text{(clamped)} \tag{33}$$

$$\begin{gathered} y(\xi,t) = 0, \\ y_{xx}(\xi,t) = -I\, y_{xtt}(\xi,t) - \kappa y_x(\xi,t) - \beta y_{xt}(\xi,t); \end{gathered} \qquad \text{(pinned)} \tag{34}$$

$$\begin{gathered} y_{xx}(\xi,t) = -I\, y_{xtt}(\xi,t) - \kappa y_x(\xi,t) - \beta y_{xt}(\xi,t), \\ y_{xxx}(\xi,t) = M\, y_{tt}(\xi,t) + k y(\xi,t) + b y_t(\xi,t); \end{gathered} \qquad \text{(free)} \tag{35}$$

$$\begin{gathered} y_x(\xi,t) = 0, \\ y_{xxx}(\xi,t) = M\, y_{tt}(\xi,t) + k y(\xi,t) + b y_t(\xi,t); \end{gathered} \qquad \text{(sliding)} \tag{36}$$

where ξ is the nondimensional boundary point $x = 0$ or $x = 1$. The boundary conditions include linear lumped elements as well, such as translational masses M, rotary inertias I, translational and rotary springs k and κ, and translational and rotary viscous dampers b and β, respectively. Exogenous inputs at the boundaries can be incorporated into the analysis, but are neglected here for simplicity.

The Lyapunov design method is applied to the beam vibration problem by first constructing an energy functional that describes the total system energy. For the beam described by Eq. (32), this is given by

$$
\begin{aligned}
J(t) = {} & \frac{1}{2}\int_0^1 \left[(y_{xx})^2 + (y_t)^2\right] dx + \frac{1}{2}M\left[y_t(\xi,t)\right]^2 \\
& + \frac{1}{2}I\left[y_{xt}(\xi,t)\right]^2 + \frac{1}{2}\kappa\left[y_x(\xi,t)\right]^2 + \frac{1}{2}k\left[y(\xi,t)\right]^2.
\end{aligned} \tag{37}
$$

This functional represents the sum of the beam's strain and kinetic energies, as well as any energy storage mechanisms of the linear lumped elements which appear at either or both of the boundaries. Vibration damping is then based upon total energy considerations. This approach has the advantage over conventional methods by dealing directly with the distributed plant model without resorting to approximations, save any approximations implicit in the use of a Bernoulli-Euler model. Consequently, conditions sufficient for asymptotic stability can be derived using the present distributed parameter formulation [22,23]. This is in sharp contrast to the modally truncated plant models generally used in control design, as described in section I.

The time derivative of the functional (37) is combined with the nondimensional Bernoulli-Euler beam governing Eq. (32) and *any* combination of the boundary conditions (33) - (36) to yield an expression for the power flow from the beam,

$$
\dot{J}(t) = \int_0^1 y_t(x,t)\, u(x,t)\, dx - \beta\left[y_{xt}(\xi,t)\right]^2 - b\left[y_t(\xi,t)\right]^2. \tag{38}
$$

Note that the control input appears only within the spatial integral. The boundary lumped viscous damping terms, which are decoupled from the control input, always extract energy from the system (by convention, negative power means power flow *out* of the system), hence they are stabilizing. They will henceforth be dropped from Eq. (38).

The vibration control design reduces to finding a distributed feedback control input $u(x,t)$ that ensures $\dot{J}$ is negative semi-definite for all time t. Essentially, one requires distributed rate feedback to stabilize the beam by always removing energy. Depending upon the type(s) of actuator(s) used, Eq. (38) can be manipulated into various forms to suggest the requisite control and sensed variables and their spatial distribution.

As an example of how Eq. (38) is used to analyze a spatial control distribution, consider a cantilevered beam with a spatially-uniform piezoelectric film actuator distribution bonded to one face. The film distribution takes the form (28), hence the distributed input $u(x,t)$ becomes

$$u(x,t) = V_{xx} = \left[\langle x\rangle^{-2} - \langle x-1\rangle^{-2}\right] u(t). \tag{39}$$

The film actuator provides two opposed moments at the ends of the distribution. Utilizing the clamped boundary condition at the root end $x = 0$, Eq. (38) becomes

$$\dot{J}(t) = -u(t)\, y_{xt}(1,t), \tag{40}$$

where $x = 1$ is the free end. To maximize $\dot{J}(t)$ and extract the most energy from the beam, one chooses the exogenous control time signal $u(t)$ to be

$$u(t) = \text{sgn}[y_{xt}(1,t)]. \tag{41}$$

This requires a measurement of the angular velocity at the free end $x = 1$. Consequently,

$$\dot{J}(t) = -\text{sgn}[y_{xt}(1,t)]\, y_{xt}(1,t), \tag{42}$$

which shows that the switching control law (41), combined with the uniform film actuator distribution, removes energy from the beam *for all modes* [16]. Naturally, other control temporal functions $u(t)$ can be chosen, such as angular rate feedback; the switching control law, however, provides a quasi-optimal solution.

This particular control design has been successfully implemented experimentally by Plump *et al* [24]. The PVF_2 damper developed in this experiment was tested under single mode conditions as a proof-of-concept demonstration of the uniformly distributed piezo-film actuator. Table III defines the specific piezo-film and cantilever beam parameters used in the experiment. The Lyapunov control was applied to the first bending mode of this beam. The beam was fitted with a tip mass which included a 0.5 gram piezoresistive linear accelerometer. The acceleration signal was then integrated to produce a velocity signal. For a single mode, linear velocity is proportional to angular velocity by a constant factor, and hence linear velocity may be used as input to the controller. The first cantilever bending mode was approximately $6Hz$. Active damping tests were performed by initially displacing the tip of the beam $2cm$, releasing it, and observing the decay envelope of the resulting vibration. These vibration test were performed with switching control amplitudes of 100, 200, 300, 400 and 500 volts. PVF_2 film can support voltages as high as 2000 volts before dielectric breakdown occurs. Although relatively large voltages are needed for control, the power level required is very small.

Table III. Cantilever beam/film static properties.

Parameter	Value
Length	1.22 m
Width	0.152 m
Beam Thickness	3.18×10^{-3} m
Film Thickness	2.3×10^{-5} m
Beam Modulus	7.6×10^{10} Pa
Film Modulus	2.0×10^{9} Pa
Beam Density	2.84×10^{3} kg/m^3
Film Density	1.3×10^{3} kg/m^3
Film d_{31}	2.2×10^{-11} m/V

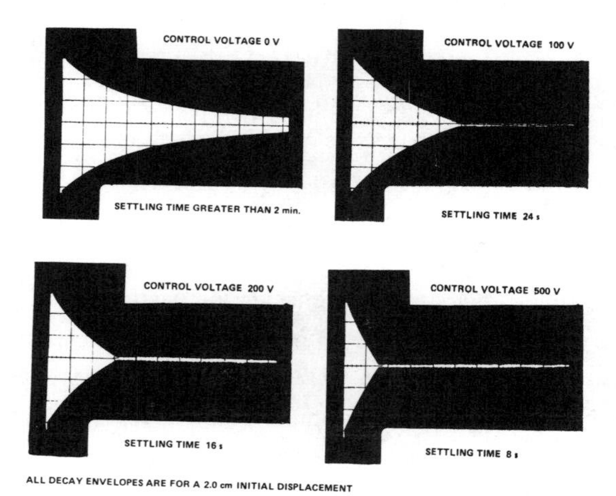

Fig. 6. Uncontrolled and controlled cantilever beam tip displacement time histories.

Figure 6 shows the displacement time histories obtained from the active damping experiment described above. These traces show the effectiveness of the active damper, with a uniform spatially distributed actuator using the control law suggested by Eq. (41), as the control voltage is varies from 0 to 500 volts. The settling time is decreased from a free decay time of over 90 seconds to a controlled settling time of 8 seconds. Due to the nonlinear control law, the damping achieved yields a linear decay envelope indicating a change in effective loss factor which increases as the vibration amplitude decreases. Even though the amount of energy dissipated per cycle is decreasing, the amount of energy in the system is decreasing at a faster rate as the vibration amplitude decreases.

In addition to boundary control problems defined by spatially-uniform control distributions, Eq. (38) facilitates the examination of more general spatial control distributions. It will be shown that a spatially-uniform film actuator distribution is ineffective for certain beam boundary condition combinations, and that Eq. (38) can be used to gain insight into the design of appropriate *spatially varying* distributed controls.

A.2. LIMITATIONS OF CONTROL DESIGNS BASED UPON SPATIALLY UNIFORM FILM ACTUATOR DISTRIBUTIONS

The uniformly-distributed film actuator results of the previous section will now be extended to more clearly understand the implications of Eq. (38). For the ensuing discussion, Eq. (38) will be written

$$\dot{J}(t) = V(x,t)\, y_{xt}(x,t)\Big|_0^1 - V_x(x,t)\, y_t(x,t)\Big|_0^1 + \int_0^1 y_t(x,t)\, V_{xx}(x,t)\, dx. \tag{43}$$

A spatially uniform control distribution is sketched in Fig. 2, with its loading [Eq. (39)] depicted in Fig. 7. This distribution will only produce boundary control terms in Eq. (43) in terms of linear and angular velocities at $x = 0$ and $x = 1$, since the uniform distribution provides a pair of point moments at the boundaries. Thus, a beam with clamped-clamped boundary conditions is *not controllable* using a spatially-uniform film control.

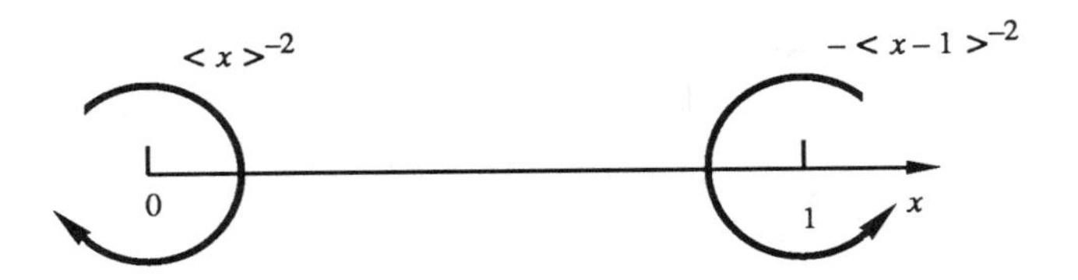

Fig. 7. Uniform spatial film distribution loading.

Beams with clamped ends, where the boundary conditions (32) hold, will not display boundary control at those ends in terms of linear and/or angular velocities, because y_x and y must vanish there. Thus, a beam with clamped-sliding boundary conditions will not be controllable with a uniform distribution; the sliding end of the beam, like the root, must have $y_x = 0$. However, a beam clamped at one end, with a pinned boundary condition at the other end *is* controllable with this spatial distribution. In this case Eq. (43) takes the form

$$\dot{J}(t) = u(t)\, y_{xt}(1,t), \tag{44}$$

where $x = 1$ is the pinned beam boundary.

As a further example, consider a pinned beam with a spatially uniform piezoelectric film control distribution. For this configuration, Eq. (43) becomes

$$\dot{J}(t) = u(t)[\, y_{xt}(1,t) - y_{xt}(0,t)\,]. \tag{45}$$

The beam is controllable for motions where the slopes at its ends are not equal. Physically, this corresponds to odd-numbered (symmetric about mid-span) modes of the pinned beam. However, even-numbered (asymmetric about mid-span) modes have slopes at the boundaries that are equal, hence $\dot{J}(t) = 0$ for these modes. Since the time derivative of the energy functional corresponds to instantaneous power flow, Eq. (45) shows no energy can be removed from or added to the system for asymmetric vibrational modes. Parenthetically, one notes that for even-numbered modes and the uniform control distribution, the integrands in Eqs. (38) and (43) are products of odd and even (about mid-span) functions; the integral must vanish over the length of the beam for the pinned beam.

The spatially uniform film actuator distribution, essentially a portion of a square wave in space, can be expressed mathematically in a Fourier sine series containing only *odd* harmonics. When this series is substituted into the spatial integral in Eq. (38), one obtains a restatement of modal orthogonality for the pinned beam, *for odd-numbered modes*. As a result, only symmetric modes lead to non-zero contributions to the beam's vibration control. The distribution demonstrates this property because it can be decomposed into the same set of orthogonal functions as the Laplacian of the system eigenfunctions. These observations will be generalized in an ensuing section to develop techniques for targeting specific modes or modal subsets for control.

Similarly, a free-free beam with uniform spatial control yields the same results as equations (43); its asymmetric modes are not controllable using counter-opposed moments at the boundaries. Thus, in spite of the utility and simplicity of the uniform film spatial control distribution for most boundary condition combinations, there exist configurations for which it is inappropriate. Non-uniform spatial control distributions will be considered next for the vibration damping of systems having both symmetric and asymmetric eigenfunctions.

A.3. APPLICATION OF SPATIALLY VARYING CONTROL DISTRIBUTIONS: THE PINNED BEAM

The pinned beam will be discussed in detail as a model problem for spatially varying control distributions. While these results can also be applied to a broad class of distributed control transducers, piezoelectric film actuators are utilized in the exposition. First, the beam's displacement field $y(x,t)$ is expressed as a product of spatial and temporal functions,

$$y(x,t) = \psi(x)\varphi(t). \tag{46}$$

Substituting Eq. (46) and the control input representation (11) into Eq. (38) gives

$$\dot{J}(t) = u(t)\varphi'(t)\int_0^1 \psi''(x)\Lambda(x)\, dx. \tag{47}$$

As before, the notation $(\cdot)'$ denotes ordinary differentiation with respect to the argument. Owing to the separation in Eq. (47), the control design can now be divided into a spatial part and a temporal part.

The integral in Eq. (47) could be maximized with a constant amplitude spatial control distribution that "switched" in space at zeroes of the beam curvature. This "spatial bang-bang" controller, however, would require an inordinate amount of associated instrumentation to implement, and would necessitate segmenting the film over the beam's span at zeroes of the beam's curvature for some number of modes n, thus imposing a bandwidth limit on the effectiveness of the control. The scheme would require an infinite number of segments to control all modes. However, noting that the pinned beam's even-numbered modes have odd symmetry about $x = 1/2$, and odd-numbered modes have even symmetry, one is led to choose a spatially varying film control distribution having both even and odd spatial symmetry of the form

$$\Lambda(x) = \langle x \rangle^0 - \langle x \rangle^1 + \langle x - 1 \rangle^1 . \tag{48}$$

This distribution is shown in Fig. 3. The control can be implemented in practice, for example, by varying the electrode plating on the film layer over the beam's length, by varying the film's thickness or the bond layer thickness (subject to the constraint, in the present analysis, that such a variation does not affect the homogeneity assumption), or by varying the film's constitutive piezoelectric properties over space during manufacture. The film distribution's loading on the beam, given by its Laplacian, can be deduced using the table of singularity functions as

$$\Lambda''(x) = \langle x \rangle^{-2} - \langle x \rangle^{-1} + \langle x - 1 \rangle^{-1} . \tag{49}$$

This distribution's loading is sketched in Fig. 8. One notes that this spatially weighted actuator aperture provides a concentrated moment at $x = 0$, and opposed point forces at the two ends; since the film is a self-reacting actuator, these contributions must sum to zero to satisfy static equilibrium. Substituting the expression for the varying spatial distribution, Eq. (48), into (47) and integrating by parts gives an expression for the power flow from the beam/film composite,

$$\begin{aligned} \dot{J}(t) &= -u(t)\, \varphi'(t)\, \psi'(0) \\ &= -u(t)\, y_{xt}(0,t). \end{aligned} \tag{50}$$

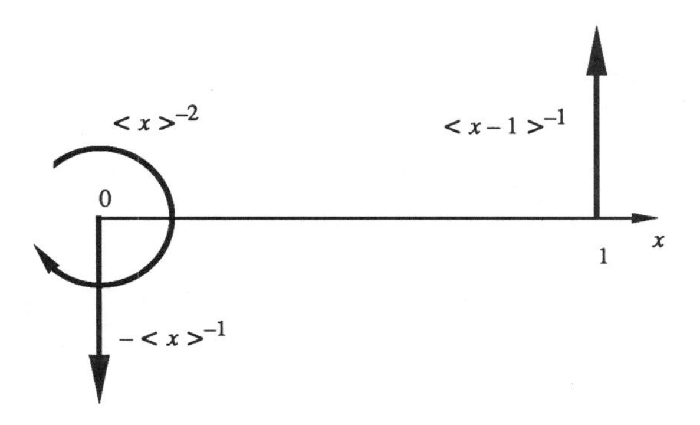

Fig. 8. Ramp spatial film distribution loading.

To extract the most energy from the pinned beam system for vibration damping for the given spatial distribution, the time portion of the control $u(t)$ is chosen to be

$$u(t) = \text{sgn}\,[y_{xt}(0,t)]. \tag{51}$$

Naturally, other stabilizing temporal compensators can be chosen, such as rate feedback.

The physical significance of the control distribution described by Eq. (48) is most easily seen by examining the inhomogeneous governing equation Eq. (9). The forcing term in Eq. (9) due to the chosen distribution will involve point forces and a point moment, as shown in Eq. (49), and illustrated in Fig. 8. Since translational motion is prohibited by the pinned boundary conditions, terms involving the delta functions will vanish in the time derivative of the energy functional. Thus, the spatially-varying film control appears to the system as a boundary controller *at one end*. Forcing the distribution to zero "smoothly" (e.g. without a discontinuity in amplitude) at $x = 1$ does not give rise to a point moment there, precluding the result for the uniform control distribution. As noted in section II, discontinuities in amplitude give rise to point moments, while discontinuities in slope give rise to point forces. This is analogous to the interpretation of static loading distributions on beams in elementary structural mechanics texts; the film distribution $\Lambda(x)$ is precisely the moment distribution for the pinned beam corresponding to the loading given by Eq. (49).

Finally, note that the Fourier sine series representation of the "ramp" distribution, Eq. (48), is expressible as a superposition of sinusoids with both odd *and* even harmonic components. Thus, this actuator distribution has Fourier components coincident with the *all* modal components of the beam's curvature. Using this insight, one is tempted to design actuator distributions that are expressed as superpositions of the beam's curvature functions for only certain modes, so as to control only those modes. General modal concepts and interpretations will be developed later in this section.

An experimental demonstration of the concepts presented in this section has been performed by Burke and Hubbard [25]. The experimental structure consisting of a steel beam with knife edges machined into its ends to provide pinned supports. The static properties of the beam are presented in Table II. The temporal control function implemented was that described by Eq. (51). This control law requires a measurement of angular velocity at $x = 0$. A 0.5 gram linear accelerometer was mounted near (but not at) this beam boundary, as shown in Fig. 9. As a result of this placement the linear accelerometer had the same sense as an angular (acceleration) measurement for "low order" modes where no node exists between the accelerometer and the boundary.

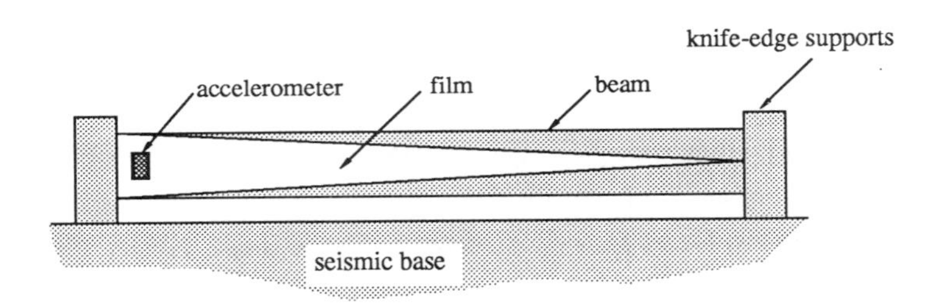

Fig. 9. Pinned beam experimental configuration.

A spatially uniform layer of PVF_2 film was bonded to the beam for actuation. The maximum voltage applied to the film for control was 250 volts. Since this film has a high input impedance ($> 30M\Omega$), very little current is required by the controller, and hence little electrical power. In order to provide the structure with a repeatable initial condition, the controller was driven with positive feedback. Once the desired initial condition was established, the controller was either turned off to provide a free decay measurement or was switched according to Eq. (51) to control the beam. Experimental results for the first two modes of the structure are shown in Figs. 10 and 11. The controller is quite effective for the first mode, increasing the effective damping ratio ζ

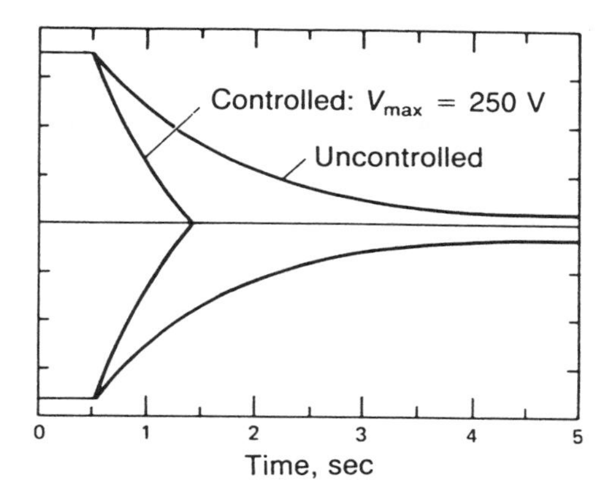

Fig. 10. Pinned beam mode 1 response with uniform film actuator distribution.

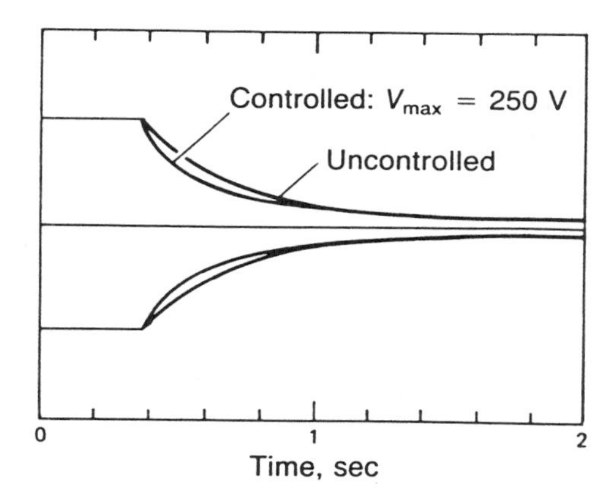

Fig. 11. Pinned beam mode 2 response with uniform film actuator distribution.

by a factor of up to 5 versus the inherent passive structural damping. Mode 2 is primarily unaffected by the controller. The initial amplitude condition imparted to the beam for mode 2 is much smaller as well; even though the uniform distribution seems to weakly excite the beam's second mode, its effectiveness as a controller is minimal. This serves to uphold the contentions of the analysis presented above: a spatially uniform control distribution will not control even-order asymmetric modes of a simply supported beam.

A linearly varying control distribution was also applied to the beam. Identical test to those described above for the uniform distribution were then performed. Plots of the corresponding decay envelopes are shown in Figs. 12 and 13 for the first two structural modes. Unlike the previous results for the uniform distribution, both symmetric and asymmetric modes of the structure are effectively controlled by the linear distribution. The effective damping ratio is increased by up to a factor of 4.5.

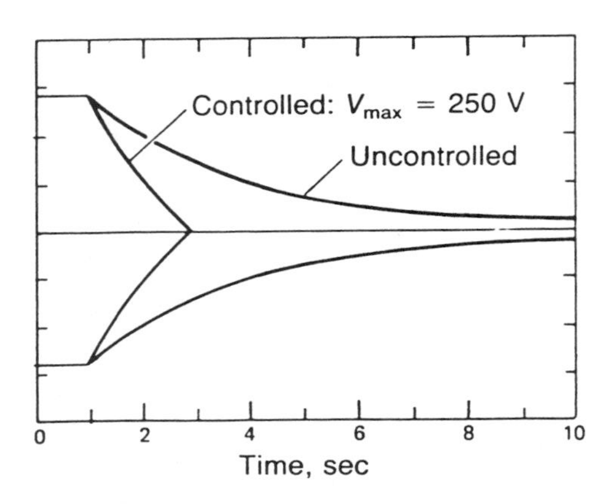

Fig. 12. Pinned beam mode 1 response with ramp film actuator distribution.

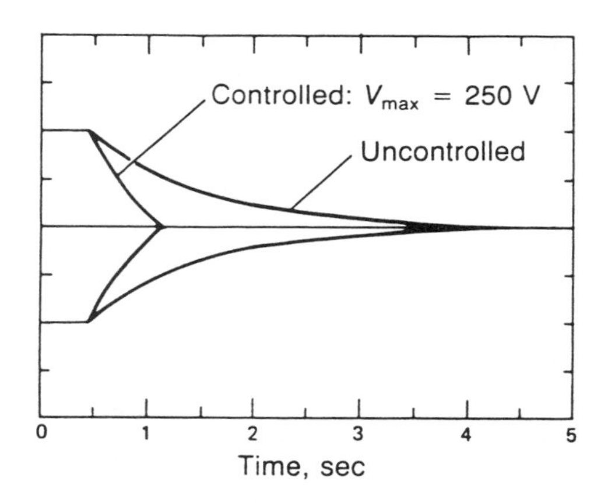

Fig. 13. Pinned beam mode 2 response with ramp film actuator distribution.

Lyapunov design has been shown to yield simple compensators that can be readily implemented to stabilize (in the Lyapunov sense of stability) structural components. The examples presented herein can be extended to develop spatially-varying control distributions for other boundary condition combinations [17]. The technique has also been applied analytically to the vibration control of thin plates [26] and the Timoshenko beam model [27]. Experimental implementations of Lyapunov designs include the vibration damping of cantilever beams [16,24], the aforementioned pinned beam [25], and a more complex interconnected beam structure [28].

One of the limitations of these Lyapunov designs, however, is that they deal only with global energy flow; alternative temporal compensation schemes must be employed to achieve more specific performance measures. The application of impedance control to beams using distributed actuators, presented in the following section, provides one realizable approach.

B. CONTROL DESIGN BASED UPON IMPEDANCE METHODS

In the analysis that follows, a controller for the active vibration damping of a continuous beam is designed using another technique which does not require model truncation. This method has an advantage over the Lyapunov approach in that specific performance goals can be achieved and explicitly addressed in the formulation. The method, based upon the analyses found in [29,30,31], develops a state space model of the continuous, infinite-dimensional beam, and then appends a boundary controller which has the same functional form as the characteristic impedance of bending waves within the beam. This characteristic impedance termination nulls any reflections from the boundary, so that the beam appears to be semi-infinite and no standing waves are produced.

This approach is analogous to well known electrical transmission line termination methods [32]. The end point impedance controller is not designed explicitly to damp modes of a beam, but instead to *absorb waves travelling along the beam*, and therefore suppress them before standing waves are established. As will be shown, the strategy requires the application of a collocated shear force and bending moment at the beam's boundary. This requirement is well suited to the types of degenerate actuators and sensors discussed in section II.

The analysis utilizes the Bernoulli-Euler beam model. Four variables are required to completely describe the state of this system: the beam's transverse displacement y, angular rotation θ, bending moment M, and transverse shear force Q. These parameters are related via the constitutive relations

$$M = EI\, y_{xx}, \tag{52}$$

$$Q_x = -\,\rho A\, y_{tt}, \tag{53}$$

$$Q = M_x, \tag{54}$$

$$\theta = y_x . \tag{55}$$

Equations (52) through (55) are combined using dynamic equilibrium constraints to yield the Bernoulli-Euler beam governing equation in dimensional form

$$a^2 y_{xxxx} + y_{tt} = 0\,, \tag{56}$$

where

$$a^2 \equiv \frac{EI}{\rho A}\,. \tag{57}$$

The Laplace transform of the governing equation Eq. (56), assuming homogeneous initial conditions, provides an ordinary differential equation in the spatial and Laplace transform variables x and s,

$$a^2 \frac{d^4 y(x,s)}{dx^4} + s^2\, y\,(x,s) = 0\,, \tag{58}$$

where

$$y\,(x,s) \equiv L\{y\,(x,t)\} \tag{59}$$

for the Laplace transform operator $L\{\cdot\}$. The constitutive relations Eqs. (52) – (55) can then be Laplace transformed, combined with the governing equation Eq. (56), and expressed in a matrix state space form as

$$\frac{d}{dx}\begin{bmatrix} v \\ \omega \\ m \\ q \end{bmatrix} = \begin{bmatrix} 0 & 1 & 0 & 0 \\ 0 & 0 & \left(\frac{s}{a}\right) & 0 \\ 0 & 0 & 0 & 1 \\ -\left(\frac{s}{a}\right) & 0 & 0 & 0 \end{bmatrix}\begin{bmatrix} v \\ \omega \\ m \\ q \end{bmatrix}, \tag{60}$$

where the vector of states is defined as

$$\begin{bmatrix} v \\ \omega \\ m \\ q \end{bmatrix} \equiv \begin{bmatrix} s\,y(x,s) \\ s\,\theta(x,s) \\ \frac{a}{EI}M \\ \frac{a}{EI}Q \end{bmatrix}. \tag{61}$$

The states m and q have been scaled to have the same dimensions as v and ω, respectively.

Equation (60) may be written compactly as

$$\frac{d\mathbf{y}}{dx} = \mathbf{A}\,\mathbf{y}. \tag{62}$$

The system eigenvectors and eigenvalues are found by solving

$$\left|\,\lambda\mathbf{I} - \mathbf{A}\right| = \lambda^4 + \left(\frac{s}{a}\right)^2 = 0. \tag{63}$$

The homogeneous solution of Eq. (62) takes the form

$$\mathbf{y}(x,s) = \mathbf{H}(x - x_0,s)\,\mathbf{y}(x_0,s), \tag{64}$$

where $\mathbf{H}(x,s)$ is the fundamental matrix, given by

$$\begin{aligned} \mathbf{H}(x,s) &= e^{\mathbf{A}(s)x} = \mathbf{I} + \mathbf{A}(s)x + [\mathbf{A}(s)]^2\frac{x^2}{2!} + \cdots \\ &= \mathbf{T}(s)\,e^{\Gamma(s)x}\,\mathbf{T}^{-1}(s). \end{aligned} \tag{65}$$

The matrix $\Gamma(s)$ is diagnol and contains the system eigenvalues, while $\mathbf{T}(s)$ is a transition matrix whose columns are the eigenvectors of the system, as determined from Eqs. (62) and (63). This formulation, based on the Caley-Hamilton theorem [33], is analogous to the evaluation of the state transition matrix for lumped parameter systems. The homogeneous solution (64) may now be written as

$$\mathbf{y}(x,s) = \mathbf{T}(s)\, e^{\Gamma(s)x}\, \mathbf{T}^{-1}(s)\, \mathbf{y}(x_0,s). \tag{66}$$

By defining a new transition matrix $\mathbf{U}(x,s)$, vis

$$\mathbf{U}(x,s) \equiv \mathbf{T}^{-1}(s)\, \mathbf{y}(x,s), \tag{67}$$

Eq. (66) is transformed to

$$\mathbf{U}(x,s) = e^{\Gamma(s)x}\, \mathbf{U}(x_0,s). \tag{68}$$

Since $\Gamma(s)$ is a diagnol matrix of system eigenvalues $\lambda_i(s)$, the matrix $e^{\Gamma(s)x}$ is also a diagnol matrix of exponentials $e^{\lambda_i(s)x}$. In order for the system variables to remain finite for any positive x, no positive eigenvalues can appear in the matrix $e^{\Gamma(s)x}$. This also implies that the components of $\mathbf{U}(x,s)$ corresponding to any positive eigenvalues must be zero in a semi-infinite system. For example, consider a system in which the first m eigenvalues of a kth order system are positive. Therefore,

$$\mathbf{U}(x,s) = \left[\begin{array}{c} \mathbf{U}_1(x,s) \\ \vdots \\ \mathbf{U}_m(x,s) \\ \hline \mathbf{U}_{m+1}(x,s) \\ \vdots \\ \mathbf{U}_k(x,s) \end{array}\right] = \left[\begin{array}{c} 0 \\ \hline \mathbf{U}(x,s) \end{array}\right]. \tag{69}$$

However,

$$\mathbf{U}(x,s) = \mathbf{T}^{-1}(s)\, \mathbf{y}(x,s) = \left[\begin{array}{c} \mathbf{W}_1(s) \\ \hline \mathbf{W}_2(s) \end{array}\right] \mathbf{y}(s,x) = \left[\begin{array}{c} 0 \\ \hline \mathbf{U}(x,s) \end{array}\right], \tag{70}$$

where $\mathbf{W}_1(s)$ is an $m \times k$ matrix whose columns are the first m reciprocal or left-eigenvectors. Therefore, in a "physical semi-infinite system", the dynamic system variables are related through the *equations of constraint*

$$\mathbf{W}_1(s)\,\mathbf{y}(x,s) \equiv 0 \quad \text{for } all\ x. \tag{71}$$

The equations of constraint define a relationship between the dynamic variables of a semi-infinite, one dimensional distributed parameter system. This relationship must hold at all points along the semi-infinite system, *or along any finite system before any traveling waves have reached a boundary or end point where there may be reflections.* The equation of constraint can be used to define a control law for end point control such that the characteristic constraint is enforced at the end point, making the system appear to be semi-infinite. In this case, any waves will not be reflected at the end point, but instead will be absorbed by the controller.

The functional form of the end point controller can be derived by first determining the transition matrix in terms of the system eigenvalues, i.e.

$$\mathbf{T} = \frac{1}{2}\begin{bmatrix} \left(\frac{s}{a}\right) & \left(\frac{s}{a}\right) & \left(\frac{s}{a}\right) & \left(\frac{s}{a}\right) \\ \left(\frac{s}{a}\right)\lambda_1 & \left(\frac{s}{a}\right)\lambda_2 & \left(\frac{s}{a}\right)\lambda_3 & \left(\frac{s}{a}\right)\lambda_4 \\ \lambda_1^2 & \lambda_2^2 & \lambda_3^2 & \lambda_4^2 \\ \lambda_1^3 & \lambda_2^3 & \lambda_3^3 & \lambda_4^3 \end{bmatrix}. \tag{72}$$

Because of the relationship between the eigenvalues of the system in the present implementation, the inverse transformation takes on the special form

$$\mathbf{T}^{-1} = \frac{1}{2}\begin{bmatrix} \left(\frac{s}{a}\right)^{-1} & \left(\frac{s}{a}\lambda_1\right)^{-1} & \lambda_1^{-2} & \lambda_1^{-3} \\ \left(\frac{s}{a}\right)^{-1} & \left(\frac{s}{a}\lambda_2\right)^{-1} & \lambda_2^{-2} & \lambda_2^{-3} \\ \left(\frac{s}{a}\right)^{-1} & \left(\frac{s}{a}\lambda_3\right)^{-1} & \lambda_3^{-2} & \lambda_3^{-3} \\ \left(\frac{s}{a}\right)^{-1} & \left(\frac{s}{a}\lambda_4\right)^{-1} & \lambda_4^{-2} & \lambda_4^{-3} \end{bmatrix}. \tag{73}$$

Inspection of Eq. (70) reveals that

$$\mathbf{W}_1 = \frac{1}{2}\begin{bmatrix} \left(\frac{s}{a}\right)^{-1} & \left(\frac{s}{a}\lambda_1\right)^{-1} & \lambda_1^{-2} & \lambda_1^{-3} \\ \left(\frac{s}{a}\right)^{-1} & \left(\frac{s}{a}\lambda_2\right)^{-1} & \lambda_2^{-2} & \lambda_2^{-3} \end{bmatrix}. \tag{74}$$

The *equation of constraint for a Bernoulli-Euler beam* is therefore

$$\begin{bmatrix} \left(\frac{s}{a}\right)^{-1} & \left[\left(\frac{s}{a}\right)\lambda_1\right]^{-1} & \lambda_1^{-2} & \lambda_1^{-3} \\ \left(\frac{s}{a}\right)^{-1} & \left[\left(\frac{s}{a}\right)\lambda_2\right]^{-1} & \lambda_2^{-2} & \lambda_2^{-3} \end{bmatrix} \begin{bmatrix} v \\ \omega \\ m \\ q \end{bmatrix} = \begin{bmatrix} 0 \\ 0 \end{bmatrix}. \tag{75}$$

Equation (75) can be expressed as a system of complex algebraic equations

$$\begin{aligned} v + \left[\frac{1-j}{\sqrt{2\left(\frac{s}{a}\right)}}\right]\omega - jm - \left[\frac{1+j}{\sqrt{2\left(\frac{s}{a}\right)}}\right] q = 0, \\ v + \left[\frac{1+j}{\sqrt{2\left(\frac{s}{a}\right)}}\right]\omega + jm - \left[\frac{1-j}{\sqrt{2\left(\frac{s}{a}\right)}}\right] q = 0, \end{aligned} \tag{76}$$

which are solved to form a characteristic impedance matrix

$$\begin{bmatrix} m \\ q \end{bmatrix} = \begin{bmatrix} -1 & -\sqrt{2\left(\frac{a}{s}\right)} \\ \sqrt{2\left(\frac{s}{a}\right)} & 1 \end{bmatrix} \begin{bmatrix} v \\ \omega \end{bmatrix}. \tag{77}$$

Equation (77) can be used as a control law at one end of a finite beam to effectively create a system that behaves as if it were semi-infinite, i.e. a beam in which traveling waves are not reflected when they reach the end of the beam. The necessary sensed parameters at the end are the translational and angular velocities v and ω, while the required control inputs are the moment m and force q. A more general form of the suggested control law is given by

$$\begin{bmatrix} m \\ q \end{bmatrix}_b = \begin{bmatrix} -G_v & -G_\omega\sqrt{2\left(\frac{a}{s}\right)} \\ K_v\sqrt{2\left(\frac{s}{a}\right)} & K_\omega \end{bmatrix} \begin{bmatrix} v - v^* \\ \omega - \omega^* \end{bmatrix}_b, \tag{78}$$

where G_v, G_ω, K_v, and K_ω represent the appropriate scalar gains of the controller, and the $(\cdot)^*$ variables are reference values. In particular, if the gains are chosen such that $G_v = G_\omega = K_v = K_\omega = 1$, then the condition (77) is achieved exactly, and no waves are reflected from the end of the beam.

The form of the control law (78) requires sensing both translational and angular velocity at the end point of the controlled beam, as well as the application of both a bending moment and shear force at the same end point location. This is readily achieved using the transducer design techniques outlined in section II of this chapter. The required point moment and point force actuators have already been presented and discussed in this section. It has been shown that the choice of a "ramp" spatial piezoelectric film actuator distribution $\langle x - a \rangle^1$ results in an effective loading that is equivalent to a time modulated *point force* at $x = a$ due to the action of the spatial derivative. Similarly, a time modulated *point moment* is achieved using a uniform film actuator spatial distribution $\langle x - a \rangle^0$.

An experimental implementation of the impedance controller, combined with the degenerate distributed actuators described above has been performed by Procopio and Hubbard [34]. The relevant parameters for the experimental beam are presented in Table IV. Motion of the tip of the experimental beam was measured through the use of both linear and angular accelerometers. A schematic of the experimental set-up is shown in Fig. 14. Recall that the control gains needed to match the impedance of the beam are $G_v = G_\omega = K_v = K_\omega = 1$. The gains used in the experiment described here were limited to $G_v = 0.1282$, $G_\omega = 166.2 \times 10^{-6}$, $K_v = 0.0071$, $K_\omega = 287.35 \times 10^{-6}$ because of the small control authority provided by the PVF_2 film actuators. The control voltages supplied to the film actuators was limited in magnitude to 400 volts. The experiments illustrated here were performed to demonstrate the utility of using degenerate distributed transducers with the technique of end point impedance control. While the results can be applied to a broad class of distributed transducers, PVF_2 film actuators were used here as a matter of convenience.

Table IV. Cantilever beam static properties.

Parameter	Value
Length	52 in
Cross-section Area	0.75 in^2
Elastic Modulus	11×10^6 psi
Density	0.266×10^{-3} lb-sec^2/in^4

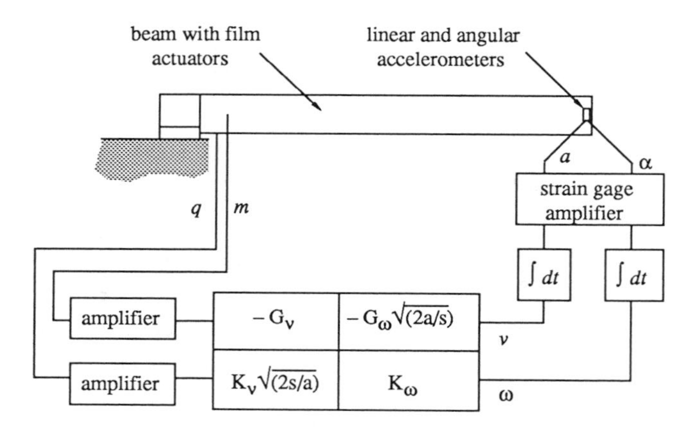

Fig. 14. Impedance control experimental schematic.

The experiments were performed by displacing the tip of the beam 1 inch and then releasing it. This type of input chiefly excited the first structural mode. Figure 15 shows the linear tip acceleration time history produced for a free decay with the controller unattached. The end point impedance controller was then activated to apply both a control moment and colocated force at the free end of the beam. Figure 16 shows the linear tip acceleration with the controller energized. Active control of multiple modes was also demonstrated by exciting the structure with an impact hammer. Figures 17 and 18 illustrate the resulting structural vibrations with the controller off and on, respectively.

Figures 15 and 16 shows an effective damping of first mode vibrations even though the control gains are very small. The settling time of the structure is reduced from well over 200 seconds to 50 seconds. Figures 17 and 18 illustrate the damping of both the first and second structural modes simultaneously since the end point impedance controller absorbs traveling waves and does not simply target individual modes of the system. The first two modes of the beam occurred at $1.35Hz$ and $8.65Hz$, respectively. The controller was implemented in hardware such that the $\sqrt{s}$ propagation operator required by the controller had a break frequency of $18Hz$, hence the higher frequency modes were not controlled. Other boundary conditions and controller forms were evaluated in [34] using distributed transducers.

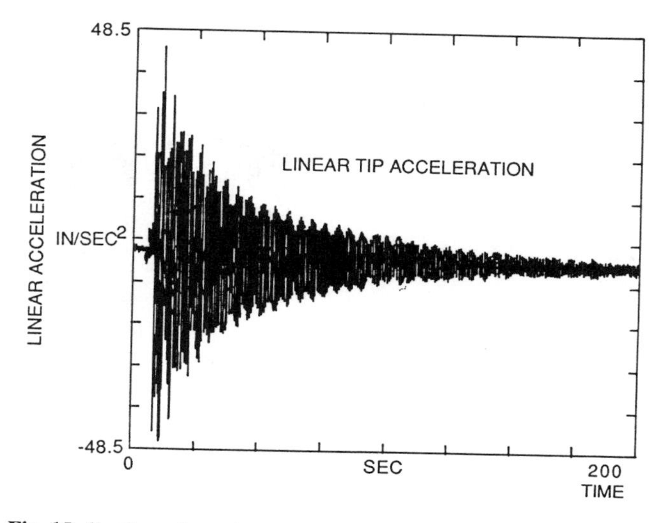

Fig. 15. Cantilever beam initial condition response with control turned off.

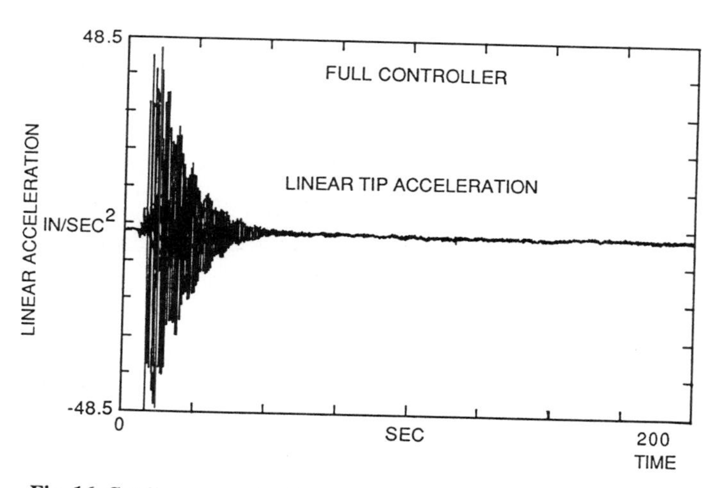

Fig. 16. Cantilever beam initial condition response with impedance control.

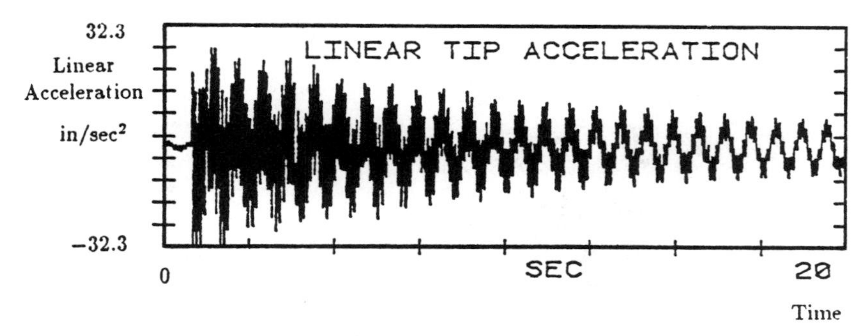

Fig. 17. Cantilever beam impact response with control turned off.

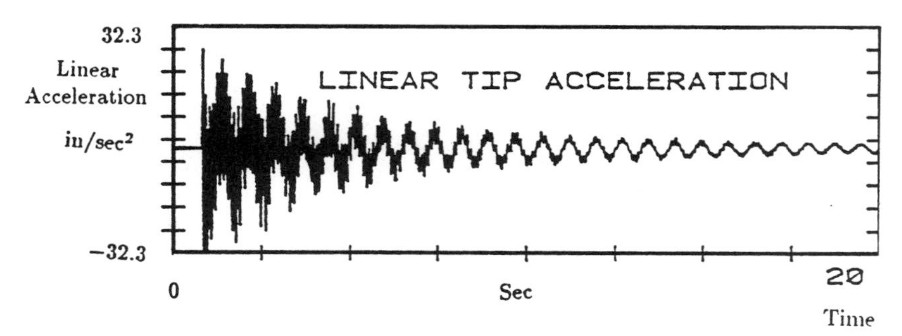

Fig. 18. Cantilever beam impact response with impedance control.

The uniform and ramp distributions can be combined in a unique manner to form a single "impedance actuator", as illustrated in Fig. 19. Note the spatial variation of the electrode deposition. Electrode I is used to provide the ramp distribution, while electrodes I and II are used in combination to provide the uniform distribution; one merely exploits linear superposition in the compensator implementation. This single actuator provides both force and moment control loads which are collocated at the free end, clearly indicating an advantage that is offered by the distributed actuator synthesis techniques of the previous section for the development of self-reacting, *physically realizable* devices. This same choice of spatial distributions may be used to synthesize an "impedance sensor" which will measure both linear and angular velocity at the free end.

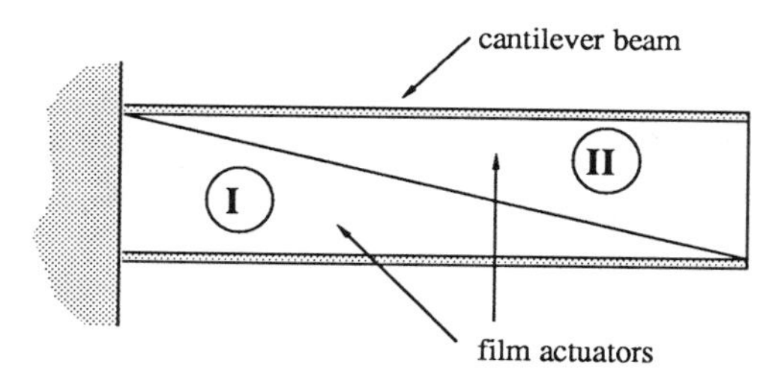

Fig. 19. "Impedance actuator" configuration.

C. MODAL INTERPRETATIONS FOR DISTRIBUTED TRANSDUCERS

Eigenfunction expansions provide a convenient means of assessing the effectiveness of spatially discrete or distributed transducers. A transducer distribution's "modal effectiveness" is based upon the completeness of its modal expansion. This result can be exploited in the design of distributed transducers that target specific modes or modal subsets. While the results are generic to any control design methodology, the Lyapunov method is used here to provide the requisite mathematical foundation for the actuator problem. In particular, eigenfunction expansions provide a modal interpretation of the non-modally based Lyapunov control designs. The sensor problem is treated later using similar analytical techniques. The modal decomposition is used to verify the contentions of the previous control designs, where distributed transducers were synthesized to control *all* modes of structural components.

The Bernoulli-Euler beam governing equation Eq. (32) and boundary conditions (33) - (36) constitute a Sturm-Liouville problem. Consequently, the equation's solution can be expressed as an expansion in terms of a complete set of orthogonal eigenfunctions $\{ \psi_i(x) \}$;

$$y(x,t) = \sum_{i=1}^{\infty} c_i(t)\, \psi_i(x) . \tag{79}$$

The eigenfunctions are normalized, so that

$$\int_0^1 \psi_i(x)\,\psi_j(x)\,dx = \delta_{ij}. \tag{80}$$

The control input $u(x,t)$ is assumed to be separable,

$$u(x,t) = \Lambda(x)\,u(t), \tag{81}$$

where the actuator distribution function $\Lambda(x)$ is square integrable over $x \in [0,1]$. The actuator distribution admits an eigenfunction expansion as well of the form

$$\Lambda(x) = \sum_{j=1}^{\infty} \lambda_j\,\psi_j(x). \tag{82}$$

Substituting Eqs. (79), (81), and (82) into the time derivative of the Lyapunov functional (38) yields

$$\dot{J}(t) = \int_0^1 \left[\sum_{i=1}^{\infty} \dot{c}_i(t)\,\psi_i(x)\right]\left[\sum_{j=1}^{\infty} \lambda_j\,\psi_j(x)\right] u(t)\,dx. \tag{83}$$

Using the orthonormality property of the eignefunctions (80), this reduces to a modal form for the power flow from the beam system,

$$\dot{J}(t) = \left[\sum_{i=1}^{\infty} \lambda_i\,\dot{c}_i(t)\right] u(t). \tag{84}$$

The expansion coefficients c_i are the modal amplitudes, and their time derivatives $\dot{c}_i$ are the modal velocities of the beam response. Since the actuator expansion coefficients λ_i are defined [using Eqs. (82) and (80)] by

$$\lambda_i = \int_0^1 \Lambda(x)\,\psi_i(x)\,dx, \tag{85}$$

all modes of the beam will be controllable if the eigenfunctions are complete with respect to the actuator distribution $\Lambda(x)$, such that $|\lambda_i| > 0$ for *all* modes $i = 1,\ldots,\infty$.

This quantifies the intuitive results developed above. The Lyapunov control design reduces to choosing an actuator spatial distribution to control all modes of interest by ensuring the actuator distribution's modal coefficients are nonzero for those modes, then choosing a control time dependence $u(t)$ to guarantee asymptotic stability, i.e. generalized modal control. In principle, one can then synthesize a spatial actuator distribution not only to control, say, odd- or even-order modes (as discussed for the pinned beam in section II), but *individual* modes, using Eqs. (82) and (85). However, the Lyapunov design can yield a *single* actuator distribution for the control of all beam modes. This contrasts with the Independent Mode Space Control (IMSC) distributed parameter control design technique [35], which requires a dedicated actuator for each mode to be controlled. However, IMSC is capable of independently targeting and controlling individual modes given these additional degrees of freedom.

As an illustration of the modal interpretation, consider the beam/film control system designed by Bailey and Hubbard [16] used to damp the vibrations of a thin cantilever beam. For this problem a spatially uniform piezoelectric film actuator distribution was applied over the entire length of the beam. The eigenfunctions of the cantilever beam take the form [36]

$$\psi_i(x) = A[cos(\mu_i x) - cosh(\mu_i x)] + B[sin(\mu_i x) - sinh(\mu_i x)]. \tag{86}$$

The eigenvalues μ_i are defined as the positive definite roots of the characteristic equation

$$cos(\mu_i)cosh(\mu_i) = -1. \tag{87}$$

All of the beam modes will be controllable given the uniform film actuator distribution if all of the Fourier coefficients of the actuator distribution's expansion (85) are nonzero. Using Eq. (86), these coefficients are given by

$$\lambda_i = \mu_i A[sin(\mu_i) + sinh(\mu_i)] + \mu_i B[cosh(\mu_i) - cos(\mu_i)]. \tag{88}$$

The contribution from the point moment at $x = 0$ has vanished during the evaluation of the integral (85). This is because the eigenfunctions satisfy the boundary conditions at the clamped end.

Since the eigenvalues μ_i must satisfy (87), there is no eigenvalue that also sets the actuator expansion coefficients λ_i defined in Eq. (88) to zero; one can easily see this by testing the first few roots of (88), then examining the plots of sin(μ_i), cos(μ_i),

$\sinh(\mu_i)$, and $\cosh(\mu_i)$. Thus, the entire set of cantilever beam functions is complete with respect to the film actuator control distribution. This means that the distributed piezoelectric film controller can damp *all* modes. A similar result holds for the pinned beam example discussed above. Consequently, the modal expansion provides a convenient interpretation of the Lyapunov damper results, as well as an affirmation of the claim that all modes are controllable for this actuator distribution and beam configuration.

A similar analysis can be developed for the structural measurement problem. Degenerate linear sensors have output voltages $V_0(t)$ proportional to a weighted integral of the structural response $y(x,t)$ [27],

$$V_0(t) = \int_0^1 b(x)\mathrm{L}[y(x,t)]dx. \tag{89}$$

The sensor distribution is parameterized by $b(x)$, as developed in section II, and is assumed to be square integrable. The operator $\mathrm{L}[\cdot]$ is a linear differential operator that models the operation of the sensor; for example, a strain gage would have $\mathrm{L} \equiv (\cdot)_{xx}$, while an accelerometer placed at $x = a$ would have $\mathrm{L}[y] \equiv y_{tt}$ and a distribution $b(x) \equiv \langle x - a\rangle^{-1}$. The form (89) can also be recovered from the distributed piezoelectric film sensor model, Eq. (27) after two integrations by parts, as the Laplacian operator is self-adjoint.

All spatial information about the sensor distribution is contained within the integral (89). Since the expansion of the beam displacement field (79) is separable, the linear operator L can be divided into a spatial operator L_x and a temporal operator L_t. Combining the expansion (79) with the composition integral (89) then yields a decomposition of the sensor output in the system eigenfunctions,

$$V_0(t) = \sum_{i=1}^{\infty} \mathrm{L}_t[c_i(t)] \int_0^1 b(x)\mathrm{L}_x[\psi_i(x)]dx, \tag{90}$$

where the uniform convergence of the expansion permits moving the modal coefficients $c_i(t)$ outside the spatial integral. Then, consistent with the constraints imposed upon $b(x)$ in the previous section, Eq. (89) can be rewritten as

$$V_0(t) = \sum_{i=1}^{\infty} \mathrm{L}_t[c_i(t)] \int_0^1 \mathrm{L}_x[b(x)]\, \psi_i(x)dx. \tag{91}$$

The modified sensor distribution $L_x[b(x)]$ is assumed to have a eigenfunction decomposition

$$L_x[b(x)] = \sum_{i=1}^{\infty} \beta_i \psi_i(x) \tag{92}$$

analogous to the actuator distribution expansion (82). Thus, the sensor output (91) can be written as a modal expansion,

$$V_0(t) = \sum_{i=1}^{\infty} L_t[c_i(t)] \beta_i, \tag{93}$$

where the sensor modal coefficients β_i are defined using Eqs. (92) and (80) as

$$\beta_i = \int_0^1 L_x[b(x)] \, \psi_i(x) \, dx. \tag{94}$$

All modes of the beam will be observable if the eigenfunctions are complete with respect to the modified sensor distribution $L_x[b(x)]$, such that $|\beta_i| > 0$ for *all* modes $i = 1,\ldots,\infty$. Sensor design reduces to choosing a sensor spatial distribution to measure all modes of interest by ensuring the sensor distribution's modal coefficients are nonzero for those modes. In principle, one can synthesize a spatial sensor distribution to target *individual* modes, using Eqs. (92) and (94).

As an example of this analysis we once again consider the cantilever beam problem, but now with a spatially uniform piezoelectric film sensing layer bonded to one face. This distribution, defined in Eq. (28), has a modified distribution defined by its Laplacian,

$$L_x[b(x)] = b_{xx}(x) = \langle x \rangle^{-2} - \langle x - 1 \rangle^{-2}. \tag{95}$$

All of the beam modes will be observable given the uniform film sensor distribution if all of the sensor modal coefficients are nonzero. Here, these modal coefficients β_i are given by

$$\beta_i = \mu_i A[sin(\mu_i) + sinh(\mu_i)] + \mu_i B[cosh(\mu_i) - cos(\mu_i)]. \tag{96}$$

As for the actuator problem, the contribution from the end $x = 0$ has vanished. This is because the eigenfunctions must satisfy the zero slope boundary condition at the clamped end. Since the eigenvalues μ_i must satisfy the characteristic equation Eq. (87), there is no eigenvalue that also sets the sensor modal coefficients β_i defined in Eq. (96) to zero; Eq. (96) has the *exact* same form as Eq. (88) for the actuator example. Thus, the entire set of cantilever beam functions is complete with respect to this spatially uniform film sensor distribution. This means that the distributed piezoelectric film sensor can observe the slope at the cantilever beam's free end for *all* modes. Interestingly, this is the time integral of the sensed parameter required to implement a Lyapunov design active vibration controller for the cantilever beam; this result shall be exploited in the ensuing discussion of collocation and smart structures. As for the actuator problem, the modal expansion provides a convenient interpretation of the sensor distributions.

D. COLOCATED DISTRIBUTED SENSORS AND ACTUATORS FOR SMART STRUCTURAL CONTROL

Using the nondimensional Lyapunov formulation, it is a straight-forward exercise to demonstrate that colocated piezoelectric sensor and actuator distributions can be used to construct stabilizing vibration controllers for Bernoulli-Euler beams. This is a generalization of the well known result that colocated spatially discrete actuators and rate sensors can be used to generate stabilizing controllers. The advantage of distributed piezoelectric transducers is that they can be embedded within beam structural components to autonomously sense and control vibrations. More complex superstructures can then be built from these "*smart*" structural components to provide decentralized active vibration control based upon autonomous, low-order control compensators.

The development here is based upon the Lyapunov beam analysis, and commences by rewriting the expression (38) for power flow from a Bernoulli-Euler beam having distributed piezoelectric transducers bonded to its faces,

$$\dot{J}(t) = u(t)\frac{d}{dt}\left[\int_0^1 \Lambda(x) y_{xx}(x,t)\,dx\right]. \tag{97}$$

The sensor and actuator distributions $b(x)$ and $\Lambda(x)$ are chosen to be *identical*, with this colocated distribution denoted by $\tilde{\Lambda}(x)$. If the time portion of the exogenous control signal $u(t)$ is chosen to be proportional to the time derivative of the sensing distribution's voltage output,

$$u(t) = -k^2 \frac{d}{dt}\left[\int_0^1 \tilde{\Lambda}(x) y_{xx}(x,t)\, dx\right], \tag{98}$$

where k is a nonzero constant, then the energy flow expression (97) takes the form

$$\dot{J}(t) = -k^2 \left\{\frac{d}{dt}\left[\int_0^1 \tilde{\Lambda}(x) y_{xx}(x,t)\, dx\right]\right\}^2 . \tag{99}$$

The power expression (99) is negative semi-definite *for all time t,* given the colocated sensor and actuator distributions. Consequently, no matter what the sensor and actuator distributions bonded to the beam are, if they are *identical*, and the two are interconnected via rate feedback as per Eq. (98), then the resulting closed loop system is stable in the sense of Lyapunov. This result is true for any spatial distribution, including those that control/sense only certain modal components. The result also holds for any distributed transducer that operates like PVF_2 film, such as piezoceramic crystal sensors and actuators.

If a mode is not controllable with the chosen film actuator distribution, then it is also not observable using the same film sensing distribution. The Bernoulli-Euler beam is open-loop stable, hence all modes are stabilizable/detectable with respect to any distribution. However, if the actuator distribution is appropriately chosen such that all modes are controllable, then utilizing a distributed sensor with the same spatial distribution permits one to simply construct a closed-loop vibration controller *for all modes*. Further, transducer distributions can be developed for tasks other than vibration damping, but still utilize the resultant actuator distribution for some level of vibration reduction (this result has been exploited experimentally in [37]). Mode targeting interpretations for specific configurations can be deduced using the eigenfunction representations developed in the previous section.

The colocation result has important practical application for complex interconnected aerospace structures. The distributed transducers illustrated here are self-reacting, hence they need no fixed inertial framework to react against. The devices can be incorporated into individual structural components during manufacture, with the corresponding compensators implemented on board the component. Since the compensators are local, and guaranteed to always extract energy from the individual beam elements, an interconnected system of such beams will be stable; the component pieces merely appear to have higher inherent damping. This strategy has been successfully tested in an interconnected experimental beam structure [28].

The colocation result can be extended to degenerate devices other than distributed film transducers. However, restrictions on the types of sensors and actuators must be satisfied. To see this, the energy flow expression (97) is written generically for the Bernoulli-Euler beam model as

$$\dot{J}(t) = u(t)\frac{d}{dt}\left[\int_0^1 y(x,t)\Lambda(x)dx\right]. \tag{100}$$

Similarly, the voltage output from a general, degenerate linear sensor with high temporal bandwidth is expressible as

$$V_0(t) = \int_0^1 y(x,t)\mathrm{L}_x[b(x)]dx. \tag{101}$$

If the modified sensor distribution $\mathrm{L}_x[b(x)]$ is equal to the actuator distribution $\Lambda(x)$, then the time derivative of the sensor output can be fed back as the control signal $u(t)$ to construct a stabilizing controller. This requires the transducers to not only have the same physical distribution over the beam, but to have the same *spatial derivative order* $\mathrm{L}_x[\cdot]$. For example, point moment actuators cannot be colocated with point linear velocity sensors and interconnected via rate feedback to construct a stabilizing controller in the sense of Lyapunov, but they *can* be colocated with point angular velocity sensors. This qualifies the colocation argument as applied to distributed parameter systems for distributed transducers.

IV. DESIGN IMPLICATIONS AND CONSTRAINTS

The transducers models presented in this chapter have broad applicability to a wide range of vibration measurement and control problems, even though they represent a small subset of distributed transducers because they are degenerate. A unified design approach has been summarized which permits the performance of both distributed sensors and actuators to be evaluated and described using a compact notation in the form of singularity functions. While specific applications and experimental results were presented using a piezoelectric polymer film transducer (PVF_2), the design models and control techniques are readily exploited using any of the popular sensors and actuators found in modern control systems. The models and techniques may be used, for example, to spatially weight the distribution of a fiber optic sensor. These and other transducers can then be used for mode targeting, shape measurement, structural identification, or non-destructive testing [38,39,40]. They may also be used to spatially weight the control gains to an array of discrete actuators and/or sensors such that specific parameters of interest may be measured and controlled [41].

The Lyapunov and impedance design methodologies have been shown to produce simple, effective, and physically realizable active structural control systems. The control designs were physically motivated, as the synthesis methods were based upon intuitive energy and power flow considerations. The *a priori* incorporation of distributed transducers in the design stage permitted the construction of control systems with a minimum of input/output channels. These techniques can be applied equally well to spatially discrete transducers. The associated compensators were of very low order. Extensions to smart structural components led to the design of structural elements with "built in" actuation, sensing, and control compensation.

Concomitant with this simplicity and appeal to physical insight, the design approaches developed herein only deal with *global* stability and damping. The Lyapunov synthesis method as presented here is only concerned with asymptotic stability of the entire plant, rather than any specific performance goal, such as settling time, steady-state tracking error, or overshoot. The control compensators developed here were designed to remove energy from the *entire system* at all times. Consequently, vibrational energy in such an actively-damped interconnected structure can "move" between various resonant modes (as there is no independent mode targeting) or between the structural components. This behavior was noted in multi-modal tests of the interconnected beam smart structural control experiment discussed in

[28], as well as the cantilever beam impedance control impact response experiments presented in [34]. Nevertheless, for many structural control applications this type of performance is more than adequate, and provides a realizable approach to an important class of challenging control problems.

Other control analysis and design methods are required to address specific performance goals, and to develop compensators for applications such as tracking or shape control. Recent developments in multivariable control theory permit the design of multi-input/multi-output compensators for achieving prescribed damping and tracking performance using LQG, LQG/LTR, H_∞, and μ-synthesis techniques [42,43,44]. In fact, distributed parameter control system performance can be directly assessed using suitable performance measures developed in both temporal and *spatial* domains that are extensions of these techniques [45,46]. LQG/LTR design has been successfully utilized in a structural beam dynamic shape control experiment [27,37]. However, there are limits to achievable performance, robustness, and system realization that must be considered when using degenerate transducers and lumped parameter design techniques for the control of distributed parameter systems.

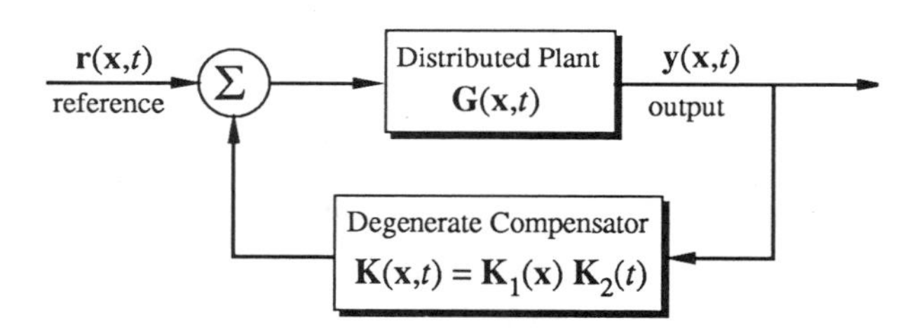

Fig. 20. Degenerate control compensator topology.

The aforementioned design methods have been developed primarily for lumped parameter models. Their application to more complex structural control problems suffers because of the complexities of interconnected elastic systems: an extremely large (and theoretically infinite) number of modes that lead to high-order compensators, high modal densities, and continuum (spatial) robustness issues relating to transducer placement. The transducers and plants discussed here can be represented in the degenerate control system topology depicted in Fig. 20. It is important to note that the distributed parameter control designer is not only charged with developing and implementing the temporal portion of the controller, but the spatial portion as well. This compensation form has been studied at length by Butkovskiy [7]. One can consider the transducers and their placement/distribution as the spatial portion of a degenerate distributed parameter compensator matrix $\mathbf{K}_1(\mathbf{x})$. The poles and zeros of the closed loop system are affected by the designer's choice of the temporal and spatial portion of the compensator matrix. Zero placement may be achieved by proper spatial weighting of transducers [47], while pole placement is fixed by the temporal compensator design. Most importantly, the overall stability of the distributed parameter control system will be profoundly affected by the choice of these spatial and temporal weightings.

The control designer must consider the robustness of a control system's performance and stability in the presence of modelling uncertainties. Complex structural systems often have plant models that are imperfectly specified. The Lyapunov and impedance control synthesis approaches developed herein do not address robustness explicitly. However, since they do not rely upon precise *modal* representations of the underlying plants, they are in many ways insensitive to variations in a structure's dynamics, such as its resonance frequencies. And while the designs do require specification of a plant's boundary conditions for deducing an appropriate transducer distribution to assure controllability and observability, they can accomodate variations. This "spatial robustness" is guaranteed if one exploits the colocation results developed in section III. If the boundary conditions differ from those of the design plant model, a transducer distribution designed for the nominal plant will only suffer from variations in the controllability and observability of certain modes; their colocation and rate feedback interconnection guarantees closed-loop stability in the sense of Lyapunov. All that will suffer is performance.

Explicit stability-robustness [48] and performance-robustness [42] measures have been developed for multivariable lumped parameter control design. Recent developments in sliding mode control [49] have led to design techniques that provide performance guarantees in the presence of plant model variations and bounded disturbance environments for lumped systems. The application and extension of these and other techniques to active structural control using distributed transducers is a topic of current research. The recent interest in extending H_∞ control to explicitly address distributed parameter control, as well as developments in infinite dimensional linear systems theory [50], may provide tractable solutions to these problems.

V. REFERENCES

1. M. Balas and J. Canavin, "An Active Modal Control System Philosophy for a Class of Large Space Structures", *AIAA Symposium on Dynamics and Control of Large Flexible Spacecraft*, Blacksburg VA (1977).

2. M. Athans, P. Kapasouris, E. Kappos, and H. Spang, "Linear-Quadratic Gaussian with Loop-Transfer Recovery Methodology for the F-100 Engine", *AIAA Journal of Guidance and Control* **9**(1), pp. 45-52 (1986).

3. M. Balas, "Active Control of Flexible Systems", *Journal of Optimization Theory and Applications* **25**, pp. 415-436 (1978).

4. M. Balas, "Feedback Control of Flexible Systems", *IEEE Transactions on Automatic Control* **AC-23**, pp. 673-679 (1978).

5. R. Haftka, Z. Martinovic, W. Hallauer, and G. Schamel, "Sensitivity of Optimized Control Systems to Minor Structural Modifications", *Proceedings of the AIAA/ASME/ASCE/AHS 26th Structures, Structural Dynamics and Materials Conference*, Palm Springs CA (1985).

6. M. Balas, "Modal Control of Certain Flexible Dynamic Systems", *Proceedings of the 2nd IFAC Symposium on Distributed Parameter Control*, Coventry, England (1977).

7. A. Butkovskiy, *Structural Theory of Distributed Parameter Systems*, Ellis Horwood, Chichester, England (1983).

8. S. Crandall, *An Introduction to the Mechanics of Solids*, McGraw-Hill, New York, N.Y. (1972).

9. W. Macauley, "Note on the Deflection of Beams", *Mes. Math.* **48**, pp. 170-174 (1964).

10. W. Pilkey, "Clesbsch's Method for Beam Deflections", *Journal of Engineering Education* **54**, pp. 170-174 (1964).

11. W. Kaplan, *Operational Methods for Linear Systems*, Addison-Wesley, Reading, MA, pp. 44-63 (1962).

12. M. Lighthill, *An Introduction to Fourier Analysis and Generalized Functions*, Cambridge University Press, Cambridge, England (1958).

13. J. Hubbard, Jr., "Method and Apparatus Using a Piezoelectric Film for Active Control of Vibrations", United States Patent No. 4,565,940 (1986).

14. Pennwalt Inc., *Technical Manual for Kynar Film* (1985).

15. J. Hubbard, Jr., "Distributed Sensors and Actuators for Vibration Control in Elastic Components", *Proceedings of Noise-Con 87*, State College, PA (1987).

16. T. Bailey and J. Hubbard, Jr., "Distributed Piezoelectric Polymer Active Vibration Control of a Cantilever Beam", *AIAA Journal of Guidance and Control* **8**(5), pp. 605-610 (1985).

17. S. Burke and J. Hubbard, Jr., "Distributed Actuator Control Design for Flexible Beams", *Automatica* **24**(5), pp. 619-627 (1988).

18. S. Miller and J. Hubbard, Jr., "Observability of a Bernoulli-Euler Beam Using PVF_2 as a Distributed Sensor", *Proceedings of the Sixth VPI&SU/AIAA Symposium on Dynamics and Control of Large Structures*, Blacksburg, VA (1987).

19. J. Hubbard, Jr., "Distributed Transducers for Smart Structural Components", *Proceedings of the 6th International Modal Analysis Conference*, Kissimee, FL, pp. 856-862 (1988).

20. R. Kalman and J. Bertram, "Control Systems Analysis and Design via the 'Second' Method of Lyapunov", *Transactions of the ASME, Journal of Basic Engineering*, pp. 371-400 (1960).

21. D. Franke, "Control of Bilinear Distributed Parameter Systems", in I. Hartmann (Ed.), *Advances in Control Systems and Signal Processing*, Vol. 1, Friedr. Vieweq. & Sohn, Weisbaden, West Germany (1980).

22. P. Wang, "Stability Analysis of Elastic and Aeroelastic Systems via Lyapunov's Direct Method", *Journal of the Franklin Institute* **281**, pp. 51-72 (1966).

23. M. Slemrod, "An Application of Maximal Dissipative Sets in Control Theory", *Journal of Mathematical Analysis and Applications* **46**, pp. 369-387 (1974).

24. J. Plump, J. Hubbard, Jr., and T. Bailey, "Nonlinear Control of a Distributed System: Simulation and Experimental Results", *Transactions of the ASME, Journal of Dynamic Systems, Measurement, and Control* **109**, pp. 133-139 (1987).

25. S. Burke and J. Hubbard, Jr., "Active Vibration Control of a Simply Supported Beam Using a Spatially Distributed Actuator", *IEEE Control Systems Magazine*, pp. 25-30 (August 1987).

26. S. Burke and J. Hubbard, Jr., "Limitations of Degenerate Distributed Actuator Control Design for Thin Plates", SPIE OE/Aerospace Sensing '90, Orlando, FL (1990).

27. S. Burke, "Shape and Vibration Control of Distributed Parameter Systems – Extension of Multivariable Concepts Using Spatial Transforms", C.S. Draper Laboratory Report CSDL-T-1017 (1989).

28. S. Miller and J. Hubbard, Jr., "Smart Components for Structural Vibration and Control", *Proceedings of the 1988 ACC*, Vol. 3, Atlanta, GA, pp. 1897-1902 (1988).

29. D. Vaughan, "Application of Distributed Parameter Concepts to Dynamic Analysis and Control of Bending Vibrations", *Transactions of the ASME, Journal of Basic Engineering*, pp. 157-166 (1968).

30. F. Brown, "On the Dynamics of Distributed Systems", *Applied Mechanics Reviews* **17**(5) (1964).

31. J. Van de Vegte, "The Reflection Matrix in Beam Vibration Control", *Transactions of the ASME, Journal of Dynamic Systems, Measurement, and Control*, pp. 94-101 (1971).

32. R. Schwarz and B. Friedland, *Linear Systems*, McGraw-Hill, New York, N.Y. (1965).

33. K. Ogata, *Modern Control Engineering*, Prentice-Hall, Englewood Cliffs, N.J. (1970).

34. G. Procopio and J. Hubbard, Jr., "Active Damping of a Bernoulli-Euler Beam via End Point Impedance Control Using Distributed Parameter Techniques", in D. Inman (Ed.), *Vibration Control and Active Vibration Suppression*, DE-Vol. 4, 11th Biennial Conference on Mechanical Vibration and Noise, Boston, MA, pp. 35-46 (1987).

35. L. Meirovitch and H. Bauh, "Nonlinear Natural Control of an Experimental Beam", *AIAA Journal of Guidance and Control* **7**, pp. 437-442 (1984).

36. K. Graff, *Wave Motion in Elastic Solids*, Ohio State University Press (1975).

37. S. Burke and J. Hubbard, Jr., "Closed-Loop Dynamic Shape Control Of Structural Systems", SPIE OE/Aerospace Sensing '90, Orlando FL (1990).

38. F. Berkman and D. Karnopp, "Complete Response of Distributed Systems Controlled by a Finite Number of Linear Feedback Loops", *Proceedings of the ASME, Journal of Engineering for Industry*, pp. 1063-1068 (November 1969).

39. R. Griffiths and R. Lamson, "Adaptation of an Electro-Optic Monitoring System to Aerospace Structures", SPIE E/O Fibers '87, Paper 838-54, San Diego, CA (1987).

40. H. Gallantree, "Ultrasonic Applications of PVDF Transducers", *The Marconi Review*, First Quarter, pp. 50-64 (1982).

41. J. Creedon and A. Lindgren, "Control of the Optical Surface of a Thin, Deformable Primary Mirror with Application to an Orbiting Astronomical Observatory", *Automatica* **6**, pp. 643-660 (1970).

42. J. Doyle and G. Stein, "Multivariable feedback design: Concepts for a classical-modern synthesis", *IEEE Transactions on Automatic Control*, **AC-26**, pp. 4-16 (1981).

43. G. Stein and M. Athans, "The LQG/LTR procedure for multivariable feedback control design", *IEEE Transactions on Automatic Control*, **AC-32**, pp. 105-114 (1987).

44. J. Doyle, K. Lenz, and A. Packard, "Design examples using μ-synthesis: space shuttle lateral axis FCS during reentry", *Proceedings of the 25th Conference on Decision and Control* (December 1986).

45. S. Burke and J. Hubbard, Jr., "Performance measures for distributed parameter control systems", *Proceedings of the IMACS/IFAC International Symposium on Modelling and Simulation of Distributed Parameter Systems*, Hiroshima Japan, pp. 559-565 (1987).

46. S. Burke and J. Hubbard Jr., "Spatial filtering concepts in distributed parameter control", accepted for publication in the *Transactions of the ASME, Journal of Dynamics, Measurement, and Control* (1990).

47. S. Burke and J. Hubbard, Jr., "Distributed Transducers, Colocation, and Smart Structural Control", SPIE OE/Aerospace Sensing '90, Orlando, FL (1990).

48. N. Lehtomaki, N. Sandell Jr., and M. Athans, "Robustness results in linear-quadratic Gaussian based multivariable control designs", *IEEE Transactions on Automatic Control*, **AC-26**(1), pp. 75-93 (1981).

49. G. Young and S. Rao, "Robust sliding mode control of a nonlinear process with uncertainty and delay", *Journal of Dynamic Systems, Measurement, and Control* **109** (September 1987).

50. R. Curtain and A. Pritchard, "Infinite Dimensional Linear Systems Theory", LCIS - Vol. 8, Springer-Verlag, Berlin, West Germany (1978).

ROBUST ADAPTIVE IDENTIFICATION AND CONTROL DESIGN TECHNIQUES

GARY A. MCGRAW

The Aerospace Corporation
El Segundo, California 90009-2957

I. INTRODUCTION

Robustness of adaptive identification and control algorithms to disturbances and unmodeled dynamics has been recognized as an important issue for several years. This chapter develops techniques for designing robust identification and control algorithms and applies these techniques to a flexible beam laboratory experiment. The problems of modal identification, active damping, and command tracking control of the beam are considered.

A. CONTROL AND IDENTIFICATION DESIGN PROBLEM

The basic problem considered is identification and control of a discrete-time plant given by

$$y(t) \quad = \quad G_0(q)u(t) + v(t), \ t = 1, 2, \ldots, \tag{1}$$

where $u(t)$ and $y(t)$ are the input and output of the system, respectively, and $v(t)$ is a disturbance (noise). The transfer function $G_0(q)$ is a function of the delay operator q^{-1} ($q^{-k}x(t) = x(t-k)$) and can be decomposed into two parts: $G_0(q) = G(q, \theta^*) + \Delta G$. The operator $G(q, \theta^*)$ is derived from a nominal mathematical model of the plant. The vector θ^* represents parameters such as gains and modal frequencies in the true system, and is often referred to as the "tuned parameter" [1,2]. The operator ΔG is the *unmodeled dynamics* such as neglected or unknown high frequency and nonlinear dynamics.

The model used for control design is $G(q, \hat{\theta})$. The error $\hat{\theta} - \theta^*$ is the *parametric* or *structured uncertainty.* In the context of control design, the goal of system identification is to minimize the structured uncertainty. The presence of disturbances and unmodeled dynamics often lead to biased transfer function estimates and techniques to minimize this bias are presented here.

There are two basic approaches to coping with uncertainty in control system design. One approach is robust control, in which a single controller is designed to operate over all possible parameter variations and unmodeled dynamics. The other approach is adaptive control, in which the structured parameters are estimated and the controller is updated accordingly.

Adaptive controllers offer the potential for improved performance (e.g. higher control bandwidth) over robust designs, because of the reduction in the structured uncertainty of the system. However in many applications this performance improvement is slight unless there is considerable uncertainty in the plant parameters. Another trade-off is that adaptive controllers are inherently nonlinear and time-varying, so they are more complex than robust controllers. For these reasons adaptive control should only be considered after a fixed robust design has been shown not to meet the design objectives. The effect of unmodeled dynamics in adaptive control must also be considered as was observed in the early 80's when many standard adaptive control algorithms in the literature were shown to be unstable in the presence of minor unmodeled dynamics and disturbances [3,4].

B. OVERVIEW

Section II describes the laboratory setup and modeling. Section III discusses the prediction error method (PEM) of system identification. Then, following the development of [5,6], frequency domain analysis techniques for system identification design are considered. An overview of various identification design issues, such as sampling rate, model type, known dynamics, and prefilter design is presented in Section IV. In Section V a laboratory identification design based on modal decomposition of the model is developed. A set of control design rules for the laboratory experiment are derived in Section VI. These rules are based on classical frequency domain control design techniques. The resulting controllers are robust to unmodeled high frequency dynamics. The identification and controller designs are integrated into an adaptive control strategy in Section VII, where the feasibility of this strategy is demonstrated. Means for enhancing the transient response of the identification and control algorithms are also considered in those sections.

II. IDENTIFICATION AND CONTROL LABORATORY EXPERIMENT

Much of the research discussed in this chapter was motivated by experiences with a flexible beam laboratory experiment. In this section, the laboratory setup, measurements, and modeling will be described. Detailed discussion of the laboratory apparatus and modeling may be found in [7,8].

A. LABORATORY EXPERIMENT SETUP

1. EXPERIMENT HARDWARE

The laboratory setup is pictured in Fig. 1. The experimental setup consists of a thin, flexible fiberglass beam hung from a fixture. Roller bearings permit the fixture to slide on a set of rails. The beam is four feet long, two inches wide, and one-quarter inch thick, and is constrained to move in a vertical plane perpendicular to its width. The beam bends easily in the direction parallel to the motion of the sliding fixture. The vertical orientation of the beam helps damp-out torsional modes; side-to-side bending is small. Five evenly-spaced holes were drilled along the beam so that weights can be attached to alter the modal characteristics of the beam. (For reference purposes, these hole locations will be numbered 1–5; hole 1 at the top of the beam, hole 5 at the bottom.) An accelerometer mounted on the bottom of the beam senses motion there. A linear variable differential transformer (LVDT) attached to the sliding fixture measures position at the top of the beam. A linear DC motor at the top of the beam is the control actuator. The control objective is to damp-out oscillations at the bottom of the beam due to position commands at the top of the beam and disturbances applied along the beam.

2. EXPERIMENT ELECTRONICS

Fig. 2, is a block diagram of the setup which illustrates the electronic equipment used. A high bandwidth analog proportional-integral-derivative (PID) controller using the LVDT as a feedback sensor controls the position of the motor. This

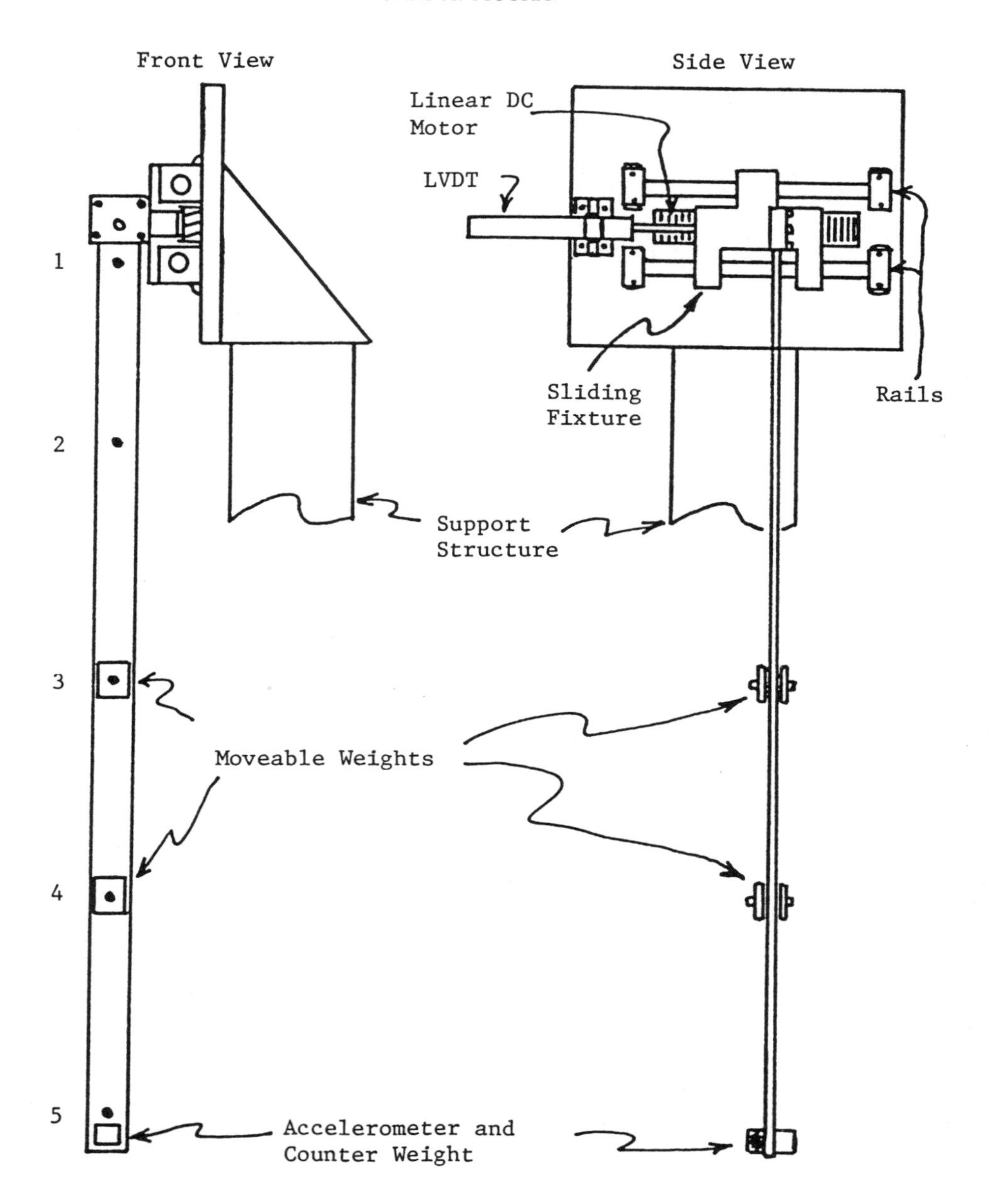

Figure 1: Laboratory experimental setup.

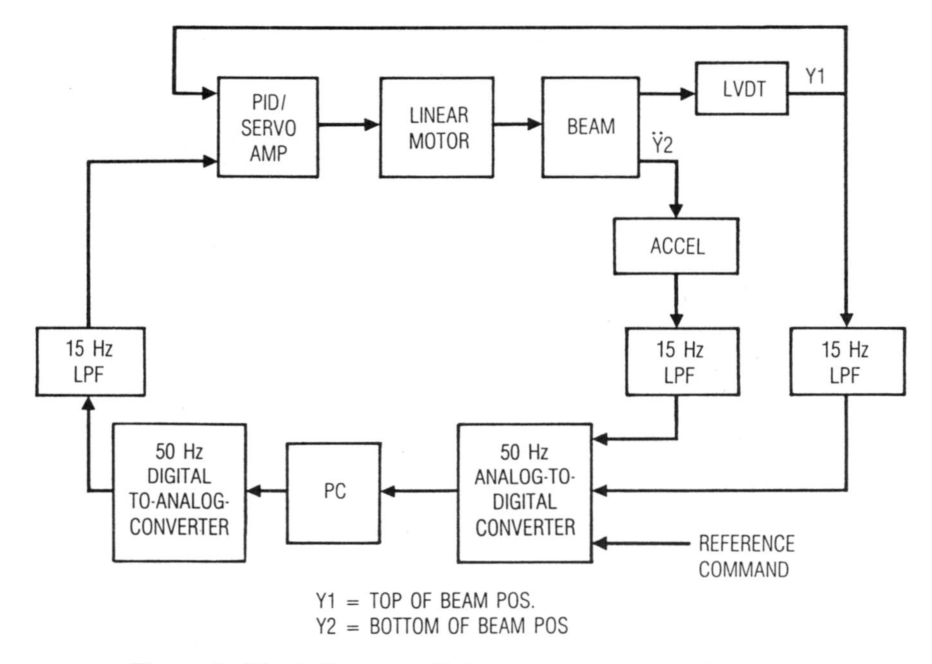

Figure 2: Block diagram of laboratory experimental setup.

loop prevents the motor from drifting, helps reduce the effects of motor nonlinearities, and is unaffected by the variations in the weight mounted on the beam. The dynamics of the PID controller, motor, and beam are lumped together for the purposes of developing a mathematical model.

Other electronic components in the setup included analog filters, a personal computer (PC), and data acquisition electronics. The PC is an 8086-based, 4.77 MHz clone of an IBM PC-XT. A sampling rate of 50 Hz was selected as compromise between system identification and control design performance objectives. Two 15 Hz first-order Butterworth lowpass filters are used for antialiasing filtering of the accelerometer and LVDT signals prior to sampling by the 12 bit analog-to-digital converter (ADC). A third ADC channel is used for a reference signal input. A 15 Hz analog filter on the 12 bit digital-to-analog converter (DAC) output attenuates high frequency components of the zero-order-hold output from the DAC, to minimize excitation of unmodeled bending modes.

The PC was not capable of processing measurements for system identification in real-time. Therefore true adaptive control experiments could not be conducted. However, identification and control could be performed separately, and experience with laboratory implementation greatly enhanced the adaptive control simulation studies that were performed.

B. LABORATORY EXPERIMENT MODELING

Modeling of the laboratory experiment was a two-stage process consisting of experimental measurements of the system transfer functions and derivation of an analytical lumped-mass model of the setup. Parameters in this analytical model were adjusted to match the experimental measurements.

1. EXPERIMENTAL MEASUREMENTS

A spectrum analyzer was used to measure the transfer functions from the PID input to the accelerometer and LVDT outputs for six weight configurations. The input was wideband noise. Only the first two modal frequencies in the accelerometer transfer function were determined experimentally. The weight configurations and the corresponding modal and gain measurements are summarized in Table I.

Fig. 3 shows a typical measured Bode plot (for case 2). Note that the damping on the flexible modes in the accelerometer transfer functions is very small. The first mode varies in a range of approximately 1.1–1.8 Hz; the second mode 8.4–12.5 Hz. Higher frequency modes are not explicitly included in the control system designs, but are modeled in simulations to ensure that the algorithms are robust to these dynamics. The LVDT transfer functions are basically flat, except at the second mode frequency, indicating that the PID controller works as intended.

2. ANALYTICAL MODELS

A lumped-mass model of the laboratory setup was the basis for analytical models. The details of the derivation of the dynamic equations may be found in [7,8]. For each output, the derivation yields a four-mode transfer function model from the PID input of the form

$$Y(s) \quad = \quad \sum_{k=1}^{4} \left(\frac{(c_k s + d_k)\omega_k^2}{s^2 + 2\zeta_k \omega_k s + \omega_k^2} \right) U(s), \tag{2}$$

where ω_k are the modal frequencies (in rad/sec), ζ_k the damping ratios, and c_k and d_k the modal gains. Measured values for the first two modes were used in Eq. (2); analytical values were used for the higher frequency modes and gains. This model serves as the truth model for the purposes of evaluating designs.

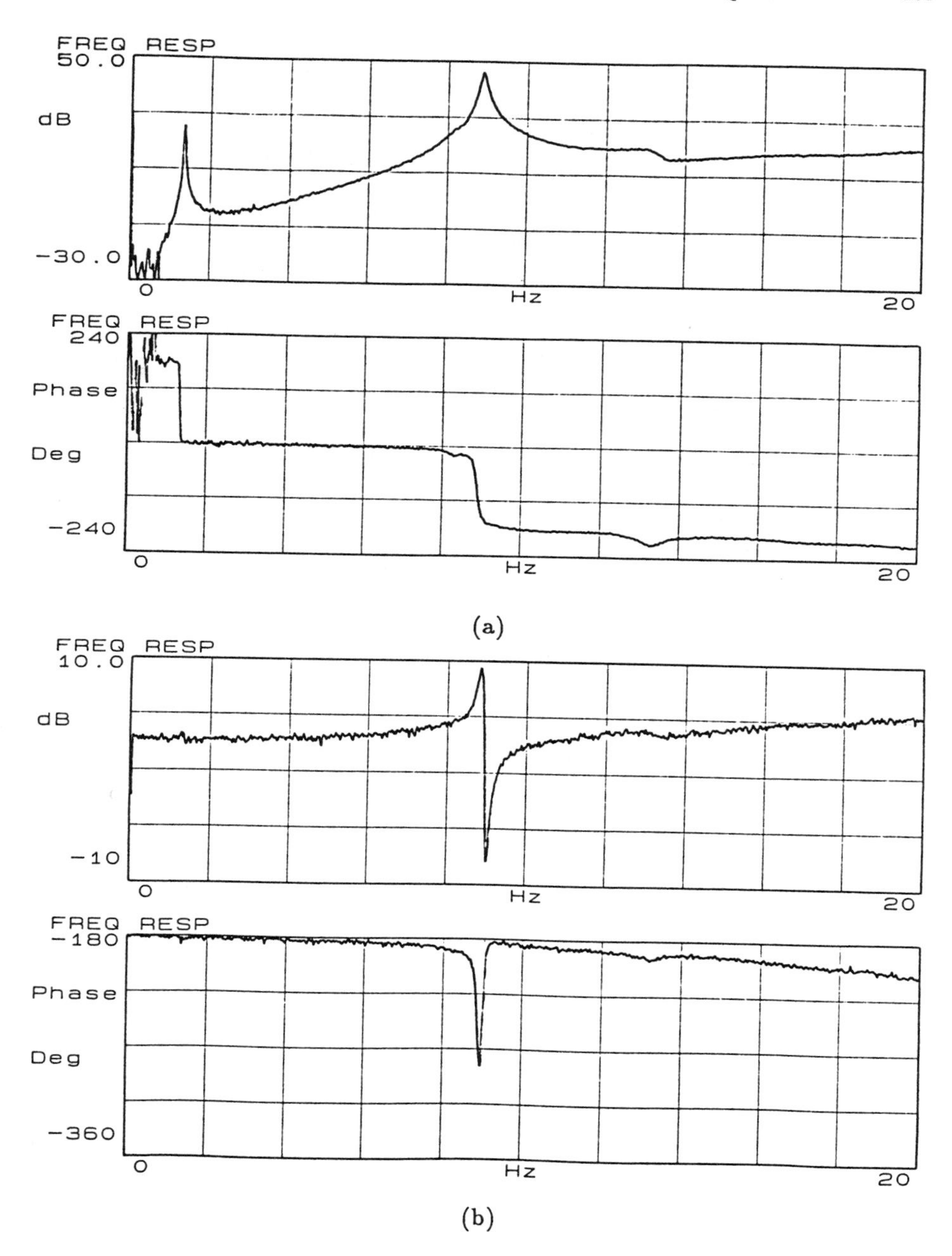

Figure 3: Measured laboratory transfer functions for case 2. (a) accelerometer output transfer function, magnitude and phase. (b) LVDT output transfer function, magnitude and phase.

Case	Weight Config. (hole # - weight)	1st Mode (freq., Hz/ damping)	2nd Mode (freq., Hz/ damping)	DC Gain (V/V)
1	5 - A & B	1.175 0.0008	12.4 0.0026	1.372
2	5 - A 3 - B	1.346 0.0011	8.879 0.0013	1.369
3	2 - A 3 - B	1.706 0.0012	8.48 0.0015	1.382
4	2 - A 4 - B	1.558 0.0012	9.857 0.0034	1.389
5	1- A & B	1.774 0.0009	12.32 0.014	1.371
6	no add'l weights	1.773 0.0067	12.42 0.014	1.413

Table I: Beam experiment configurations and measured modal frequencies and damping, and DC gains. Weight A = 371.8 g; weight B = 373.8 g.

Reduced order models were extracted from Eq. (2) for the purposes of control and identification design. The transfer function from the PID input to the accelerometer output is

$$\frac{\ddot{Y}_2(s)}{U(s)} = \frac{K_a s 2(s_z - s)(s_z + s)}{(s^2 + 2\zeta_1\omega_1 s + \omega_1^2)(s^2 + 2\zeta_2\omega_2 s + \omega_2^2)}. \tag{3}$$

The LVDT transfer function is modeled as a single gain, i.e., $Y_1(s) = K_p U(s)$. Because the actuator and accelerometer are not collocated, the zeros in Eq. (3) are real, with $\omega_2 > s_z > \omega_1$. The nonminimum phase (right-half plane) zero places limits on achievable control performance.

The analytical model of Eq. (2) was implemented along with models for the various electronics in two simulations. A PC-Matlab based simulation was used for design and analysis. A time domain FORTRAN simulation implemented on an Apollo workstation was used for the adaptive control simulations.

III. ANALYSIS TECHNIQUES FOR SYSTEM IDENTIFICATION DESIGN

This section develops the PEM identification methods and discusses frequency domain analysis techniques for designing robust system identification algorithms. The model class of primary concern is autoregressive-exogenous-input (ARX) models, since these models are predominant in on-line applications. The techniques discussed here are applied to the laboratory system identification design in Section V.

A. TRANSFER FUNCTION MODELS

A design model of the true system of Eq. (1) is given by the general transfer function model

$$y(t) \quad = \quad G(q,\theta)u(t) + H(q,\theta)e(t), \; t = 0,1,2,\ldots, \tag{4}$$

where $e(t)$ is mean-zero, white Gaussian noise (WGN) with variance λ. The transfer functions $G(q,\theta)$ and $H(q,\theta)$ have Laurent series expansions $G(q,\theta) = \sum_{k=1}^{\infty} g_k q^{-k}$ and $H(q,\theta) = \sum_{k=0}^{\infty} h_k q^{-k}$. The noise transfer function is assumed to be monic, i.e., $h_0 = 1$. (A non-unity value can be absorbed into the noise variance, λ, which may also be a parameter to be estimated.) In addition, $H(q,\theta)$ and $1/H(q,\theta)$ are assumed to be stable.

The frequency response of a discrete time transfer function $G(q)$ is computed by evaluating $G(q)$ around the unit circle in the complex plane, $q = e^{j\omega T}$, $-\pi/T \leq \omega \leq \pi/T$, where T is the sampling interval. Examples of commonly-used models included in the transfer function model class Eq. (4) will now be considered.

1. ARX MODELS

The autoregressive-exogenous-input (ARX) model is given by

$$\begin{aligned} A(q,\theta)y(t) &= B(q,\theta)u(t) + e(t), \; t = 0,1,2,\ldots, \\ A(q,\theta) &= 1 + a_1 q^{-1} + \cdots + a_n q^{-n}, \\ B(q,\theta) &= b_0 q^{-d} + \cdots + b_m q^{-m-d}, \\ \theta &= \left[\begin{array}{cccccc} a_1, \ldots, & a_n, & b_0, \ldots, & b_m \end{array} \right]^T. \end{aligned} \tag{5}$$

The corresponding transfer function model is

$$\begin{aligned} G(q,\theta) &= A(q,\theta)^{-1}B(q,\theta), \\ H(q,\theta) &= A(q,\theta)^{-1}. \end{aligned}$$

The notation ARX(n,m,d) will be used to indicate the orders of the polynomials and the input delay in Eq. (5).

2. ARMAX MODELS

The autoregressive-moving-average-exogenous-input (ARMAX) model is given by

$$\begin{aligned} A(q,\theta)y(t) &= B(q,\theta)u(t) + C(q,\theta)e(t),\ t = 0,1,2,\ldots, \\ A(q,\theta) &= 1 + a_1 q^{-1} + \cdots + a_n q^{-n}, \\ B(q,\theta) &= b_0 q^{-d} + \cdots + b_m q^{-m-d}, \\ C(q,\theta) &= 1 + c_1 q^{-1} + \cdots + c_p q^{-p}, \\ \theta &= \begin{bmatrix} a_1,\ldots, & a_n, & b_0,\ldots, & b_m, & c_1,\ldots, & c_p \end{bmatrix}^T. \end{aligned} \tag{6}$$

The corresponding transfer function model is

$$\begin{aligned} G(q,\theta) &= A(q,\theta)^{-1}B(q,\theta), \\ H(q,\theta) &= A(q,\theta)^{-1}C(q,\theta). \end{aligned}$$

B. PREDICTION ERROR METHOD IDENTIFICATION

For control design, the typical objective of the identification procedure is to determine the parameter θ so that $G(q,\theta) \approx G_0(q)$ over the frequency range of interest. In general, this model matching problem is difficult to solve on-line. Most on-line identification algorithms are based on criteria like least squares which are computationally more tractable but often lead to bias in the asymptotic transfer function estimates. The designer can minimize this bias by careful choice of the design parameters, such as sampling rate, and data prefilters. Prefiltering of system identification data is particularly useful for altering the transfer function bias distribution since other design parameters are often constrained by other objectives, such as control performance. Here, $F(q)$ will denote a filter transfer function to used on both the input and output data.

1. ONE-STEP PREDICTION

Prediction error identification methods are based on a criterion which minimizes the mean square one-step prediction error between the model and the true system. For a given model the optimal one-step predictor of $y(t)$ (in the mean square sense) based on the past observations $\{y(t-1), u(t-1), y(t-2), u(t-2), \ldots\}$

is [9]

$$\hat{y}(t/t-1;\theta) = (1 - H(q,\theta)^{-1})F(q)y(t) + H(q,\theta)^{-1}G(q,\theta)F(q)u(t). \quad (7)$$

The predictor Eq. (7) minimizes the mean squared prediction error for a given model. This prediction may be quite poor if the true system transfer function differs substantially from the model. Other predictors are also possible. For example, a k-step prediction horizon can be used [5]. The filtered one-step prediction error is defined to be $\epsilon(t/t-1;\theta) = F(q)y(t) - \hat{y}(t/t-1;\theta)$.

2. LEAST SQUARES CRITERION

The most common PEM criterion is least squares:

$$\hat{\theta}_N = \arg\min_{\theta \in \mathcal{M}} V_N(\theta),$$

$$V_N(\theta) = \frac{1}{N}\sum_{t=1}^{N} \epsilon(t/t-1;\theta)^T \epsilon(t/t-1;\theta). \quad (8)$$

where the arg min function is the value of θ which minimizes the cost function and $\mathcal{M}$ is the set of parameters which yield permissible models. (For example, the model may required to be of a given order and stable.)

The least squares criterion has the advantages that conditions for the minimum PEM cost are easily formulated, and a unique, global minimum always exists. However, criterion Eq. (8) is seldom of interest itself, but is only a mathematical convenience. L. Ljung and his co-workers have analyzed prediction error methods in terms of how well the estimates fit the true system transfer function [5,10,11,6]. *This analysis is valid even when the true system is not in the model class.* The transfer function fit paradigm will be developed below, after a brief discussion of on-line implementation of least squares algorithms.

3. ON-LINE IMPLEMENTATION FOR ARX MODELS

The ARX model Eq. (5) can be written as a linear regression:

$$y(t) = \phi(t)^T\theta + e(t), \quad (9)$$

where $\phi(t)$ is the regressor, which for the scalar ARX(n,m,d) model is

$$\phi(t)^T = \left[\begin{array}{cccccc} y(t-1), \ldots, & y(t-n), & u(t-d), \ldots, & u(t-d-m) \end{array}\right].$$

The parameter estimate which minimizes the least squares criterion Eq. (8) is given by the normal equations [12,13]:

$$\left(\sum_{k=1}^{t} \phi(k)\phi(k)^T\right)\hat{\theta} = \sum_{k=1}^{t} \phi(k)y(k). \quad (10)$$

A recursive form of Eq. (10) can be derived [12,13], yielding the so-called recursive least squares (RLS) algorithm:

$$\hat{\theta}(t+1) = \hat{\theta}(t) + K(t)[y(t) - \phi(t)\hat{\theta}(t)] \tag{11}$$

$$K(t) = P(t)\phi(t)[R(t) + \phi(t)^T P(t)\phi(t)]^{-1} \tag{12}$$

$$P(t+1) = P(t) + Q(t) - P(t)\phi(t)[R(t) + \phi(t)^T P(t)\phi(t)]^{-1}\phi(t)^T P(t), \tag{13}$$

where $R(t)$ is the sensor noise covariance, and the covariance matrix $Q(t)$ represents a random walk model for the parameter vector that prevents the algorithm gain $K(t)$ from going to zero. This enables the algorithm to track slow parameter variations.

The RLS algorithm is similar to the standard Kalman filter algorithm and suffers from the same well-known numerical problems. Bierman's U-D factorization algorithm [14,13] is a numerically stable implementation of RLS (and Kalman filter). A different approach is lattice least squares (LLS) [13,15], which is recursive in both time and model order. LLS yields models of all orders up to some desired maximum, a feature which can be very useful in adaptive identification. In addition, the signal processing for the LLS yields a numerically stable identification algorithm, which has been exploited to estimate modal frequencies of very high order systems [15]. The drawback of LLS is that it is applicable to fewer types of systems than U-D factorization.

The U-D factorization algorithm was implemented in the time domain FORTRAN simulation. The identification results computed using PC-Matlab presented later in were generated using the standard RLS algorithm. For a given data set, the two simulations yield nearly identical parameter estimates and covariances. For low order models, the standard RLS algorithm often yields satisfactory results. However, the operation count for the U-D factorization algorithm is approximately the same as the regular RLS algorithm [13], therefore it is recommended that the U-D factorization be used whenever there is any doubt as to the numerical stability of an estimation problem.

C. FREQUENCY DOMAIN ANALYSIS OF PEM IDENTIFICATION

1. CONVERGENCE OF PEM ESTIMATES

It will be assumed that the disturbance $v(t)$ in the true system Eq. (1) is stationary with spectrum $\Phi_v(\omega)$. In addition, the known system input $u(t)$ is assumed to be generated by

$$u(t) = -R(q)y(t) + r(t), \tag{14}$$

where $r(t)$ is a known stationary signal with spectrum $\Phi_r(\omega)$, independent of $v(t)$, and $R(q)$ is chosen such that the closed loop system is stable. More precise conditions for convergence analysis may be found in [6,7,8].

Under the above conditions, the PEM cost function approaches the following limit:

$$\lim_{N\to\infty} V_N(\theta) = \bar{V}(\theta),$$

where

$$\bar{V}(\theta) = \lim_{N\to\infty} \frac{1}{N} \sum_{1}^{N} E\{\epsilon(t/t-1;\theta)^T \epsilon(t/t-1;\theta)\}. \tag{15}$$

It can be shown [6] that PEM estimators possess the following convergence property:

$$\hat{\theta}_N \rightarrow D_c, \text{ as } N \rightarrow \infty,$$
$$D_c = \{\theta^* \in \mathcal{M} \mid \bar{V}(\theta^*) \leq \bar{V}(\theta),\ \theta \in \mathcal{M}\}.$$

This condition shows that the PEM estimate converges to the best possible estimate in the model set $\mathcal{M}$ for any given data set. When the model set and input signals permit convergence to a unique parameter estimate, the limiting estimate is given by

$$\theta^* = \arg\min_{\theta\in\mathcal{M}} \bar{V}(\theta). \tag{16}$$

2. FREQUENCY DOMAIN EXPRESSIONS

The above formulation and convergence results for the PEM estimator are valid for both scalar and multivariable systems. To simplify the following developments, the focus will be restricted to single-input, single-output systems.

The asymptotic least squares criterion, Eqs. (15) and (16), can be written in the frequency domain as [7,8]

$$\bar{V}(\theta) = \frac{T}{2\pi} \int_{-\pi/T}^{\pi/T} \Phi_\epsilon(\omega)\, d\omega, \tag{17}$$

where $\Phi_\epsilon(\omega)$ is the spectrum of the one-step prediction error. Equation 17 shows how the PEM cost is distributed in frequency. The specific form of the frequency dependence will depend on details of the identification procedure, which will now be explored.

The prediction error for the filtered one-step predictor, Eq. (7), is

$$\begin{aligned}\epsilon(t/t-1;\theta) &= F(q)y(t) - \hat{y}(t/t-1) \\ &= H(q,\theta)^{-1}F(q)y(t) - H(q,\theta)^{-1}G(q,\theta)F(q)u(t) \\ &= H(q,\theta)^{-1}F(q)\{[G_0(q) - G(q,\theta)]u(t) + v(t)\}.\end{aligned}$$

It can be seen that the main function of the prefilter $F(q)$ is to modify the noise transfer function, $H(q,\theta)$.

For $u(t)$ generated as in Eq. (14), the prediction error becomes

$$\begin{aligned}\epsilon(t/t-1;\theta) &= H(q,\theta)^{-1}F(q)\left\{\frac{[G_0(q) - G(q,\theta)]}{1+G_0(q)R(q)}r(t)\right. \\ &\quad \left. + \frac{1+G(q,\theta)R(q)}{1+G_0(q)R(q)}v(t)\right\}. \end{aligned} \tag{18}$$

Computing the spectrum of $\epsilon(t/t-1;\theta)$, and inserting the result into Eq. (17) yields

$$\begin{aligned}\bar{V}(\theta) &= \frac{T}{2\pi}\int_{-\pi/T}^{\pi/T}\{W_B(\omega)|G_0(e^{j\omega T}) - G(e^{j\omega T},\theta)|^2 \\ &\quad + W_V(\omega)\Phi_v(\omega)\}\, d\omega,\end{aligned} \tag{19}$$

$$W_B(\omega) = \frac{|F(e^{j\omega T})|^2\Phi_r(\omega)}{|H(e^{j\omega T},\theta)|^2|1+G_0(e^{j\omega T})R(e^{j\omega T})|^2}, \tag{20}$$

$$W_V(\omega) = \frac{|F(e^{j\omega T})|^2|1+G(e^{j\omega T},\theta)R(e^{j\omega T})|^2}{|H(e^{j\omega T},\theta)|^2|1+G_0(e^{j\omega T})R(e^{j\omega T})|^2}. \tag{21}$$

There are two terms in the asymptotic cost function: a weighted fit error between the true and estimated transfer functions, and a term involving the noise. Suppose the parameter θ can be separated into independent plant and noise parts, θ_1 and θ_2, i.e., $G(q,\theta) = G(q,\theta_1)$ and $H(q,\theta) = H(q,\theta_2)$. Then $\bar{V}(\theta)$ is minimized by choosing $G(q,\theta_1)$ as close to $G_0(q)$ as possible, i.e., minimize the transfer function estimation bias. The bias weighting function, $W_B(\omega)$, determines how much weight the estimation procedure will give to minimizing the bias contribution at each particular frequency. If the weight is relatively large in a particular frequency range, the algorithm will tend to yield estimates which closely fit the true system transfer function in that range. In practice, $G(q,\theta)$ can never exactly match $G_0(q)$, and $G(q,\theta)$ and $H(q,\theta)$ are often not independent. Therefore, exact conclusions about the frequency distribution of the transfer function bias may not be possible. However, this approach offers useful insights into identification performance.

An alternative interpretation to the transfer function fit criterion Eqs. (19–21) is obtained by factoring-out the true system transfer function in Eq. (19):

$$\bar{V}(\theta) = \frac{T}{2\pi}\int_{-\pi/T}^{\pi/T}\{W_R(\omega)\left|1-\frac{G(e^{j\omega T},\theta)}{G_0(e^{j\omega T})}\right|^2 + W_V(\omega)\Phi_v(\omega)\}\,d\omega, \tag{22}$$

$$W_R(\omega) = \frac{|F(e^{j\omega T})|^2\Phi_r(\omega)|G_0(e^{j\omega T})|^2}{|H(e^{j\omega T},\theta)|^2|1+G_0(e^{j\omega T})R(e^{j\omega T})|^2}. \tag{23}$$

This equivalent expression for the asymptotic PEM cost involves a weighted measure of the *relative* transfer function fit error (hence the subscript "R"). This expression is similar to the "Bode Plot interpretation" derived by Wahlberg and Ljung [5]. The utility of these expressions for analyzing and designing system identification procedures will now be discussed.

3. FREQUENCY DOMAIN ANALYSIS OF ARX MODELS

It will now be shown how the weighting functions Eqs. (20), (21), and Eq. (23) can be used to analyze system identification designs. The two interpretations, Eq. (19) and Eq. (22), of the asymptotic PEM cost function yield different weighting functions for the transfer function fit error component of the cost. The comparison between these two interpretations is more clearly demonstrated by specializing to the case of ARX models under open-loop operation ($R(q) = 0$). Denoting

$$G_0(q) = \frac{B_0(q)}{A_0(q)},$$

$$G(q,\theta) = \frac{B(q,\theta)}{A(q,\theta)},$$

$$H(q,\theta) = \frac{1}{A(q,\theta)},$$

the weighting functions become

$$W_B(\omega) = |A(e^{j\omega T},\theta)|^2|F(e^{j\omega T})|^2\Phi_r(\omega),$$

$$W_V(\omega) = |A(e^{j\omega T},\theta)|^2|F(e^{j\omega T})|^2,$$

$$W_R(\omega) = \frac{|A(e^{j\omega T},\theta)|^2|F(e^{j\omega T})|^2\Phi_r(\omega)|B_0(e^{j\omega T})|^2}{|A_0(e^{j\omega T})|^2}.$$

If $A(e^{j\omega T},\theta) \approx A_0(e^{j\omega T})$, then

$$W_R(\omega) \approx |F(e^{j\omega T})|^2\Phi_r(\omega)|B_0(e^{j\omega T})|^2. \tag{24}$$

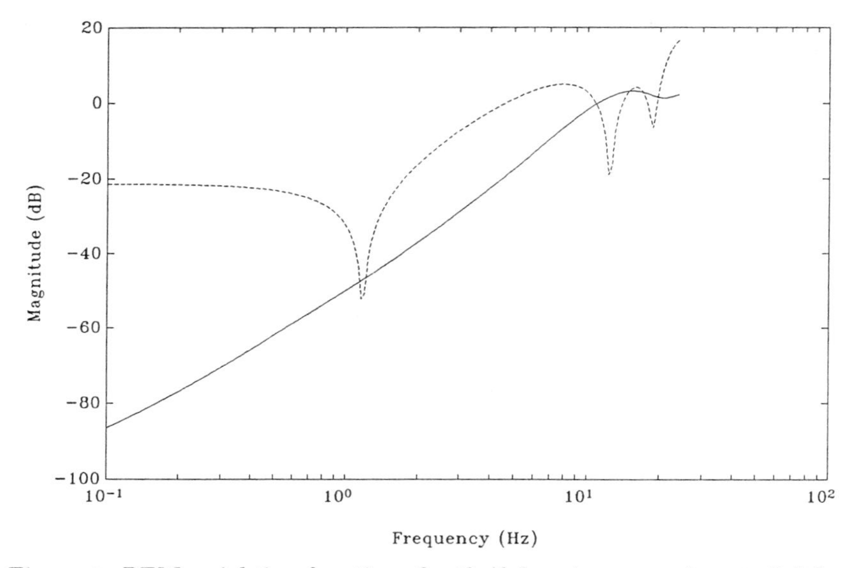

Figure 4: PEM weighting functions for the laboratory experiment. Solid: relative weighting function; dashed: bias (noise) weighting function.

Thus, the relative transfer function fit error weighting, W_R, indicates that transfer function *zeros* are what determine the frequency weighting. In contrast, the bias weighting function, W_B, indicates that the transfer function *poles* are what should be considered, as does the noise weighting function, W_V.

One general conclusion resulting from frequency domain analysis is that ARX models tend to fit high frequency ranges of the true system transfer function better than lower frequency ranges. This phenomena is due to the fact that reasonably high sampling rates cause the system poles and zeros to cluster near the $q = 1$ region of the complex plane, hence the weighting functions will usually be high-pass. The algorithm then ignores the low frequency transfer function fit and may estimate dynamics well outside the intended frequency range of the model. The highpass nature of ARX weighting functions is illustrated in Figure 4, which shows the weighting functions for the laboratory experiment. The relative weighting function increases sharply due to the presence of the accelerometer zeros. The noise weighting increases less sharply, but the frequency range around the first mode is given considerably less weight.

To reduce the weight given to high frequency dynamics a lowpass prefilter can be used. The expression for the bias weighting function suggests that the prefilter order should be chosen at least as large as the model order. However an extensive comparison of the bias and relative weighting functions presented in [7,8] shows that the relative weighting function is a better predictor of overall identification performance. This is due to the fact that W_B is insensitive to variations in several important system identification design parameters. For example, the bias weighting says that sensor type has no effect on identification performance, which is decidedly not the case [7,8]. The bias weighting also ignores sensor and actuator placement.

The utility of the frequency domain PEM cost expressions for identification analysis and design will be demonstrated in the next two sections. However, it must be stressed that these techniques are somewhat heuristic: the exact weighting functions are functions of the unknown parameters, so the weighting functions must be approximated by use of the nominal plant model. Therefore, the conclusions from this analysis must be viewed with a degree of caution. Nonetheless the frequency domain viewpoint yields many useful insights into identification practice.

IV. SYSTEM IDENTIFICATION DESIGN ISSUES

This section presents an overview of system identification design techniques. The issues considered are model type, sensor selection and placement, sampling rate, prefilter design, and known dynamics.

A. MODEL TYPE

There are two main model classes available for on-line system identification: ARX and ARMAX. For the purposes of the discussion here, the ARMAX model class may be considered to include innovation and state space models, since their asymptotic properties are similar—their transient response is not! As seen above, ARX models tend to emphasize high frequency model fit at the expense of low frequency estimates. ARMAX models avoid this because the inclusion of the MA polynomial allows some flexibility in the noise transfer function. The weighting functions for the open loop ARMAX case are

$$
\begin{aligned}
W_R(\omega) &= |H(e^{j\omega T},\theta)^{-1}G_0(e^{j\omega T})|^2\Phi_u(\omega) = \left|\frac{A(e^{j\omega T},\theta)B_0(e^{j\omega T})}{C(e^{j\omega T},\theta)A_0(e^{j\omega T})}\right|^2\Phi_u(\omega),\\
&\approx \left|\frac{B_0(e^{j\omega T})}{C(e^{j\omega T},\theta)}\right|^2\Phi_u(\omega),\\
W_V(\omega) &= |H(e^{j\omega T},\theta)|^{-2} = \left|\frac{A(e^{j\omega T},\theta)}{C(e^{j\omega T},\theta)}\right|^2.
\end{aligned}
$$

It is seen that the true system zeros still affect the relative error weighting, however the noise weighting is likely to be flatter due to the presence of the MA polynomial.

Although the frequency domain properties of ARMAX models are superior to those of ARX models, ARMAX estimates are computed using the recursive prediction error method or pseudo-linear regressions, which approximate the true solution. These techniques converge to the true solution asymptotically, provided various assumptions about the system are met [13]. These conditions will not be discussed here, but it suffices to say that convergence and computational issues make the implementation of ARMAX identification fairly complex. The addition of sensors and data prefilters can help make ARX identification robust to the presence of disturbances and noise, so these alternatives should be considered first.

B. SENSOR SELECTION

The analysis in Section III shows that rate sensors adversely affect low frequency transfer function estimates, effectively limiting the frequency range over which good estimates can be obtained. For this reason, position sensors are preferred for system identification. However, other considerations may dictate the use of gyros and accelerometers, as was the case with the laboratory experiment. An approach for using rate sensors in identification is presented in the next section.

C. SAMPLING RATE

The sampling frequency, f_s, can greatly affect identification performance. The sampling theorem says that system mode frequencies must be less than $f_s/2$ to be identified. Finite ADC precision sets a lower bound on the frequency of system modes which can be accurately identified using delay operator models. This is demonstrated by the following example. The AR model for a single, undamped mode at ω rad/s is

$$[1 - 2\cos(\omega T)q^{-1} + q^{-2}]y(t) \quad = \quad e(t).$$

For ωT small, $\cos(\omega T) \approx 1 - 1/2\omega^2 T^2$. If the ADC has b bits of resolution, then it is required that $\omega T > \sqrt{2^{-b}}$ to prevent loss of resolution in the model. For $b = 12$, the modal frequency, f, must be such that $f > f_s/200$. The effect of quantization noise can be partly mitigated by long data records, but this example shows that good identification results can be obtained only over a frequency range of about two decades. The numerical problems associated with delay operator models can be minimized by using delta-operator models [16], where $\delta = (q-1)/T$.

As observed in Section III, the identification weighting functions for ARX models tend to be highpass. This phenomena is more pronounced with high sampling rates, making identification of low frequency dynamics difficult and the sensitivity to "unmodeled" high frequency dynamics greater. If other constraints require high sampling rates (as is often the case in control design) then data prefilters can be used to reduce the influence of unmodeled dynamics.

In general, it is difficult to obtain good transfer function estimates using ARX models over a frequency range of more than a decade. This conclusion is intrinsic to the PEM approach, and is independent of numerical aspects of the identification procedure. In particular, the delta operator formulation will suffer from the same limitation.

D. PREFILTER DESIGN

The frequency domain weighting functions are the best guide for prefilter design. The prefilter will typically be low-pass or band-pass. The low-pass cutoff frequency should be chosen as low as system dynamics will permit. The low-pass filter order can be determined by the upward slope of the weighting functions at high frequencies. The low-pass filter order should roughly equal the model order. A band-pass filter is required when there are sensor biases or other low frequency disturbances. The low frequency cutoff must be not be chosen too small, otherwise the transient performance of the identification algorithm will be poor.

E. KNOWN DYNAMICS

A method for removing known dynamics from the identification process has been proposed by Clary [17,18]. Let $A_k(q)$ and $B_k(q)$ be polynomials containing the known poles and zeros. Assume that the true system is given by the transfer function model

$$y(t) = \frac{B_0(q)B_k(q)}{A_0(q)A_k(q)}u(t) + v(t).$$

The corresponding ARX model is

$$A(q,\theta)A_k(q)y(t) = B(q,\theta)B_k(q)u(t) + e(t),$$

or using prefiltered data

$$A(q,\theta)y_f(t) = B(q,\theta)u_f(t) + e(t),$$

where $y_f(t) = F(q)A_k(q)y(t)$ and $u_f(t) = F(q)B_k(q)u(t)$. The identification is carried out using y_f and u_f; only the unknown dynamics are estimated. The model order is reduced, which speeds convergence of the parameter estimates.

The performance of Clary's approach can be evaluated using the weighting function approach. The relative fit and noise weighting functions are given by

$$\begin{aligned} W_R(\omega) &= \left| \frac{F(e^{j\omega T})A(e^{j\omega T},\theta)B_k(e^{j\omega T})B_0(e^{j\omega T})}{A_0(e^{j\omega T})} \right|^2 \Phi_u(\omega) \\ W_V(\omega) &= \left| F(e^{j\omega T} A_k(e^{j\omega T})A_0(e^{j\omega T}) \right|^2 . \end{aligned}$$

The weighting functions are essentially unchanged by this procedure, therefore design requirements for $F(q)$ are unaltered. In particular, if the known dynamics are in B_k only, then those zeros still appear in the relative weighting function. In structural identification problems, Clary's method is particularly helpful for removing known rigid-body dynamics from the estimation procedure. Rigid body modes are difficult to identify accurately, and their removal from the identification problem permits faster and more accurate identification of flexible modes.

V. LABORATORY EXPERIMENT SYSTEM IDENTIFICATION DESIGN

This section discusses the system identification design for the laboratory experiment. The goal is to identify the first two modes of the beam, a frequency range of approximately 1–13 Hz. Identifying the first two modes with a single ARX model was found to be infeasible with the sensor limitations of the laboratory setup [7,8]. The frequency domain analysis technique is used to develop an identification approach which utilizes a models over different frequency ranges to determine the overall transfer function model.

1. LABORATORY DESIGN APPROACH

The sampling rate for the sensor and control signals was chosen to be 50 Hz; the example in Section III.C implies that the 12-bit resolution of the ADC will permit identification of modal frequencies down to 50/200=0.25 Hz. Therefore the lowest frequency mode of the laboratory experiment can be identified with the hardware available. The accelerometer output relative and bias weighting functions were shown in Fig. 4. The bias weighting function, which weights the noise, increases 20 dB over the 1–12 Hz frequency range and rises less steeply up to 25 Hz. The relative transfer function weighting increases 50 dB over the same frequency range, due to the presence of the accelerometer zeros. This shows why the first mode is difficult to identify.

Examination of the weighting functions show that if the frequency range over which the estimation is performed is restricted, then the change in the weighting functions will be small enough to permit accurate identification results to be obtained. Therefore, by dividing the frequency range of interest into enough pieces, accurate models can be obtained in each range. For the laboratory beam, two separate bands are sufficient to identify the first two modes. Two band-pass filters were designed for this purpose. The signals for the low frequency band estimate are filtered by a band-pass filter consisting of a second order high-pass section at 0.5 Hz cascaded with a fourth order low-pass section at 3 Hz. The high frequency band signals are produced by a fourth order high-pass section at 4 Hz cascaded with a fourth order low-pass at 14 Hz. The frequency response of these filters is shown in Fig. 5.

ARX(2,2,1) models are used for identification in each band. The estimated transfer functions for ten seconds of data for the case 3 beam configuration are shown in Fig. 6. Both modes are accurately estimated and there is reasonably good fit in the frequency range between the two modes. The success of this approach of identifying separate frequency bands can be explained by examining the relative weighting functions for this case, Fig. 7. Both weighting functions vary no more than 20–30 dB in their respective frequency ranges. In particular, the low-band weighting function is well-attenuated at the second mode frequency.

This approach of using separate estimators for the low and high frequency modes yields accurate estimates of the modal frequencies, but yields poor values for the transfer function zeros. In particular, it is important that the non-miminum phase zero between the first and second modes be estimated for the control system design. It also is desirable to connect these two ARX(2,2,1) sub-models together to form a single model for for control design. A good, simple s-plane transfer function model from the PID input to accelerometer output is

$$\begin{aligned} P(s) &= s^2 \sum_{k=1}^{2} \frac{d_k \omega_k^2}{s^2 + 2\zeta_k \omega_k s + \omega_k^2} \\ &= d_1 P_1(s) + d_2 P_2(s). \end{aligned} \tag{25}$$

Evaluating the two estimated sub-models at frequencies ω_ℓ and ω_h yields transfer function gain measurements m_ℓ and m_h, respectively. Assuming negligible damping, the modal gains d_1 and d_2 can be determined from the following set

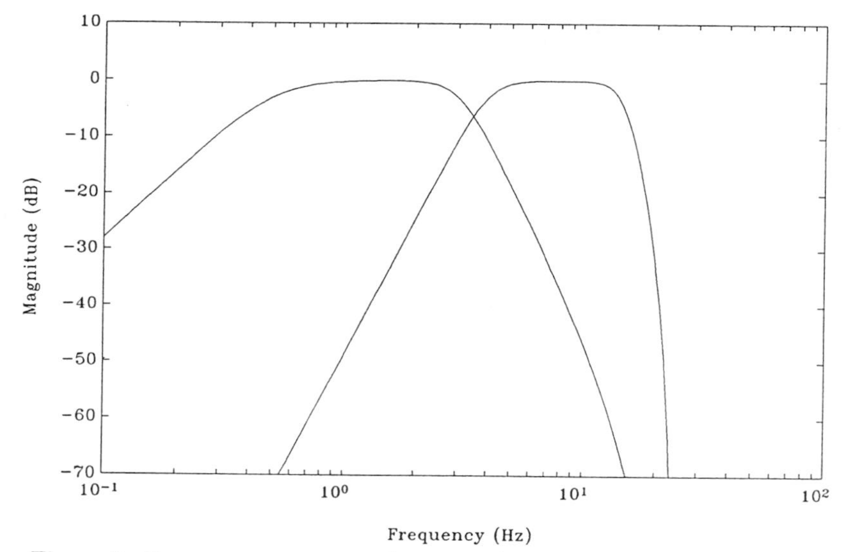

Figure 5: Frequency response of system identification bandpass prefilters.

of linear equations:

$$\begin{bmatrix} P_1(j\omega_\ell) & P_2(j\omega_\ell) \\ P_1(j\omega_h) & P_2(j\omega_h) \end{bmatrix} \begin{bmatrix} d_1 \\ d_2 \end{bmatrix} = \begin{bmatrix} m_\ell \\ m_h \end{bmatrix}. \tag{26}$$

The specific procedure implemented for the laboratory experiment is:

1. The frequencies ω_ℓ and ω_h, are chosen as

$$\begin{aligned} \omega_\ell &= \omega_1^{7/8}\omega_2^{1/8} \\ \omega_h &= \omega_1^{1/8}\omega_2^{7/8}. \end{aligned}$$

   The sub-model transfer function estimates, Fig. 6, yield accurate estimates of the true system gain at these frequencies.

2. The transfer function gain measurements, m_ℓ and m_h, are computed from the low and high ARX(2,2,1) sub-models, respectively. It may be helpful to use both sub-model estimates to determine both gain measurements, but that was not necessary in this case.

3. The modal gains are computed from Eq. (26).

4. For negligible damping, the zeros of $P(s)$ are at $s = \pm s_z$, where

$$s_z = \omega_1\omega_2\sqrt{-\frac{(d_1+d_2)}{d_1\omega_1^2 + d_2\omega_2^2}}.$$

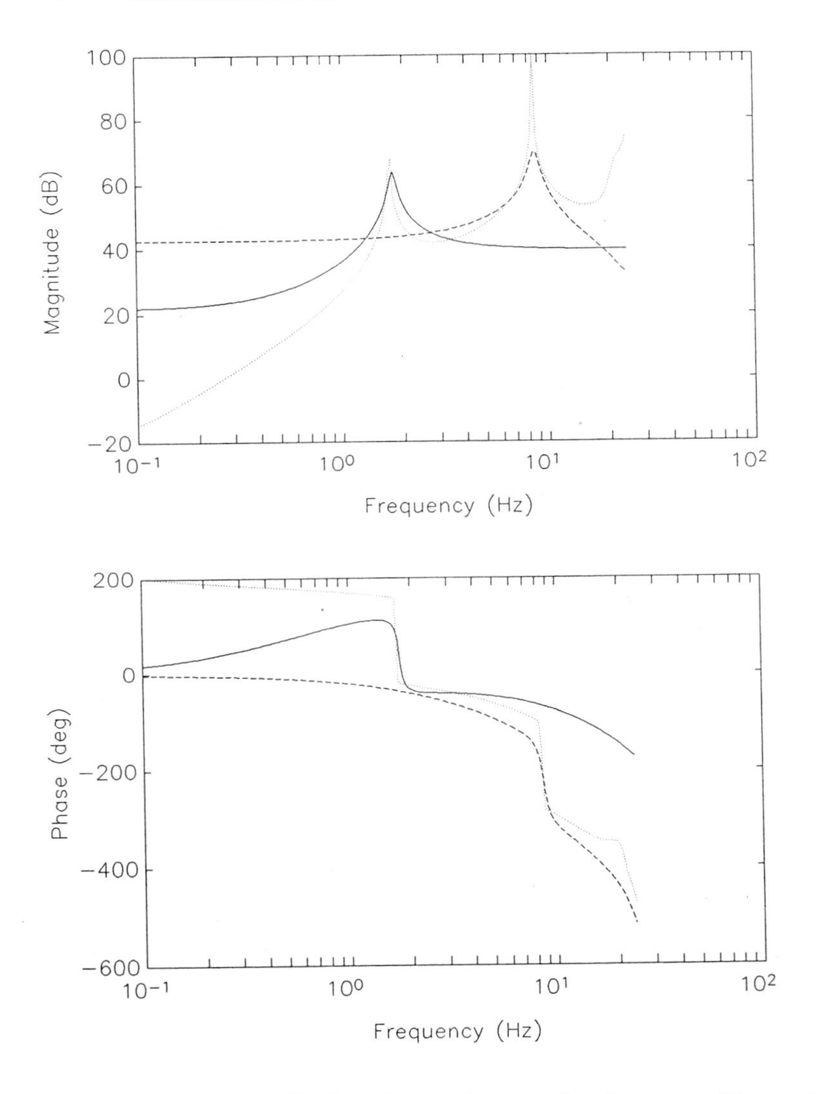

Figure 6: Estimated transfer functions using two band-pass prefilters. Solid: low-band estimate; dashed: high-band estimate; dotted: exact.

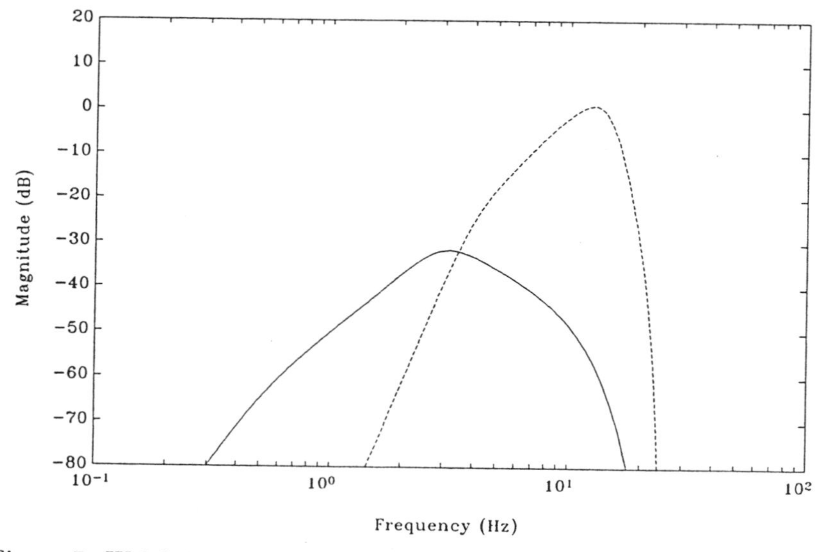

Figure 7: Weighting functions using two band-pass prefilters. Solid: low-band relative weighting function; dashed: high-band relative weighting function.

It is known *a priori* that $\pm s_z$ should be real; this provides a check to determine the validity of the estimated model.

5. The s-plane model Eq. (25) can be transformed to an ARX(4,4,1), if necessary. Note that the accelerometer zeros are explicitly included in the model, so it is not required that the ARX(2,2,1) sub-models be accurate at frequencies well below the first mode.

The estimated case 3 transfer function using the above procedure is shown in Fig. 8. The fit to the exact model is seen to be excellent. A shortcoming of this procedure is that the estimated transfer function may be accurate, but the zeros in the estimated model may not be close to those in the analytical model. The effect of this in the performance of the control designs is examined in Section VII.

2. TRANSIENT RESPONSE OF IDENTIFICATION APPROACH

The identification results for the separate frequency bands shown above were computed assuming initial covariances of $1000 \times I_5$ and intial regression vectors set to zero. With these intial values, the modal frequency estimates have a severe transient phase, as shown in Fig. 9 using case 1 data. The first mode damping ratio estimate is also negative for the first two seconds. These results

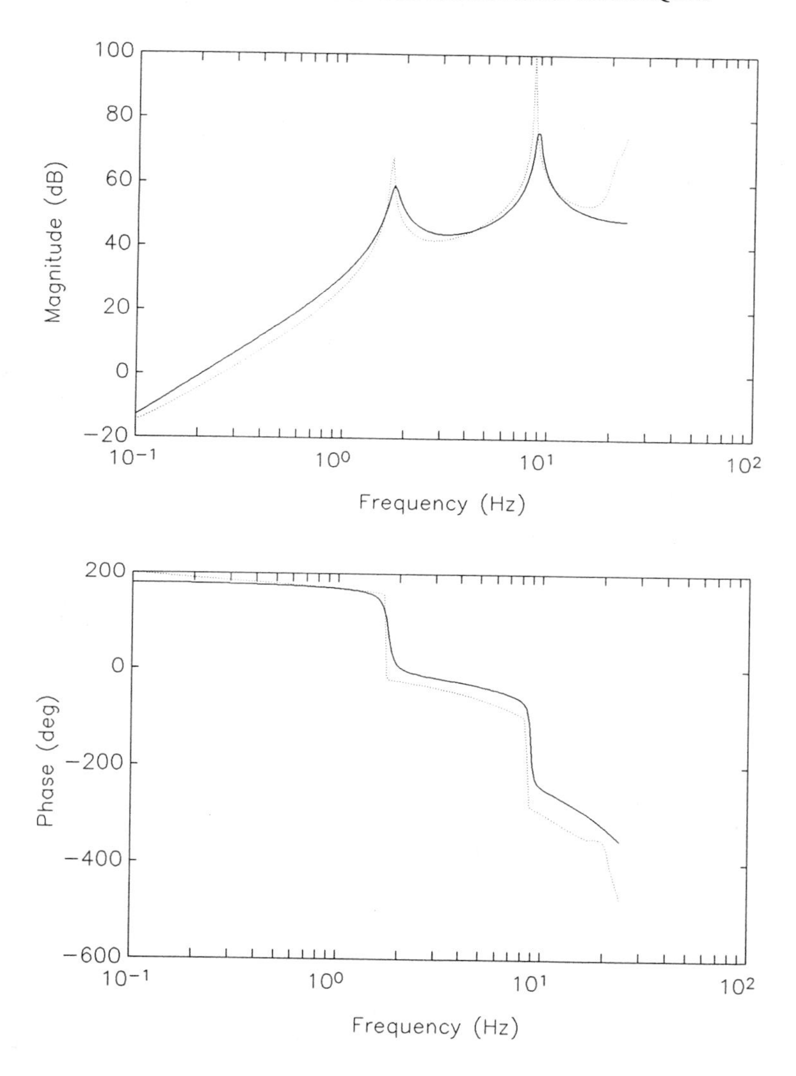

Figure 8: Estimated single transfer function using two band-pass prefilters. Solid: estimate; dotted: exact.

are essentially independent of the initial parameter vectors. The size of these transients is undesirable for on-line applications. To address this issue, the following combination of steps were found to be helpful:

1. The parameter vectors were initialized to modes at 1.5 Hz and 9 Hz for the low and high-bands, respectively. The damping ratios were both 0.02. The zero parameters for the low-band vector were initialized to two zeros at $q = 1$; the high-band vector to zeros at $q = 0.6$ and $q = 1.6$ (about 4 Hz).

2. A smaller initial covariance is used, which better reflects the *a priori* knowledge of the system. (Note that in the absence of these measures a large initial covariance is required to obtain reasonable estimates.)

3. The start of the RLS algorithm is delayed to permit the prefilter transient to decay and to obtain good initial regression vectors. A waiting period of 0.5–2.0 s gives a good response.

4. A projection facility for stabilizing the pole parameters of the parameter vectors is included. The characteristic equation for the ARX(2,2,1) model is $q^2 + \theta_1 q + \theta_2$; if $|\theta_2| \geq 1$, then the parameter and covariance update equations are modified as [13]

$$\begin{aligned} \hat{\theta}(t+1) &= \hat{\theta}(t) + gK(t)[y(t) - \phi(t)^T\hat{\theta}(t)], \\ P(t+1) &= P(t), \end{aligned}$$

   where g is chosen to be successive powers of $1/2$ until $|\theta_2| < 1$. This procedure reduces the algorithm gain, while retaining the search direction of the RLS algorithm.

With these measures, the transient response is greatly improved, as seen in Fig. 10. Standard texts in system identification [12,13] recommend using a large initial covariance for the RLS algorithm, as well as a projection algorithm. However, the parameter estimates were observed to have poor transient behavior. Instead, it appears that the initial covariance should reflect the prior knowledge of the parameters and proper initialization of the regression vector must be considered.

3. ESTIMATION OF PLANT DC GAIN

The DC gain of the lab setup is required for tracking of step commands. The measured LVDT output transfer functions are basically flat below the second

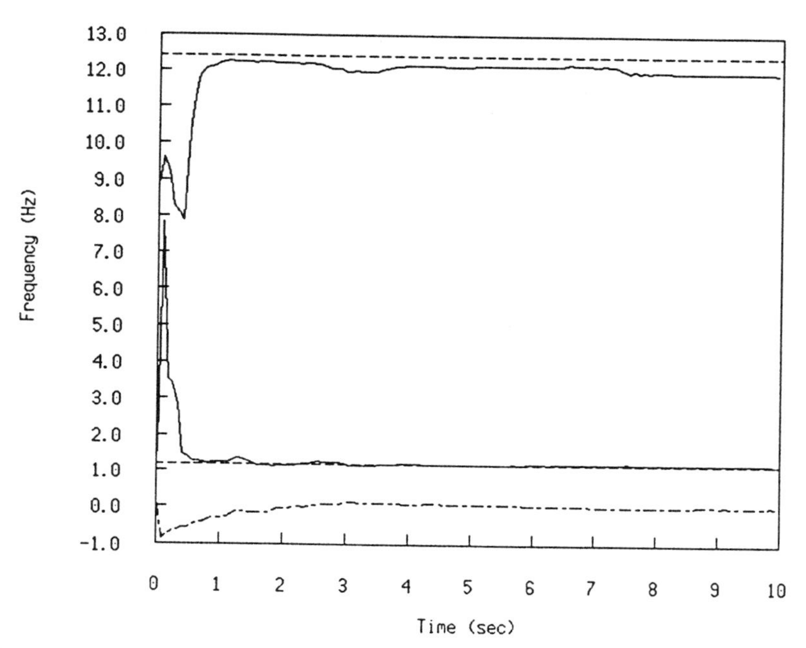

Figure 9: Transient response of RLS algorithm with poor initialization. Solid: modal frequency estimates; dashed: actual modal frequencies; dot-dashed: first mode damping ratio estimate.

mode, therefore the simple model

$$y(t) \quad = \quad \theta_p u(t-1)$$

is adequate. The signal $y(t)$ is the sampled LVDT measurement; $u(t)$ is the computer control signal. These signals are scaled so that θ_p represents the gain from the PID input to the position of the top of the beam (in cm/V). Both signals are passed through a 0.5 Hz fourth order Butterworth lowpass prefilter before the RLS estimate $\hat{\theta}_p$ is computed.

Figure 11 shows $\hat{\theta}_p$ under open loop operation with filtered noise as the input to the system. The variation in the plant gain is small, so the intial parameter covariance is set to 0.05. This value yields a compromise between speed of convergence and magnitude of initial transients. The convergence to the final value (≈ 0.38) takes about 20 seconds with this input. The convergence is much faster when step-like inputs are used.

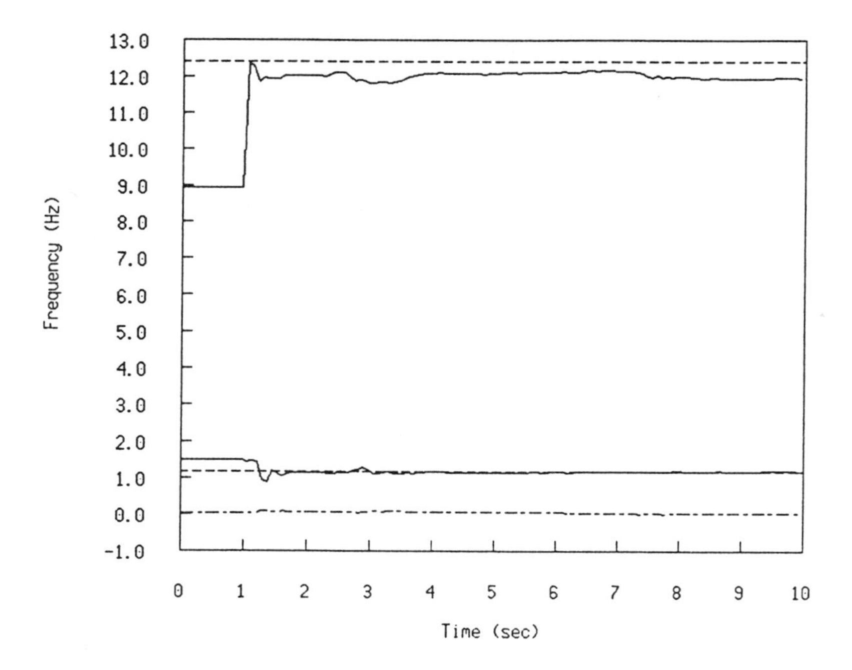

Figure 10: Transient response of RLS algorithm with good initialization. Solid: modal frequency estimates; dashed: actual modal frequencies; dot-dashed: first mode damping ratio estimate.

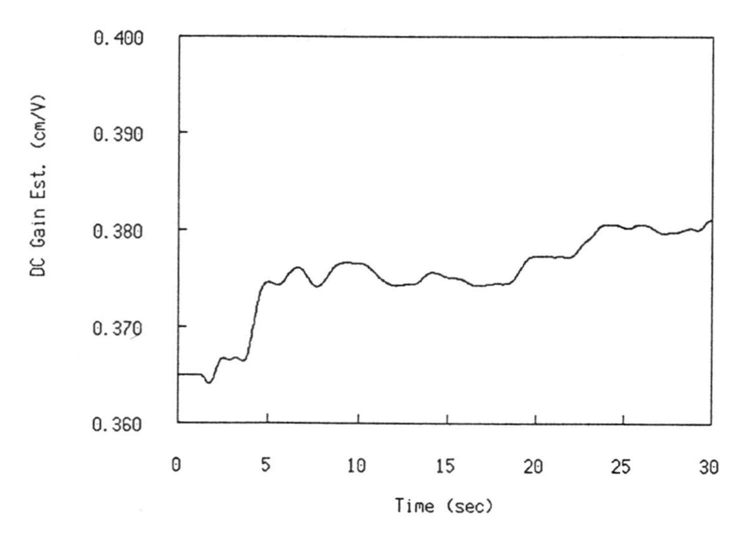

Figure 11: Estimate of Plant DC Gain

VI. LABORATORY CONTROL SYSTEM DESIGN

This section discusses the design of controllers for the laboratory experiment. The designs developed here have features that are applicable to other flexible structure control problems, but individual applications have such diverse requirements and limitations that no single control design methodology can be universally applicable. This is one reason that adaptive control system designs must be application specific.

The two-mode plant model of Eq. (3) is used for the designs, but the designs are evaluated using a model based on Eq. (2), which includes "unmodeled" dynamics. A control design rule is developed which utilizes estimates of laboratory mode frequencies and gains to determine the compensator coefficients. This rule is based on root locus arguments, in contrast with most adaptive control algorithms, which are usually based on model reference or pole placement paradigms. The approach used here yields controllers with better stability margins than pole placement designs. To demonstrate the performance improvement possible with an adaptive controller, a fixed-gain robust controller is also designed.

A. DESIGN REQUIREMENTS AND ISSUES

The following is a list of factors and requirements which the laboratory control system design addresses.

- The main objectives of the control design are to increase the damping of the flexible modes and to force the bottom of the beam to track step and step-like commands.

- No direct measurement of position at the bottom of the beam is available. The accelerometer senses vibration at the bottom of the beam and the LVDT mounted at the top of the beam provides a measurement of steady state position.

- The accelerometer has a bias which prevents an integrator from being used inside the accelerometer loop.

- There are no constant disturbances to the beam; air drafts and people bumping into the beam are the major disturbance sources.

- Only those parameters available from the system identification algorithm are to be used in the adaptive control design, namely the transfer function estimates discussed in Section V.

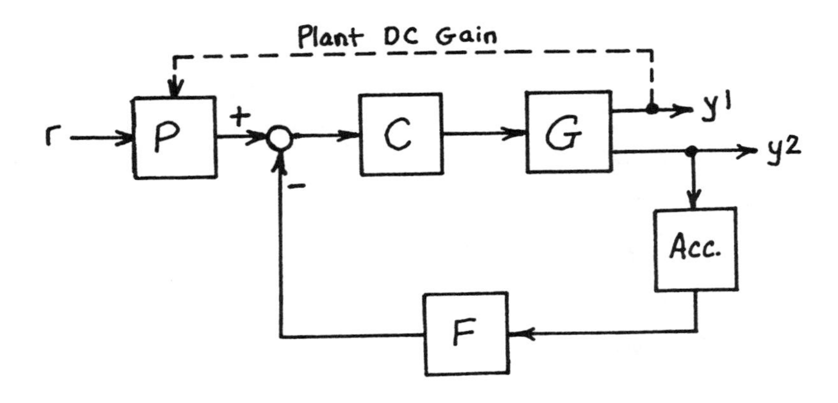

Figure 12: Adaptive controller configuration.

- The open loop transfer function should roll off as s^{-2} to guarantee robustness to high frequency dynamics.

B. ADAPTIVE CONTROL DESIGN RULE

1. CONCEPTUAL DESIGN

The conceptual design of the controllers are carried out in the s-plane, although the designs are evaluated in the z-plane to include the effect of sampling and aliasing in the control performance. The overall topology of the adaptive controller is shown in Fig. 12. A loop using the accelerometer signal is used to increase the damping of the flexible modes. A feedforward compensator scales the reference command for steady state tracking—the plant gain measurement estimate is obtained from the LVDT output.

A feedback compensator consisting of a 0.1 Hz first-order highpass filter is used to eliminate the accelerometer bias. The digital implementation is accomplished through pole-zero mapping [19]. The s-plane compensator poles and zeros are mapped to the z-plane by $q = e^{sT}$, where T is the sampling interval:

$$\frac{K_a(s-b_1)\cdots(s-b_m)}{(s-a_1)\cdots(s-a_n)} \longrightarrow \frac{K_d(q+1)^{n-m}(q-z_1)\cdots(q-z_m)}{(q-p_1)\cdots(q-p_n)}, \quad (27)$$

$$p_i = e^{a_iT},$$

$$z_i = e^{b_iT}.$$

The digital compensator has no delay from input to output; the additional zeros, if required, are placed at $q = -1$ to make the compensator gain zero at half the sampling rate. The gain, K_d, is chosen to make the passband gain equal to that of the continuous time filter.

The cascade compensator is designed to increase the damping of the modes and roll off the open loop transfer function for robustness. For the purposes of design, the beam dynamics are assumed to be

$$G(s) = \frac{Ks^2(s_z - s)(s_z + s)}{(s^2 + 2\zeta_1\omega_1 s + \omega_1^2)(s^2 + 2\zeta_2\omega_2 s + \omega_2^2)}. \tag{28}$$

The phase lag due to the antialiasing filters must also be considered in the design. The nonminimum phase zero at $s = +s_z$ sets an upper bound on the controller bandwidth. Following the advice in [20], the stable, well-damped zero at $s = -s_z$ is cancelled. Between the first and second modes, the plant effectively has four zeros and two poles. To roll off the open loop transfer function above s_z, two additional poles at $s = \omega_p$ are included in the compensator, where $0 < \omega_p < s_z$. The determination of ω_p will be discussed shortly. The conceptual design of the cascade compensator is completed with the addition of a lead-lag filter to phase stabilize the second mode. This filter consists of two zeros at a frequency of $0.8\omega_2$, damping ratio 0.4, and a pole at $s = 2\omega_2$. Therefore the cascade compensator has the form

$$C(s) = \frac{K_c[s^2 + 2(0.4)(0.8\omega_2)s + (0.8\omega)^2]}{(s + 2\omega_2)(s + s_z)(s + \omega_p)^2}. \tag{29}$$

The digital implementation of this compensator rolls-off at 40 dB/decade above the second mode because of two zeros at $q = -1$ from the pole-zero map Eq. (27).

2. DETERMINATION OF COMPENSATOR GAINS

Figure 13 shows the desired s-plane root locus for the open loop system (the feedback compensator is neglected). It is desirable to have the locus departing from the first mode move farther inside the left-half plane, then swing toward higher frequencies. This has the effect of increasing both the damping and natural frequency of the closed loop mode, resulting in faster decay in the time response. The angle of departure from the first mode pole is $\pi - \phi$ radians. The compensator poles at ω_p are chosen such that $\phi \approx 0$. Points on the root locus are determined by the rule

$$\sum(\text{angles of zeros}) - \sum(\text{angle of poles}) = \begin{cases} \pi, & \text{positive gain} \\ 0, & \text{negative gain} \end{cases} \tag{30}$$

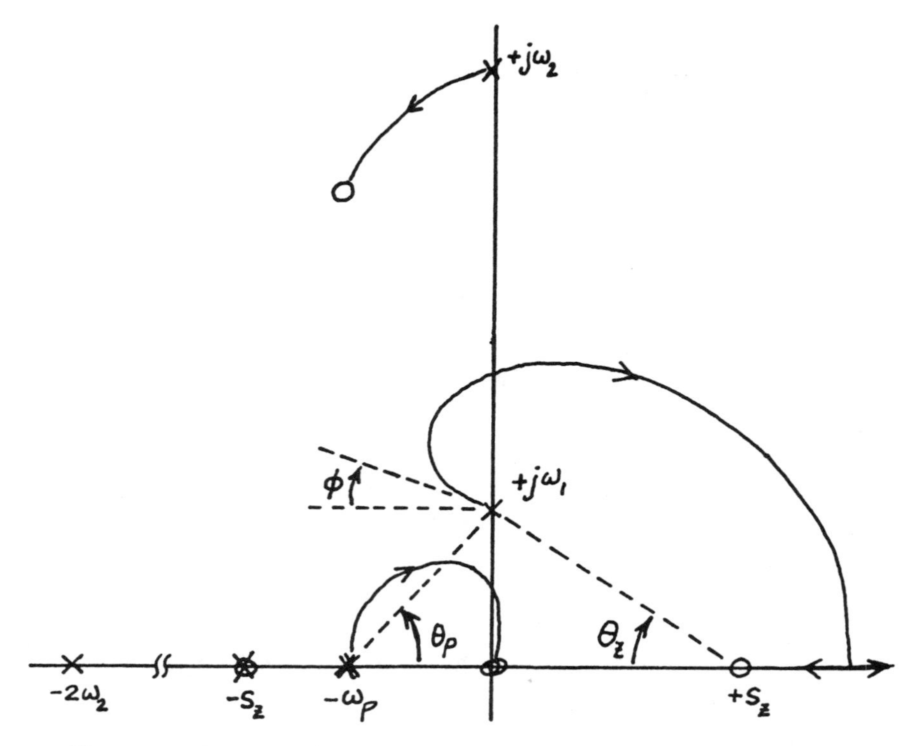

Figure 13: Conceptual s-plane root locus for accelerometer control loop.

The angle of departure from a pole can be computed by applying the rule to a point infinitesimally close to the pole. The negative gain case is used because of the nonminimum phase zero. Referring to Fig. 13, the application of Eq. (30) yields

$$2(\frac{\pi}{2}) + (\pi - \theta_z) - \frac{\pi}{2} - 2\theta_p - (\pi - \phi) \quad = \quad 0.$$

The first term is the contribution from the accelerometer zeros; the third term from the pole at $-j\omega_1$. The phase shifts from the higher frequency dynamics are ignored. Solving for θ_p yields

$$\theta_p \quad = \quad \frac{1}{2}\left[\frac{\pi}{2} + \phi - \theta_z\right], \tag{31}$$

so that

$$\omega_p \quad = \quad \frac{\omega_1}{\tan\theta_p}. \tag{32}$$

It was found that $\phi = 0.05\pi$ (9 degrees) compensated for extraneous phase lags, producing the desired root locus.

The accelerometer controller loop gain is yet to be determined. The open loop Bode plot (from the input of the cascade compensator to the feedback compensator output) for the case 3 setup is shown in Fig. 14. For stability, the loop gain must be below 0 dB at each 180° phase crossing. The $-180°$ crossing at approximately 3 Hz determines the stability of the loop. If the cross-over frequency is known, then the plant and compensator gains can be computed and the loop gain adjusted to ensure stability. A procedure for determining the cross-over frequency will now be discussed.

Let ω_c be the cross-over frequency. The phase shifts for the various plant and compensator poles and zeros at ω_c must sum to $-\pi$ radians:

$$-\pi + \pi + \left(\pi - \tan^{-1}\left(\frac{\omega_c}{s_z}\right)\right) - 2\tan^{-1}\left(\frac{\omega_c}{\omega_p}\right) - \tan^{-1}\left(\frac{\omega_c}{a_0}\right) \quad = \quad -\pi.$$

The first term is the phase shift due to the first mode; the second term is the accelerometer zeros phase shift. The third term is due to the nonminimum phase zero and the fourth term is the contribution of the compensator poles. The last term is the phase shift due to one of the antialiasing filter poles ($a_0 \approx 30\pi$ rad/s). The phase shift of the other pole at 15 Hz and the lead-lag zeros at $0.8\omega_2$ approximately cancel. Using the formula $\tan(\alpha + \beta) = (\tan\alpha + \tan\beta)/(1 + \tan\alpha \tan\beta)$, the cross-over frequency can be shown to be

$$\omega_c \quad = \quad \left[\frac{2\omega_p s_z a_0 + (a_0 + s_z)\omega_p}{2\omega_p + a_0 + s_z}\right]^{1/2}. \tag{33}$$

Let m_f, m_c, and m_p be the feedback, cascade, and plant gains at ω_c, respectively. The feedback gain multiplier is computed as

$$k_f \quad = \quad \frac{0.4}{m_f m_c m_p}, \tag{34}$$

which yields a gain margin of about 8 dB. This gain is included in the feedback compensator.

Figure 15 shows the z-plane root locus for the accelerometer loop controller. The loop is seen to be stable and the damping of the first mode is increased.

3. FEEDFORWARD COMPENSATOR

The final piece of the adaptive controller design is the feedforward compensator. This compensator has two functions. One function is modification of the reference signal for tracking. This function is accomplished by scaling the reference

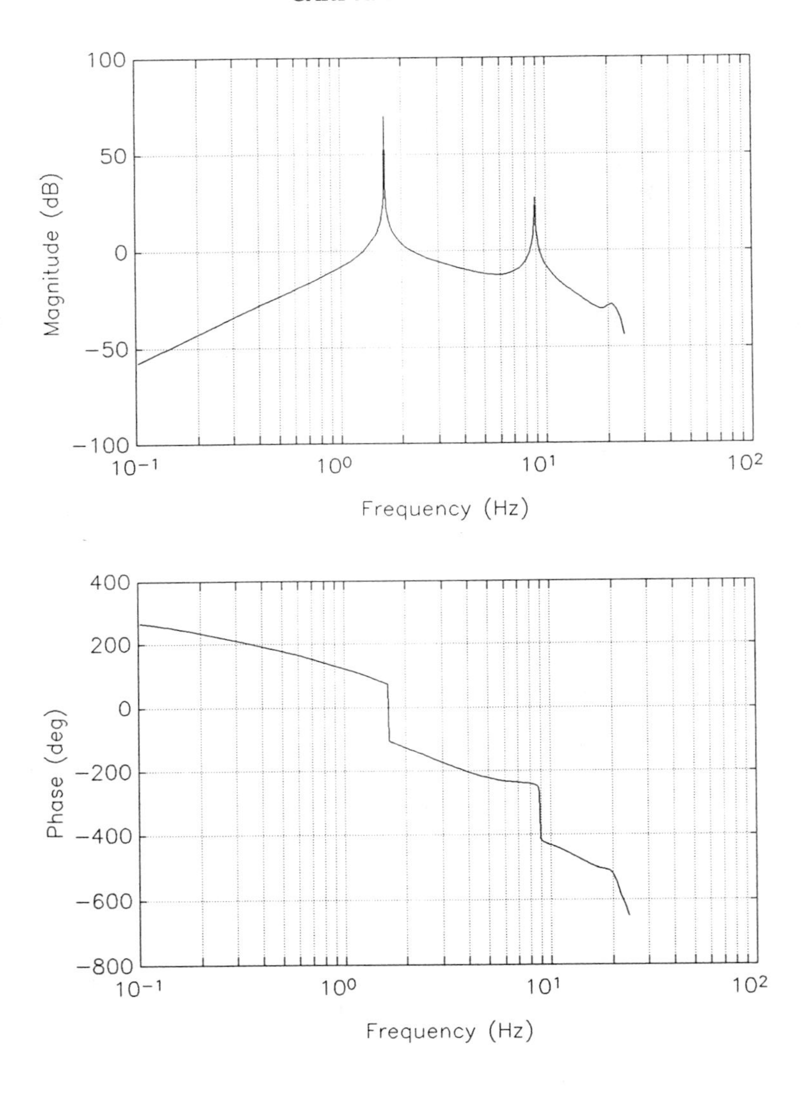

Figure 14: Open loop accelerometer control loop Bode plot.

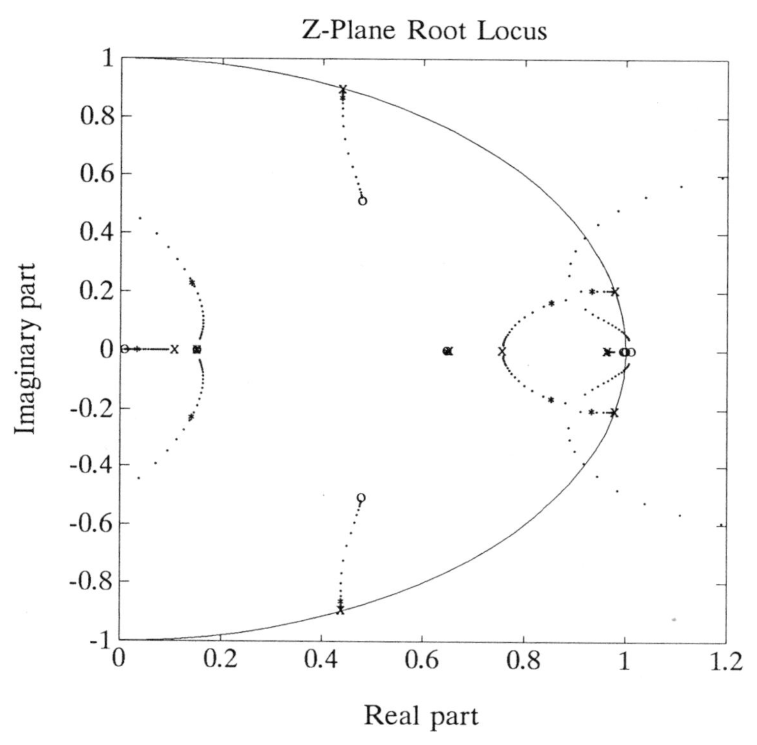

Figure 15: Accelerometer control loop root locus. ×—open loop poles; ∘—open loop zeros; ∗—closed loop poles.

and by filtering with a notch filter to prevent excessive excitation at the first mode. The s-plane form of this filter is

$$\frac{s^2 + 2(0.15)\omega_1 + \omega_1^2}{s^2 + 2\omega_1 + \omega_1^2}. \tag{35}$$

The other function of the feedforward compensator is attenuation of high frequencies in the reference signal to prevent excitation of the plant above the controller bandwidth. This is accomplished by cascading a second order Butterworth lowpass filter with Eq. (35). This filter has the form

$$\frac{\omega_c^2}{s^2 + 2(0.71)\omega_c + \omega_c^2}. \tag{36}$$

C. ROBUST CONTROLLER DESIGN

A block diagram of the robust controller is shown in Fig. 16. In contrast to the adaptive configuration, both the accelerometer and the LVDT are used in the

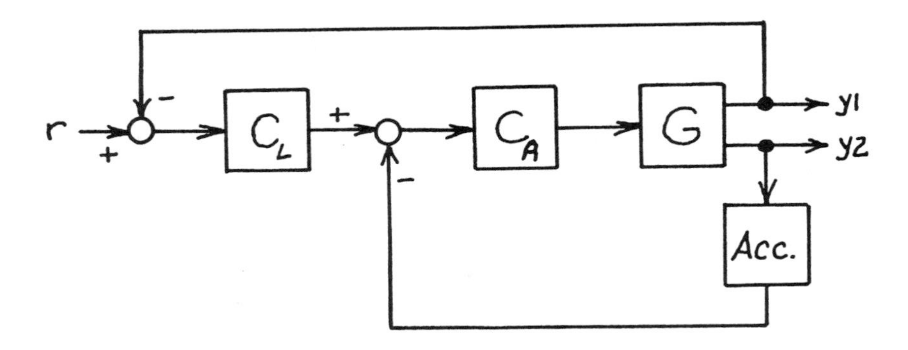

Figure 16: Robust Controller Configuration.

robust controller. Feedforward compensation of the reference command cannot be used in this case because the plant of DC gain is not assumed to be known. The robust controller consists of two loops. The inner loop compensator, C_A, uses the accelerometer for active damping of the first two beam modes. The outer loop compensator, C_L, consists of integral feedback of the LVDT error signal for command tracking. Highpass filtering of the accelerometer signal in the feedback path is not necessary because the sensor bias appears as a constant disturbance which is rejected by the integrator in the outer loop.

The design objectives of the inner loop compensator are the same as those for the adaptive rule cascade compensator. Therefore, the same compensator structure can be used. As far as stability is concerned, the plant configuration which causes the most difficulty in the design is when the first mode frequency is at its highest and the second mode frequency is at its lowest. This is troublesome because the frequency response of the plant does not roll off as much between the first and second modes. The design rules Eqs. (29–34) are applied with modal frequencies of 2 and 7 Hz and $s_z = 2\pi(4.5 \text{ Hz})$ to obtain C_A. This compensator is practically identical to the cascade compensator for the case 3 plant.

The robust inner loop compensator maintains stability for other plant configurations because the loop gain is lower than the corresponding adaptive rule controllers and does not roll off as quickly between the first and second modes. The trade-off is that damping of the modes for these other configurations is not as great.

The outer loop compensator is

$$C_L(q) = \frac{KT(q+1)}{2(q-1)},$$

which is a trapezoidal rule integrator. The gain K is set to $2\pi(1\text{ Hz})$, so there is little interaction between the inner and outer loops. Under closed loop operation the outer loop compensator acts as a lowpass filter with cutoff at K rad/s. This gain was chosen as large as possible in order to make the system response comparable to the adaptive design. As shown below, this choice results in more overshoot in some plant configurations than is desirable, but this trade-off was made to create a higher performing design.

D. EVALUATION OF CONTROLLER DESIGNS

The stability and time response of the designs will now be considered. Laboratory configurations 1 and 3 represent the extreme conditions considered here: case 1 has the widest-spaced modal frequencies and the smallest gain in-between; case 3 has the closest-spaced modal frequencies and largest midrange gain. Evaluating the controllers with these two cases does not constitute a formal proof that the adaptive design rule or robust design will meet performance objectives for all laboratory cases. However, evaluating the controllers in this manner yields many qualitative insights.

Figures 17 and 18 are Nichols charts for the accelerometer control loop for cases 1 and 3. In both cases, the adaptive rule controller with correct parameters provides about 8 dB of gain margin and 45° of phase margin at the −180° phase cross-over. Therefore the adaptive rule is insensitive to small errors in system parameter estimates. The robust controller and the case 3 adaptive rule are practically identical in terms of stability margins. When the case 3 adaptive rule control gains are applied to the case 1 plant, the closed loop system remains stable because the case 3 controller has lower gain; applying the case 1 control gains to the case 3 plant yields an unstable system. Thus, there exists enough variability in the plant parameters to cause a highly tuned controller to go unstable. This suggests that it is not supererogatory to consider adaptive control for this application. Also note that at higher frequencies (left side of the charts) the robust controller and the adaptive rule both provide significant attenuation of high frequency dynamics. The third mode is gain-stabilized by more than 20 dB in both cases. Therefore, the design goal of robustness to unmodeled dynamics has been achieved.

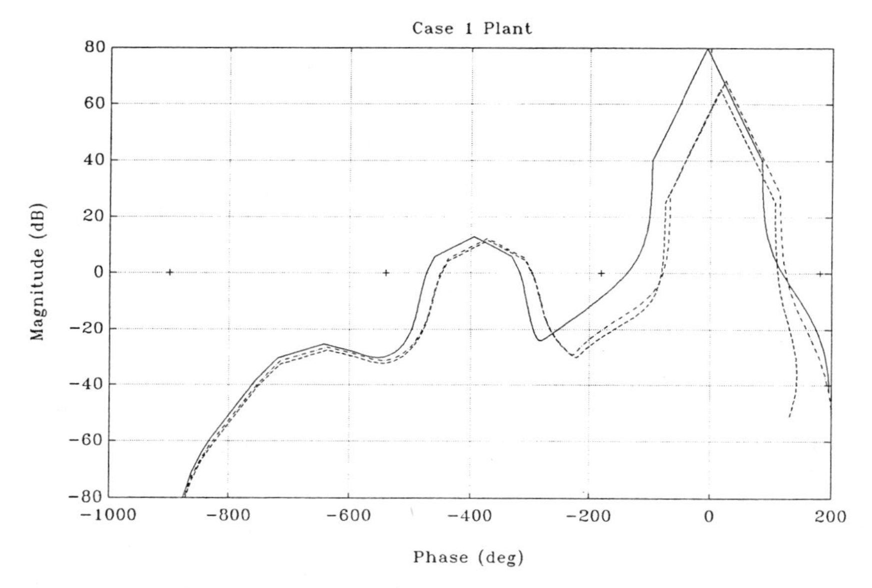

Figure 17: Accelerometer control loop Nichols chart for case 1 plant. Solid: case 1 adaptive rule; dash-dot: case 3 adaptive rule; dashed: robust controller.

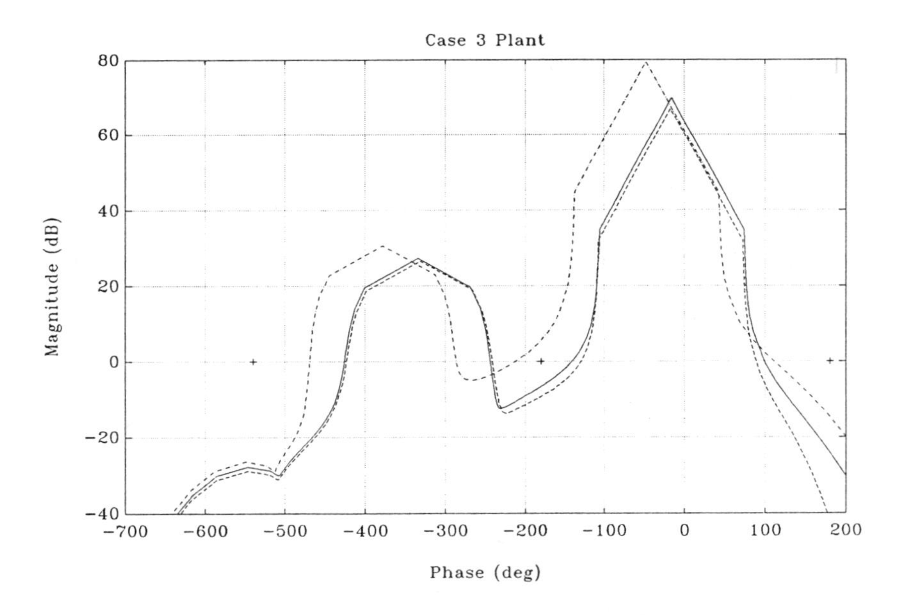

Figure 18: Accelerometer control loop Nichols chart for case 3 plant. Solid: case 3 adaptive rule; dash-dot: case 1 adaptive rule; dashed: robust controller.

Figures 19 and 20 illustrate the time responses for the controllers. These responses were generated using the FORTRAN time domain simulation, so the effects of motor and bearing nonlinearities are not included, although DAC and ADC quantization is. For the case 1 plant, Fig. 19, the adaptive rule controller settles in about half the time of the robust controller, with smaller overshoot—a big improvement in performance. As expected, the case 3 adaptive rule controller and the robust controller have nearly identical responses, as shown in Fig. 20. The adaptive rule controller has a slightly faster rise and settling time in response to a step command because of the use of feedforward compensation: the outer loop in the robust controller slows the response.

E. COMPARISON WITH ALTERNATIVE DESIGN APPROACHES

The proposed method for adapting the controller gains is non-traditional. The control designs developed here will now be compared to more standard techniques.

Figure 17 shows that, the major difference in stability between the robust and adaptive design rules is the accelerometer loop gain. If the system gain were tracked in the frequency range of the $-180°$ cross-over (about 3 Hz), then the inner loop gain of the robust design could be adjusted to yield a tighter loop for the case 1 plant. A similar idea is implemented in [21], in which a robust controller could handle all plant parameter variations except for the variation in the DC gain. Figure 21 compares the time response of a robust controller whose inner loop gain has been tuned to match the case 1 plant gain with the unmodified robust controller. The rise times are approximately the same; the settling time for the tuned controller is only slightly improved. This example illustrates that estimation of the mode frequencies is necessary to obtain better control performance.

Pole placement is a standard control design approach for indirect adaptive control [12,22]. A pole placement design was done for the plant model Eq. (28). The controller was specified to have relative degree two and the closed loop poles were assigned to locations similar to the adaptive rule. The resulting continuous time compensator was converted to discrete time by pole-zero mapping. The feedforward compensator was identical to the adaptive rule. Figure 22 is the Nichols chart for the accelerometer loop of the pole placement design; the system is unstable. The procedure is not robust to the phase lag of the antialiasing filters. When this same procedure was tried with the antialiasing filters included in the plant model, the Sylvester matrix formed as part of the solution method

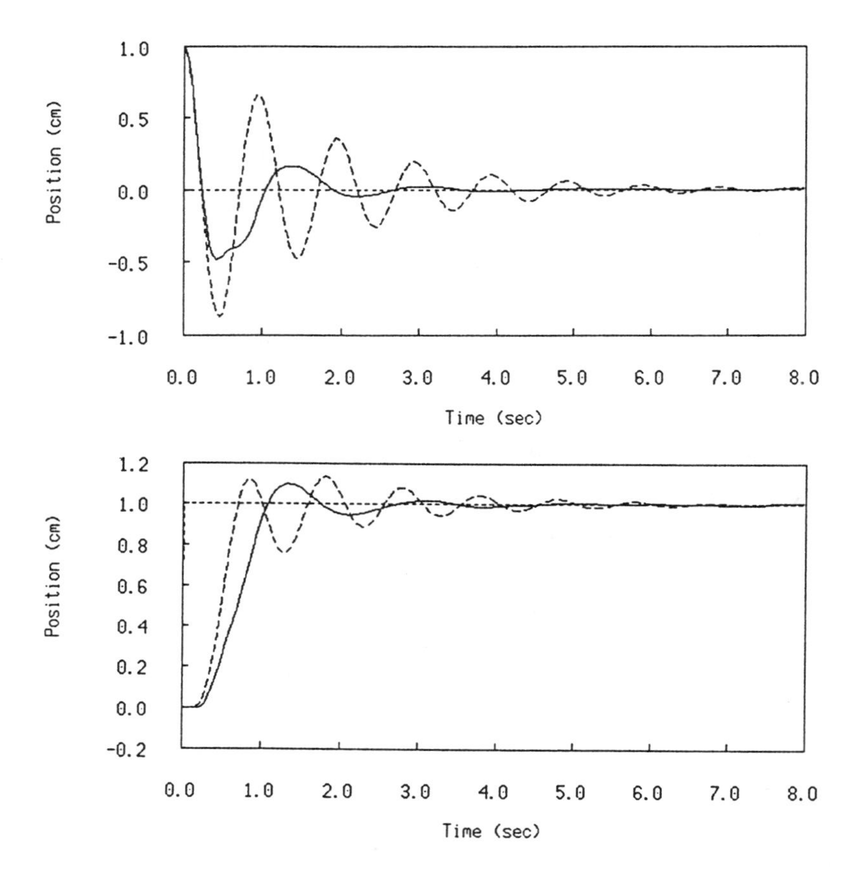

Figure 19: Time response of controllers for case 1 plant. Solid: adaptive rule; dashed: robust controller. Top: response to initial beam tip displacement; bottom: step response.

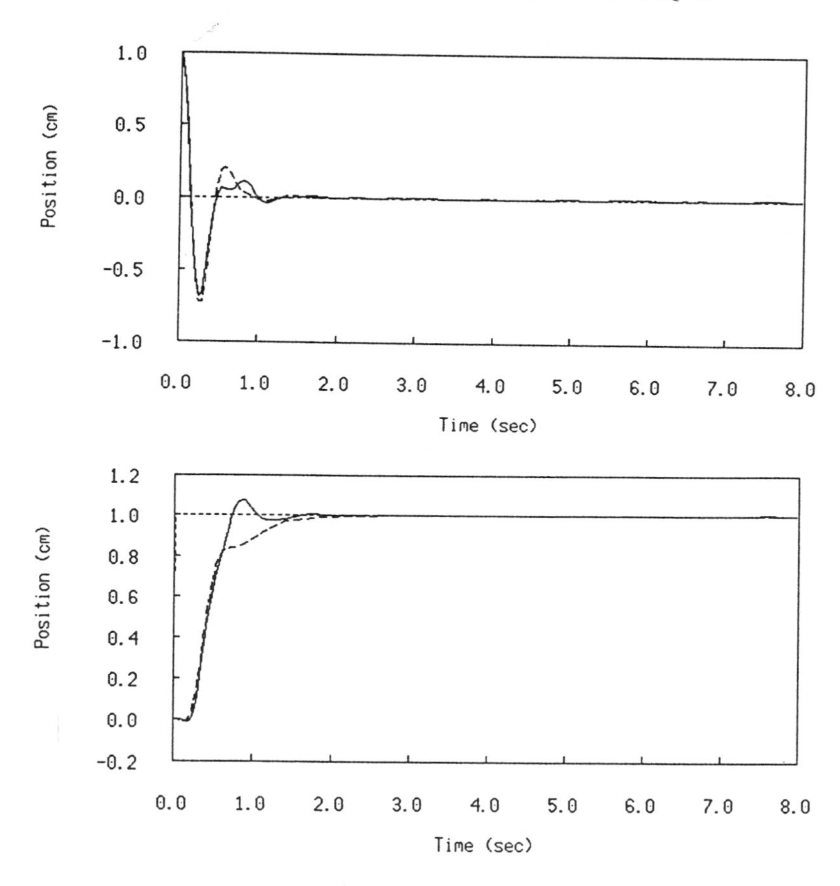

Figure 20: Time response of controllers for case 3 plant. Solid: adaptive rule; dashed: robust controller. Top: response to initial beam tip displacement; bottom: step response.

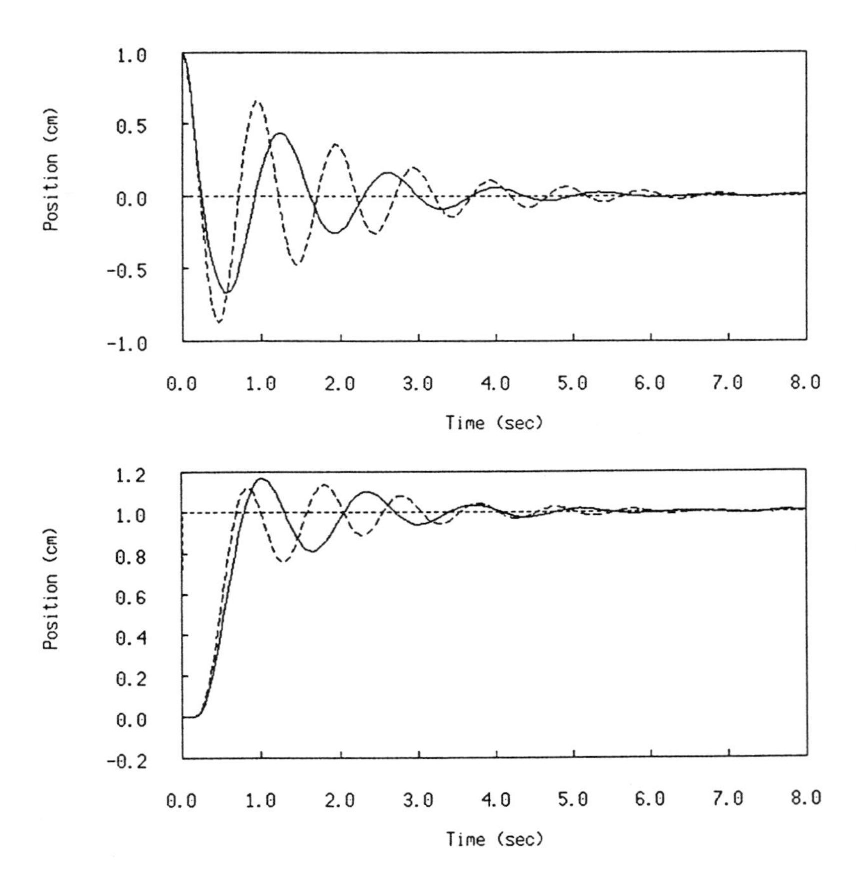

Figure 21: Comparison between robust controller and robust controller with tuned inner loop gain. Solid: tuned controller; dashed: robust controller. Top: response to initial beam tip displacement; bottom: step response.

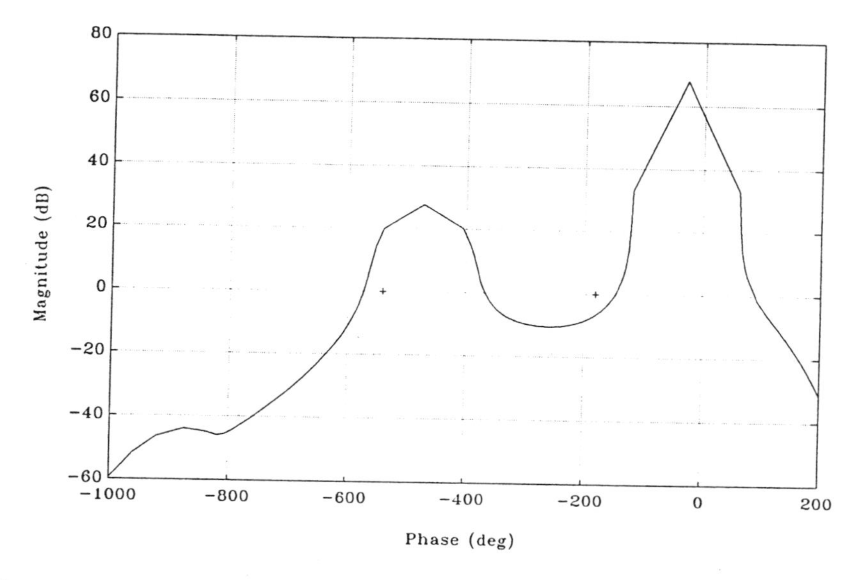

Figure 22: Accelerometer control loop Nichols chart for pole placement compensator applied to case 3 plant.

[12] was nearly singular (condition number $\approx 10^{-26}$), yielding a nonsensical result. A discrete time formulation results in a compensator that does not roll off, implying that the controller would not be robust to unmodeled dynamics. Better methods exist for solving the pole placement problem, but the complexity of the computations may not be suitable for on-line applications. Given the spread of eigenvalues and unmodeled dynamics often found in flexible structure control problems, use of pole placement design methods for on-line applications is not recommended.

VII. ADAPTIVE CONTROL DESIGN EVALUATION

This chapter evaluates three adaptive control scenarios for the laboratory model. These adaptive control schemes are constructed by integrating the identification and control designs discussed earlier. The scenarios considered in this chapter are:

Adaptive Calibration. An open loop pre-tuning phase is used to learn the system parameters, and the control is designed accordingly.

Adaptive Control (Conventional). Parameter estimation and control design update are performed throughout the control program. Nominal control gains are used initially.

Adaptive Robust Control. The control program is started with a robust design which guarantees stability for all plant variations. The controller is updated when the uncertainty in the plant parameters is sufficiently small.

The first approach yields the best controllers, but is impractical in many situations. Conventional adaptive controllers work well in many situations, but may be subject to large start-up transients if the controller is initialized poorly. Adaptive robust control is a modification of conventional adaptive control in which the control design is based on the structured uncertainty of the plant. Control performance can be improved as the uncertainty is decreased. The primary objective of this chapter is to investigate the feasibility of this approach.

A critical problem in the design of adaptive control systems is trading-off the issues of parameter variation and persistence of excitation. When using least square algorithms for parameter estimation, a forgetting factor must be included in the covariance update to keep the algorithm gain large enough to detect parameter changes. However, in the absence of proper excitation, the covariance grows and the algorithm can experience oscillations caused by noise and disturbances. The standard approach for dealing with this problem is to essentially shut off the algorithm when the excitation and prediction error is small. Methods for accomplishing this include dead zones [12,20], normalization [12,2], and σ-modification [23]. This aspect of the design is not considered here, because the emphasis is on start-up behavior of the scenarios.

A. CLOSED LOOP IDENTIFICATION

Intrinsic to adaptive control performance is the quality of estimates when the system is operating under feedback control. In [24,6] the issues of identifiability and accuracy in closed loop identification are considered. If a sufficiently rich command signal is present, then the feedback does not affect the identifiability of the system. However, feedback does significantly alter the frequency domain PEM weighting functions. Here open and closed loop identification experiments using the simulated laboratory model are compared. Reasonable modal estimates can be obtained when the robust controller is used, however additional excitation beyond the reference command may be required.

1. PERSISTENCE OF EXCITATION AND INFORMATION

The batch solution to the ARX least squares problem Eq. (10) is dependent on the existence of the matrix inverse of

$$\mathcal{I}_t = \sum_{k=1}^{t} \phi(k)\phi(k)^T.$$

If the noise variance in the ARX model is σ^2, then $\sigma^{-2}\mathcal{I}_t$ is the Fisher information matrix [25,6] for the estimation problem. If the excitation in the system is poor, then the spread in the eigenvalues of $\mathcal{I}_t$ will be large, reflecting the fact that some parameters are essentially not identifiable. Thus, there is an important connection between information and persistence of excitation. The information increases with the magnitude of the regressor, $\phi(t)$, but the excitation of some parameters may still be relatively poor, which can lead to numerical difficulties and stability problems in indirect adaptive controllers. The relative excitation of the parameters can be assessed using the matrix

$$\Phi_t = \frac{\sum_{k=1}^{t} \phi(k)\phi(k)^T}{\sum_{k=1}^{t} \phi(k)^T\phi(k)}. \qquad (37)$$

Let $U\Sigma U^T$ be the eigenvalue decomposition of Φ_t, where $\Sigma = diag\{s_1, \ldots, s_n\}$ and $U = [U_1, \cdots, U_n]$. Note $\sum_1^n s_r = 1$, independent of the magnitude of the excitation. The vector U_r indicates the linear combination of parameters which have a relative level of excitation given by s_r.

2. FREQUENCY DOMAIN ANALYSIS

The frequency domain expressions for the asymptotic PEM cost derived earlier include the effect of feedback. In general, the laboratory control signal can be written in the form

$$u(t) = M(q)r(t) + N(q)w(t) - R(q)y(t), \qquad (38)$$

where $r(t)$ is the reference command, $y(t)$ is the accelerometer output, and $w(t)$ is an extra "dither" or probing signal added to the control signal before digital-to-analog conversion. With these definitions, the relative weighting function Eq. (23) becomes

$$W_R(\omega) = \left\{ \frac{|F(e^{j\omega T})|^2 |G_0(e^{j\omega T})|^2}{|H(e^{j\omega T}, \theta)|^2 |1 + G_0(e^{j\omega T})R(e^{j\omega T})|^2} \right\} \times [|M(e^{j\omega T})|^2 \Phi_r(\omega) + |N(e^{j\omega T})|^2 \Phi_w(\omega)],$$

while the noise weighting function Eq. (21) is

$$W_V(\omega) = \frac{|F(e^{j\omega T})|^2|1+G(e^{j\omega T},\theta)R(e^{j\omega T})|^2}{|H(e^{j\omega T},\theta)|^2|1+G_0(e^{j\omega T})R(e^{j\omega T})|^2}.$$

For ARX models, making the approximation $G(q,\theta) \approx G_0(q)$, one obtains

$$W_R(\omega) = \left\{\frac{|F(e^{j\omega T})|^2|B_0(e^{j\omega T})|^2}{|1+G_0(e^{j\omega T})R(e^{j\omega T})|^2}\right\} \times [|M(e^{j\omega T})|^2\Phi_r(\omega) + |N(e^{j\omega T})|^2\Phi_w(\omega)], \tag{39}$$

$$W_V(\omega) = |F(e^{j\omega T})|^2|A_0(e^{j\omega T})|^2. \tag{40}$$

The noise weighting is essentially unchanged by feedback (see Fig. 4), but the relative weighting function is quite different from the open loop case.

The transfer functions $M(q)$, $N(q)$, and $R(q)$ in Eq. (39) are different for the robust and adaptive rule control configurations. The robust controller is considered first. Let G_1 represent the LVDT transfer function (which is modeled as a gain), and let C_A and C_L be the inner and outer loop compensators, respectively. Then the control signal is

$$u = w + C_A[C_L(r - G_1 u) - y],$$

which yields

$$M = \frac{C_A C_L}{1 + C_A C_L G_1},$$

$$N = \frac{1}{1 + C_A C_L G_1},$$

$$R = \frac{C_A}{1 + C_A C_L G_1}.$$

Fig. 23 shows the relative weighting functions for the reference and dither inputs with the robust controller and no prefilter. (The case 3 plant is shown.) Unlike the open loop case, the relative weighting function for the reference command is basically flat between the two modes, and the first mode is more heavily weighted than the second. Feedback causes a sharp notch in the weighting functions at the second mode. The high frequency fit can be expected to be relatively poor when just the reference signal is present. The dither input weighting function emphasizes high frequency fit more than the reference command, so the dither can be used to supplement the reference.

The adaptive rule controller consists of the cascade compensator, C, and the feedforward compensator, P. The feedback compensator has negligible influence on the weighting functions. The control signal in this case becomes

$$u = w + C(Pr - y),$$

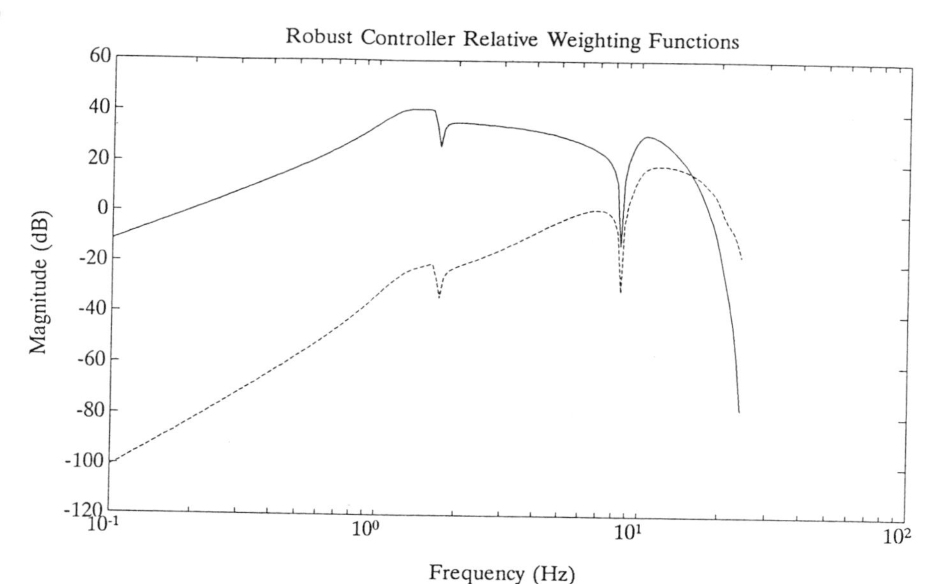

Figure 23: PEM relative weighting functions using robust controller. Solid: reference command weighting; dashed: dither signal weighting.

so that

$$
\begin{aligned}
M &= CP, \\
N &= 1, \\
R &= C.
\end{aligned}
$$

Inserting these transfer functions in Eq. (39) yields the weighting functions shown in Fig. 24. These functions are similar to the robust case, except around the first mode, where the reference command weighting function is relatively smaller for the adaptive rule case. This is due to the feedforward compensator which includes a notch filter at the first mode.

The relative flatness of the closed loop weighting functions in the frequency range between the modes suggests that a single estimator could be used, instead of the two-band estimator designed in section IV. This observation will not be pursued further here, but it does illustrate the utility of the frequency weighting approach in identification design for adaptive control.

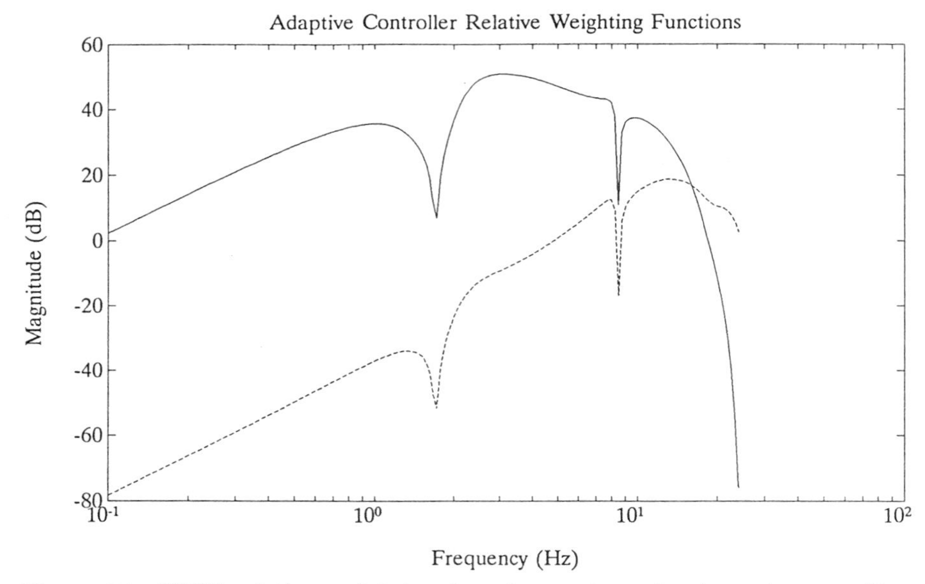

Figure 24: PEM relative weighting functions using adaptive rule controller. Solid: reference command weighting; dashed: dither signal weighting.

3. OPEN AND CLOSED LOOP IDENTIFICATION COMPARISON

Identification experiments were conducted using the time domain simulation to compare open and closed loop identification performance on the case 3 plant. The input signal for the open loop experiment was a random binary sequence (RBS) with the smallest period between set point shifts being two seconds. The amplitude of the motion at the top of the beam was was ± 0.1 cm. The duration of the experiment was 30 seconds. The closed loop experiment used the robust controller and a similar 1 cm RBS reference command. The results are summarized in Table II; the estimated Bode plots for the closed loop case are shown in Fig. 25 (compare with Fig. 6). In Table II, the subscripts ℓ and h refer to the low- and high-band estimators. The parameter vectors for the ARX(2,2,1) estimators have the form $\theta = [a_1,\ a_2,\ b_0,\ b_1,\ b_2]^T$; $P_\ell(i,i)$ is the ith parameter variance for the low-band estimator, and so forth.

Comparing the open and closed loop results, the following conclusions can be made:

- The modal frequency estimates are slightly less accurate for the closed loop case.

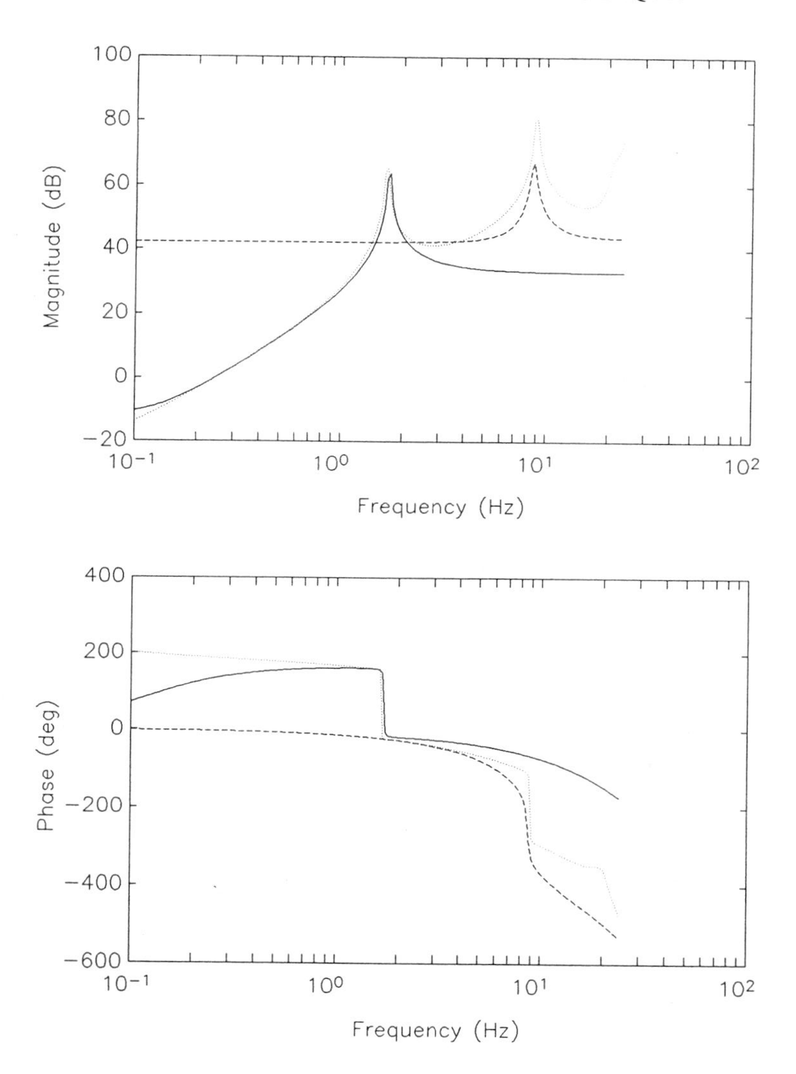

Figure 25: Transfer function estimates from closed loop identification experiment. Solid: low-band estimates; dashed: high-band estimates; dotted: true.

Parameter	Open Loop	Closed Loop	True Value
θ_p (cm/V)	0.368	0.380	0.380
f_1 (Hz)	1.657	1.701	1.648
f_2 (Hz)	8.887	8.662	8.876
s_z (rad/s)	16.8	41.8	21.4
$P_\ell(1,1)$	1.1×10^{-5}	5.5×10^{-5}	–
$P_\ell(4,4)$	17	8.0	–
$P_h(1,1)$	4.3×10^{-6}	2.0×10^{-4}	–
$P_h(4,4)$	4.9	24.8	–
$s_{\ell_{min}}/s_{\ell_{max}}$	2.1×10^{-10}	1.4×10^{-8}	–
$s_{h_{min}}/s_{h_{max}}$	4.7×10^{-9}	4.5×10^{-8}	–

Table II: Open and closed loop identification experiment comparison.

- As predicted by the weighting function analysis, the low frequency fit is greatly improved in the closed loop case, however the high frequency fit is worse. The high frequency gain is underestimated, resulting in a poor value for the real zero estimate, s_z.

- In both cases, the two largest excitation eigenvalues correspond to linear combinations of the denominator parameters; the three smaller eigenvalues correspond to numerator parameters. This helps to explain why the modal frequencies are estimated well in comparison to the transfer function magnitude.

- The spacial distribution of the excitation, given by the ratio of the minimum and maximum eigenvalues of Φ_{30} is better in the closed loop case. This indicates that the poor high frequency fit in the closed loop case is due more to the magnitude of the relative weighting function than the amplitude of the reference signal.

- The reference signal has little high frequency content; addition of a dither signal may be helpful.

B. ADAPTIVE CONTROL FEASIBILITY STUDY

Now the three adaptive control approaches will be compared. The adaptive calibration uses a random walk excitation until the parameter variances become sufficiently small. These bounds are functions of the initial parameter uncertainties:

$$\begin{aligned}(P_\ell(1,1) + P_\ell(2,2))/2 &< P_{\ell 0}(1,1)/5, \\ (P_h(a_1) + P_h(a_2))/2 &< P_{h0}(1,1)/100, \\ (P_\ell(3,3) + P_\ell(4,4) + P_\ell(5,5))/3 &< P_{\ell 0}(4,4)/2, \\ (P_h(3,3) + P_h(4,4) + P_h(5,5))/3 &< P_{h0}(4,4)/5.\end{aligned}$$

The initial covariances are

$$\begin{aligned}P_{\ell 0} &= diag\{0.001,\ 0.001,\ 100,\ 100,\ 100\}, \\ P_{h0} &= diag\{0.1,\ 0.1,\ 100,\ 100,\ 100\}.\end{aligned}$$

The bounds on the high frequency parameters are relatively tight so that the estimate of s_z will be as accurate as possible. Recall that this parameter is critical for the controller design rule. After the initial tuning phase, the controller is updated in the same manner as the conventional adaptive controller.

The conventional adaptive control uses the same initial covariances. The controller is re-designed using the rules from Section V every T_a seconds. Typically T_a is chosen to be 10–50 times the sampling interval of 0.02 s. This has the effect of reducing the gain in the adaptive algorithm, which enhances stability.

The initial re-design for the adaptive robust controller is based on the same covariance bounds as above. After that, the controller is updated in the same manner as the conventional adaptive controller. It was found that the RBS reference command alone is insufficient to reduce the initial parameter uncertainty below the required bounds (in a reasonable amount of time) when the robust controller is used. A dither signal consisting of uniform noise passed through a first order 5 Hz lowpass filter was added to the control signal during the initial tuning phase.

1. TESTS USING RBS REFERENCE COMMAND

The first test of the adaptive control schemes consists of a 1 cm RBS reference command, with two seconds being the smallest period between level shifts. In the case of the adaptive calibration approach, the RBS command is switched on after the learning phase is complete. The adaptive reset time is T_a =0.5 s. The dither signal for the robust controller is turned on after ten seconds and shut off when the control update is made. The switch from robust to adaptive rule controller involves a change in the controller configuration as well as in the compensator gains. To prevent a large transient at the initial update, a nominal feedforward compensator is run during the robust control phase, although this signal is not used. Figures 26–28 show the bottom of beam position responses.

The case 1 plant is slower to converge in the adaptive calibration case because the second mode receives less excitation than in case 3. The responses to the set point changes are similar to the adaptive rule controller step responses shown in Figs. 19 and 20. The overshoot for the case 1 plant is slightly greater because s_z is overestimated (45 rad/s instead of 31 rad/s), which causes the control bandwidth to be set too high.

The command signal is persistently exciting, which enables the conventional adaptive controller to have a small initial update transient. The adaptive robust controller update occurs at about 17 s in both cases illustrated (seven seconds after the dither signal is applied). The steady state performance of the two controllers is similar: both controllers have larger overshoot than the adaptive calibration approach. This is due to poor s_z estimates. Note that the dither signal used in the adaptive robust controller causes very little disturbance in output, but greatly speeds the parameter convergence.

2. TESTS WITH INITIAL BEAM OFFSET

The previous simulation results show that the conventional adaptive controller and the adaptive robust controller are similar when the command signal is large. In fact, the conventional adaptive controller converges faster to a well-tuned controller, because of the long learning time required by the adaptive robust approach. However, this speed of convergence comes at the price of possibly unacceptable transient behavior.

Figure 29 shows a test comparing the conventional adaptive and robust adaptive controllers using the case 3 plant with the bottom of the beam at an initial offset of 1 cm. The reference command is a 0.1 cm RBS signal. The conventional adaptive controller is initialized to the adaptive rule case 1 gains, resulting initially in an unstable system. The adaptive update rate is $T_a = 0.25$ s. Figure 29 compares the responses of the conventional and adaptive robust controllers. The conventional adaptive controller converges within two seconds, but the plant suffers several large oscillations. The robust control provides a smoother initial response. The dither signal was added to the robust control signal after two seconds, the initial control update occurs at about 12 s. The conventional adaptive control has a bad transient because of the small reference signal relative to the initial condition of the plant. In this case, the initial condition is like an unmodeled disturbance acting on the system.

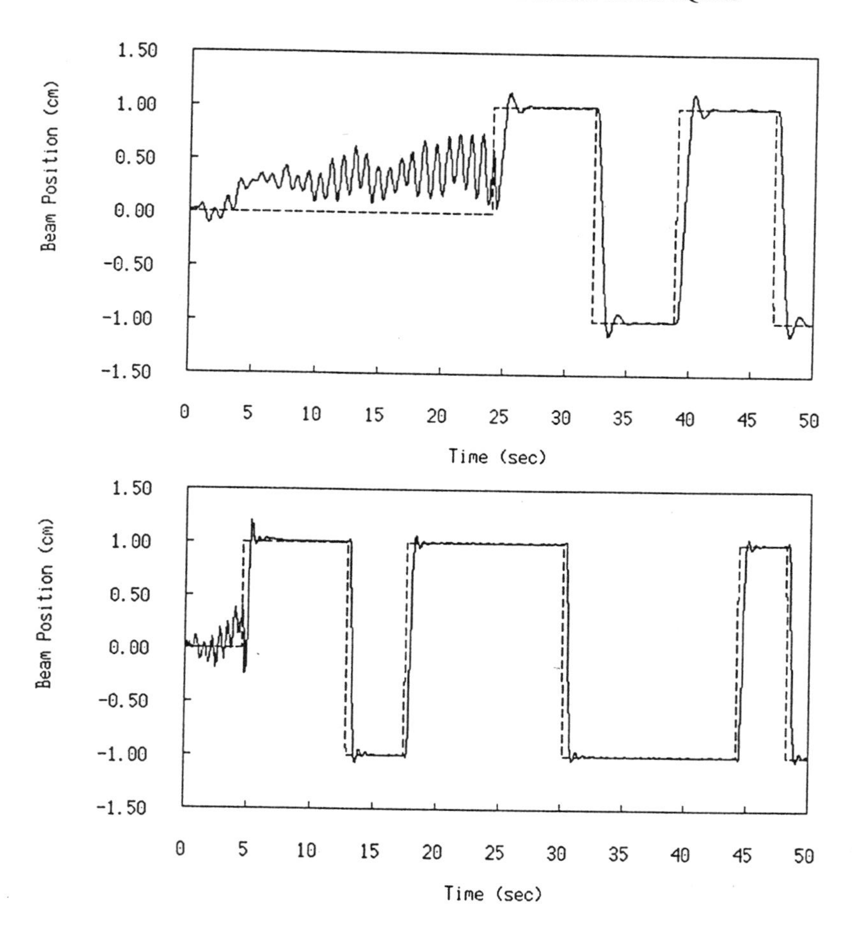

Figure 26: Adaptive calibration response to RBS command. Top: case 1 plant; Bottom: case 3. Solid: bottom of beam position; dashed: command signal.

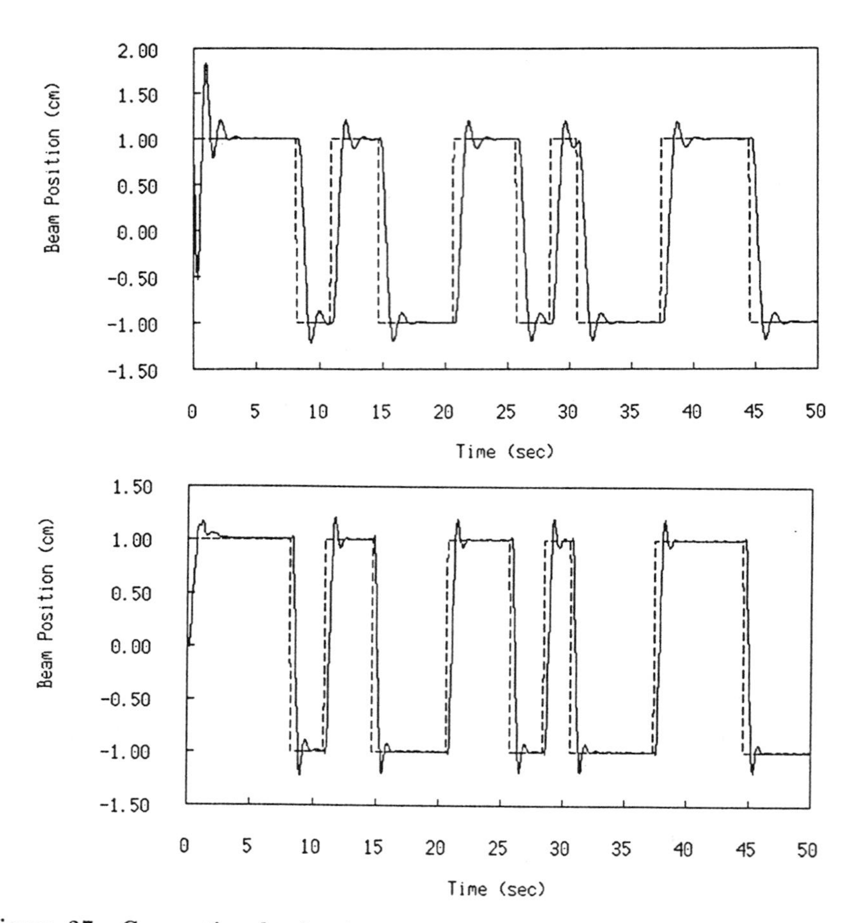

Figure 27: Conventional adaptive control response to RBS command. Top: case 1 plant; Bottom: case 3. Solid: bottom of beam position; dashed: command signal.

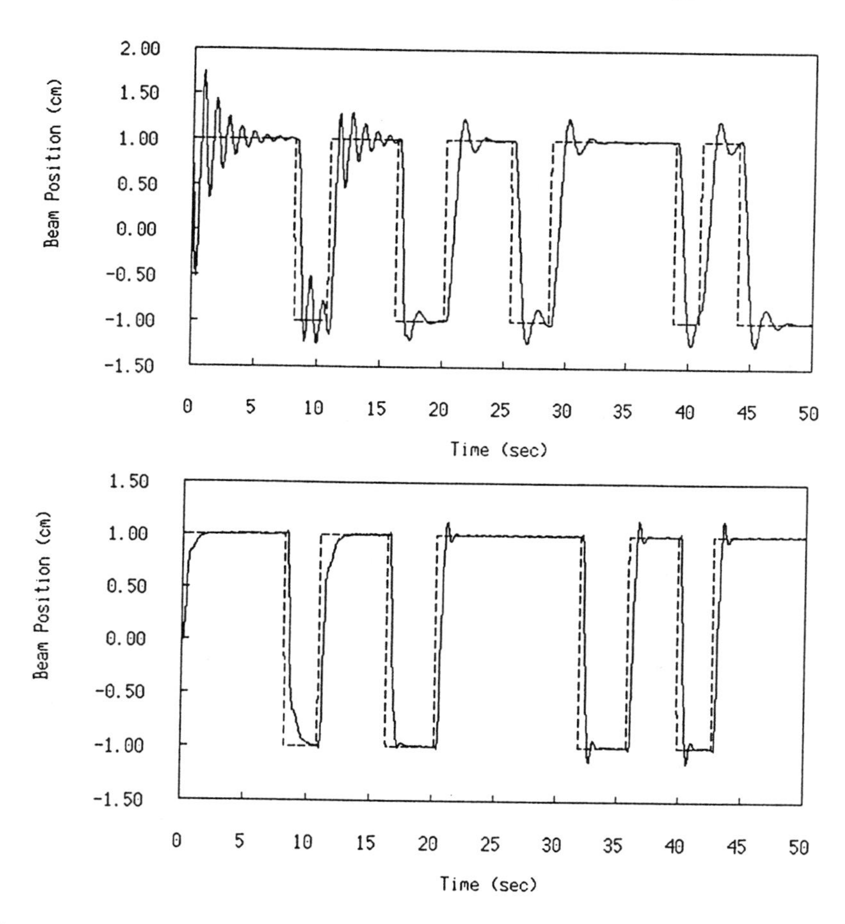

Figure 28: Adaptive robust control response to RBS command. Top: case 1 plant; Bottom: case 3. Solid: bottom of beam position; dashed: command signal.

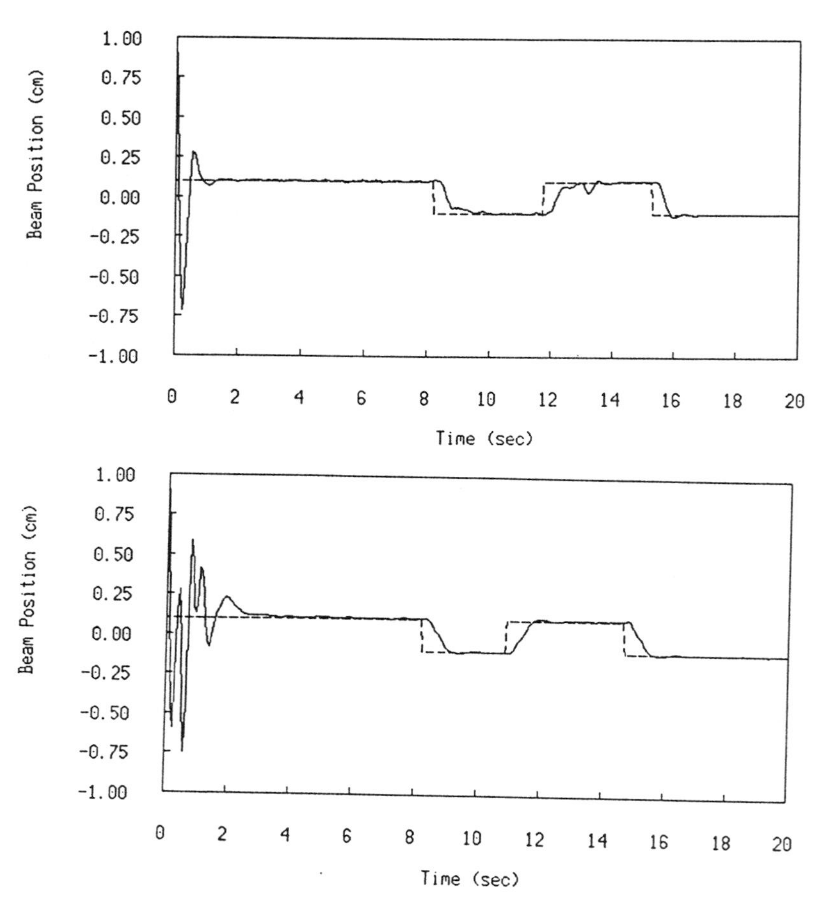

Figure 29: Adaptive controller responses to initial beam offset. Top: adaptive robust controller; Bottom: conventional adaptive controller. Solid: bottom of beam position; dashed: command signal.

VIII. SUMMARY

This chapter has considered methods for *designing* adaptive robust controllers. Contributions included (1) development of a system identification approach based on modal decomposition of the model and enhancement of the transient response of recursive least square algorithms; (2) development of robust control design rules using classical frequency domain methods; (3) study of the feasibility of adaptive robust control based on the use of parameter uncertainty to update the control design; (4) implementation of identification and control algorithms on a flexible beam experiment. The viewpoint of evaluating system identification performance in terms of transfer function fit error has been shown to be very effective for adaptive identification design.

References

[1] B. D. O. Anderson, R. R. Bitmead, P. V. Kokotvic, R. L. Kosut, I. M. Y. Mareels, L. Praly, and B. D. Riedle, *Stability of Adaptive Systems: Passivity and Averaging Analysis.* Cambridge, MA: MIT Press, 1986.

[2] S. Sastry and M. Bodson, *Adaptive Control: Stability, Convergence, and Robustness.* Englewood Cliffs, NJ: Prentice-Hall, 1989.

[3] C. E. Rohrs, L. Valavani, M. Athans, and G. Stein, "Robustness of continuous-time adaptive control in the presence of unmodeled dynamics," *IEEE Trans. Automatic Control*, vol. AC-30, pp. 881–889, Sep. 1985.

[4] P. A. Ioannou and P. V. Kokotovic, "Instability analysis and improvement of robustness of adaptive control," *Automatica*, vol. 20, no. 5, pp. 583–594, 1984.

[5] B. Wahlberg and L. Ljung, "Design variables for bias distribution in transfer function estimation," *IEEE Transactions on Automatic Control*, vol. AC-31, pp. 134–144, Feb. 1986.

[6] L. Ljung, *System Identification — Theory for the User.* Englewood Cliffs, NJ: Prentice-Hall, 1987.

[7] G. A. McGraw, *Robust Adaptive Control Design Techniques.* PhD thesis, UCLA, 1989.

[8] G. A. McGraw, "Robust adaptive control design techniques," Tech. Rep. TOR-0090(5025-05)-1, The Aerospace Corporation, 1989.

[9] K. J. Åström, *Introduction to Stochastic Control Theory.* NY: Academic, 1970.

[10] M. Gevers and L. Ljung, "Optimal experiment design with respect to the intended model application," *Automatica*, vol. 22, no. 5, pp. 543–554, 1986.

[11] L. Ljung, "Asymptotic variance expressions for identified black-box transfer function models," *IEEE Trans. Automatic Control*, vol. AC-30, pp. 834–844, Sep. 1985.

[12] G. C. Goodwin and K. S. Sin, *Adaptive Filtering, Prediction, and Control.* Englewood Cliffs, NJ: Prentice-Hall, 1984.

[13] L. Ljung and T. Söderström, *Theory and Practice of Recursive Identification.* Cambridge, MA: MIT Press, 1983.

[14] G. J. Bierman, *Factorization Methods for Discrete Sequential Estimation.* NY: Academic, 1977.

[15] F. Jabbari and J. S. Gibson, "Vector-channel lattice filters and identification of flexible structures," *IEEE Transactions on Automatic Control*, vol. AC-33, no. 5, pp. 448–456, 1988.

[16] R. H. Middleton and G. C. Goodwin, "Improved finite word length characteristics in digital control using delta operators," *IEEE Trans. Automatic Control*, vol. AC-31, pp. 1015–1021, Nov. 1986.

[17] J. P. Clary, *Robust Algorithms in Adaptive Control.* PhD thesis, Stanford University, 1984. Information Systems Laboratory Report #ISL/GFF/L-305-3.

[18] D. M. Rovner and R. H. Cannon, "Experiments toward on-line identification and control of a very flexible one-link manipulator," *Int. Journal of Robotics Research*, vol. 6, pp. 3–19, Winter 1987.

[19] G. F. Franklin and D. P. Powell, *Digital Control of Dynamic Systems.* Reading, MA: Addison-Wesley, 1980.

[20] R. H. Middleton, G. C. Goodwin, D. J. Hill, and D. Q. Mayne, "Design issues in adaptive control," *IEEE Trans. Automatic Control*, vol. AC-33, pp. 50–58, Jan. 1988.

[21] P. O. Gutman, H. Levin, L. Neumann, T. Sprecher, and E. Venczia, "Robust and adaptive control of a beam deflector," *IEEE Transactions on Automatic Control*, vol. AC-33, pp. 610–619, Jul. 1988.

[22] K. J. Åström and B. Wittenmark, *Adaptive Control.* Reading, MA: Addison-Wesley, 1988.

[23] P. A. Ioannou and K. S. Tsakalis, "A robust direct adaptive controller," *IEEE Trans. Automatic Control*, vol. AC-31, pp. 1033–1043, Nov. 1986.

[24] L. Ljung, I. Gustavsson, and T. Söderström, "Identification of linear multivariable systems operating under linear feedback control," *IEEE Transactions on Automatic Control*, vol. AC-19, pp. 836–840, Dec. 1974.

[25] G. C. Goodwin and R. L. Payne, *Dynamic System Identification: Experiment Design and Data Analysis.* NY: Academic Press, 1977.

TECHNIQUES IN THE OPTIMAL CONTROL OF DISTRIBUTED PARAMETER SYSTEMS

MUIBI ADE. SALAMI
Hughes Aircraft Company[1]
Los Angeles, California 90009

I INTRODUCTION

While the control literature shows a lot of progress in the optimal control of distributed parameter systems (D.P.S.) , a corresponding progress is yet to be noticed in the development of practical and numerical techniques for real time optimal control of D.P.S. The reason for this is not surprising, considering the complexities involved in the nature of the analytical techniques required in understanding the distributed parameter systems. However, judging from the application needs for D.P.S. in the industries (as well as our daily and practical life needs), it is apparent that this area of research and development requires as much attention as the lumped parameter systems (L.P.S.) which has enjoyed tremendous progress . Numerous examples of D.P.S can be cited in industries that need be controlled in real-time. Examples from aerospace, defence, chemical, environmental and infrastructural industries are too many to be listed . A contribution to the development of such practical and simple algorithms for optimal control of

[1]This work was done while the author was at the University of California at Los Angeles.

D.P.S. is the concern of this chapter. These algorithmic techniques are developed with a view to implementation on digital computers and in real-time.

In section II , techniques in the optimal controller synthesis are addressed for some classes of D.P.S. Linear optimal regulator problems are the focus. Optimal controller design for D.P.S. is normally infinite dimensional in nature . Correspondingly , an infinite dimensional Riccati-like Differential equation has to be solved . This yields an infinite dimensional controller which for purposes of practical implementation is prohibitive. Techniques are developed to reduce the controller to finite order controllers that can be and in some cases have been implemented in real-time. Two methods are employed in developing such finite dimensional feed-back controllers : Direct and indirect methods. The direct method synthesizes a finite order feedback controller from a reduced order dynamics for the original infinite dimensional D.P.S. . The indirect method synthesizes an infinite dimensional controller from the full order D.P.S., and then an order reduction is performed on the infinite dimensional controller to obtain a finite order feedback controller . Both results are compared.

Section III addresses the problem of developing simple algorithms for implementing the feedback controller of Section II on a digital computer. The problem turn out to be the development of algorithms for synthesizing discrete-time weighting matrices for optimal control of D.P.S. Direct and indirect methods .are also applied .

A summary of the chapter and suggestions for further research are the subject of Section IV .

II OPTIMAL CONTROL OF DISTRIBUTED PARAMETER SYSTEMS [LINEAR OPTIMAL REGULATOR PROBLEMS]

The objective of this Section is the synthesis of a finite dimensional optimal controller for D.P.S linear regulator problems. However, the development of the infinite dimensional controller will not only be instructive and valuable in the sequel, it will also form the basis for a method of synthesizing the finite dimensional controller; therefore, we will first synthesize the infinite dimensional controller. The problem statement for this section follows in subsection A. The full order (infinite dimensional) and the reduced order (finite dimensional) controller developments are taken up in subsections B and C, respectively.

A. PROBLEM STATEMENT

Given the following infinite dimensional controllable D.P.S. model as:

$$\dot{Z}(t,s) = A(s)\,Z(t,s) + B\,f(t,s) \; ; \text{ for } 0 < t_0 < T \tag{1}$$

Initial conditions: $Z(0,s) = Z_0(s)$ (2)

Boundary conditions:

$$\frac{\partial Z(t,s)}{\partial s} = -E_0\,Z(t,s) + B_0\,f_0(t) \; ; \quad \text{for } s=0 \tag{3a}$$

$$\frac{\partial Z(t,s)}{\partial s} = -E_1\,Z(t,s) + B_1\,f_1(t) \; ; \quad \text{for } s=1 \tag{3b}$$

Performance index:

$$J_D(.) = \frac{1}{2}\int_0^1\int_0^1 \langle Z(t_f : r), S_f(r,s)Z(t_f : s)\rangle \, dr\, ds \quad +$$

$$\frac{1}{2}\int_0^{t_f}\int_0^1\int_0^1 \langle Z(t:r), Q(t,r,s)Z(t:s)\rangle + \langle f(t,s), R(t:r,s)f(t,r)\rangle \, dr\, ds\, dt$$

$$+ \frac{1}{2}\int_0^{t_f} \langle f_0(t), R_0(t)f_0(t)\rangle + \langle f_1(t), R_1(t)\, f_1(t)\rangle \, dt \tag{4}$$

In general $Z(t,s)$ is a vector and $A(s)$ could be taken as :

$$A(s) = A_0 + A_1 \frac{\partial}{\partial s} + A_2 \frac{\partial^2}{\partial s^2}$$

where :

Z(t,s) = Infinite dimensional state of the D.P.S. system. $Z(t.s) \in H$, a real infinite dimensional dense separable Hilbert Space.

A(s) = Bounded linear spartial differential operator with domain $D[A(s)]$ a dense proper subset of H.

B = The control operator, finite or infinite dimensional, depending on f(.) . $B \in L(U,H)$

f(t,s) = Control function: $f(t,s) \in U$ – the control space or the set of admissible control assumed to be a class of continuous function in time and space. (5)

$Q, S_f : \in L(H,H)$ Self – adjoint nonnegative operators

$R, R_0, R_1 \in L(U,U)$ Self – adjoint positive operators

In addition, the so called weak formulation of eq. (1) will be of use in the sequel and can be stated as:

$$Z(t,s) = T(t,t_0) Z_0 + \int_{t_0}^{\tau} T(t,s) B f(t,s) dt ; \quad 0 < t_0 < t < \tau < \infty \tag{6}$$

where T(.,.) is a semigroup on H with generator A(s). It is required to find both a full order as well as a reduced order control function $f^*(t,s) \in U$ that minimizes the performance index Eq. (4).

Synthesize: (1) Full order model (infinite dimensional) controller
(2) Reduced order model (finite dimensional) controller

B. FULL ORDER INFINITE DIMENSIONAL CONTROLLER SYNTHESIS FOR DISTRIBUTED PARAMETER SYSTEM [LINEAR OPTIMAL REGULATOR PROBLEM]

Given the control problem of sub-section A, and using the Minimum principle, first we formulate the Hamiltonians corresponding to the problem. Then the necessary conditions for optimality are developed. These equations are solved to obtain the optimal controller in terms of the Riccati equation variable . The corresponding boundary and interior Riccati equations are solved likewise to obtain the corresponding boundary and interior controllers explicitly. We are interested in algorithms that are practically simple, and attractive to implement on a digital computer. Our approach falls into two categories in synthesizing the two types of controllers specified by the problem statement - direct and indirect methods .

0. NOTES :

(a) The performance index , Eq. (4) is slightly different from the conventional lumped parameter quadratic problem performance index . In particular the quadratic terms in the boundary controls are alien to the lumped parameter performance index . This calls for the synthesis of boundary optimal controllers .

(b) The notice in (a) correspondingly calls for what we term the boundary Hamiltonians and also corresponding boundary Riccati equations.

1. THE HAMILTONIANS

a. The Main or Interior Hamiltonian

$$H(.,.,.) = \frac{1}{2}\int_0^1\int_0^1 \langle Z(r,t), Q(r,s,t)\, Z(s,t) \rangle + \langle f(r,t), R(r,s,t)\, f(s,t) \rangle + \langle \lambda, [A(s)Z(s,t)+Bf(s,t)] \rangle \, dr\, ds \tag{7}$$

b. The Boundary Hamiltonians

$$H_f(t) = \frac{1}{2}\int_0^1 \langle Z(t_f:s), S_f(r,s) Z(t_f,r) \rangle \; ds \tag{7a}$$

$$H_0(t) = \frac{1}{2}\langle f_0(t), R_0(t)\, f_0(t) \rangle - \langle \lambda_0(0,t), [-E_0 Z(0,t)+B_0 f_0(t)] \rangle \tag{7b}$$

$$H_1(t) = \frac{1}{2}\langle f_1(t), R_1(t)\, f_1(t) \rangle - \langle \lambda_1(1,t), [-E_1 Z(1,t)+B_1 f_1(t)] \rangle \tag{7c}$$

2. OPTIMALITY CONDITIONS

The characterization of the extremal point in regular calculus caries over to the calculus of variations . In applications to the optimal control problems, this is known as the optimality conditions . This is parallel to the optimality conditions for the lumped parameter systems. The results are however different as we state next .

a. System Dynamics

$$\frac{\partial Z(t,s)}{\partial t} = \frac{\partial H(.)}{\partial \lambda(S,T)} = A(s)Z(s,t) + B\,f(s,t) \tag{8}$$

b. Controls Optimality Conditions

$$\frac{\partial H(.)}{\partial f(r,t)} = \int_0^1 R\,f(s,t)\,ds + B^T\lambda(r,t) = 0 \tag{9}$$

$$\frac{\partial H_0(t)}{\partial f_0(t)} = R_0(t)f_0(t) - B_0{}^T\lambda_0(0,t) = 0 \tag{10}$$

$$\frac{\partial H_1(t)}{\partial f_1(t)} = R_1(T)f_1(t) - B_1{}^T\lambda_1(1,t) = 0 \tag{11}$$

c. Adjoint Equation

$$\frac{\partial \lambda(r,t)}{\partial t} = -\frac{\partial H(.)}{\partial Z(s,t)} = \int_0^1 Q(r,s,t)\,Z(s,t)\,ds - A^T(s)\lambda(r,t) \tag{12}$$

where $\lambda(.,.)$ is the adjoint variable with the boundary condition:

$$\lambda(r,t_f) = \int_0^1 S_f(r,s)\,Z(s,t_f)\,ds \tag{13}$$

$$\text{i.e.} \quad \lambda(r,t_f) = \frac{\partial H_f(t)}{\partial Z(s,t_f)} \tag{13a}$$

Also we have:

$$D_0^T A_2^T \lambda(0,t) - A_1 \lambda(0,t) + A_2^T \frac{\partial \lambda(0,t)}{\partial s} = 0 \tag{14}$$

$$-D_1^T A_2^T \lambda(1,t) + A_1 \lambda(1,t) - A_2^T \frac{\partial \lambda(1,t)}{\partial s} = 0 \tag{15}$$

d. The Minimum Principle:

$$H\left[r,s,t:Z(s,t):f^*(s,t)\right] \le H\left[r,s,t:Z(s,t):f(s,t)\right] \tag{16}$$

From the necessary optimality conditions Eqs. (8) to (16), the following optimal controllers emerge:

e. Main or Interior Controller

$$f^*(s,t) = \int_0^1 \langle R^{-1}(r,s,t), B\lambda(s,t) \rangle \, ds \tag{17}$$

$$\text{where: } \int_0^1 \langle R^{-1}(r,s,t), R(r,s,t) \rangle \, dr = \delta(r-\rho) I \tag{18}$$

I = identity matrix

f. Boundary Controllers

$$f_0^*(t) = \langle R_0^{-1}(t), B_0 \, A_2 \lambda_0(0,t) \rangle \tag{19}$$

$$f_1^*(t) = -\langle R_1^{-1}(t), B_1 \, A_2 \lambda_1(1,t) \rangle \tag{20}$$

where the adjoint variable $\lambda(s,t)$ can be defined as :

$$\lambda(s,t) = \int_0^1 K(r,s,t) \, Z(s,t) \, ds \tag{21}$$

by Riccati transformation. $K(r,s,t)$ is the Riccati equation variable satisfying an infinite dimensional Riccati-like integro differential equation to be developed next.

3. INFINITE DIMENSIONAL RICCATI-LIKE INTEGRO-DIFFERENTIAL EQUATION

Substituting Eq. (21) into Eq. (12), we have:

$$\frac{\partial \lambda(s,t)}{\partial t} = -\left[\int_0^1 Q(r,s,t)\, Z(s,t) - A(s)\, K(r,s,t)\, Z(s,t)\, ds \right] \tag{22}$$

where we recall that A(s) is a linear spatial differential operator and we can express Eq. (22) as follows:

$$\frac{\partial \lambda(s,t)}{\partial t} = \left[\int_0^1 P(r,s,t)\, Z(s,t)\, ds \right] \tag{23}$$

where Eq. (23) is an integral operator with:

Kernel = P(r,s,t) = − Q(r,s,t) + A(s)K(r,s,t)

Using (i) Eq. (5) and (ii) integration by parts in Eq. (22) or (23) with some algebraic manipulations leads to the following infinite dimensional Riccati like integro - differential Equation :

$$\dot{K} = \frac{\partial K}{\partial t} = -K_{ss} A_2 - A_2{}^T K_{rr} + K_s A_1 + A_1{}^T K_r - K A_0 - A_0{}^T K \quad +$$

$$\int_0^1 \int_0^1 K(r,\sigma,t)\, B\, R^{-1}(\sigma,l,t)\, B^T\, K(l,s,t)\; d\sigma\, dl \quad +$$

$$K(r,l,t)\, A_2\, B_1\, R_1{}^{-1}\, B_1{}^T A_2\, K(l,s,t) + K(r,0,t)\, A_2\, B_0\, R_0{}^{-1} B_0{}^T A_2\, K(0,s,t) -$$

$$Q(r,s,t) \tag{24}$$

Similarly the boundary Riccati Equations are:

$$K_s(r,0,t)\, A_2 + K(r,0,t)\left[A_2\, E_0 - A_1 \right] = 0 \tag{25}$$

$$K_s(r,1,t)\, A_2 + K(r,1,t)\left[A_2\, E_1 - A_1 \right] = 0 \tag{26}$$

Where the subscripts "r,s,rr,ss" denote spatial derivatives with respect to the denoted variable. The terminal conditions are:

$$K(r,s,t_f) = K_F(r,s) \tag{27}$$

i.e. as $t \to \infty$, K(r,s) becomes a time constant matrix. Hence the infinite time (steady state) interior and boundary controllers become:

$$f^*(t) = \int_0^1 \int_0^1 R^{-1}(r,s,t)\, B^T\, K(\sigma,s,t)\, Z(t,\sigma)\, ds\, d\sigma \tag{28}$$

$$f_0^*(t) = R_0^{-1} B_0^T A_2^T \int_0^1 K_0(t)\, Z(s,t)\, ds \ \ (\text{at } s=0) \tag{29}$$

$$f_1^*(t) = -R_1^{-1} B_1^T A_2^T \int_0^1 K_1(t)\, Z(s,t)\, ds \ \ (\text{at } s=1) \tag{30}$$

where $s = 0$ and $s = 1$ are the boundary co-ordinates. Eqs. (17), (18) and (20) constitutes the infinite dimensional full order controllers while Eqs.. (28), (29) and (30) are the infinite-time full order controllers.

4. EXAMPLES

Example 1: Consider the class of mechanically flexible systems governed by the generalized wave equation of the form:

$$m(s)\frac{\partial^2 X(s,t)}{\partial t^2} + 2\xi A^{1/2}\frac{\partial X(s,t)}{\partial t} + A\, X(s,t) = f(s,t)$$

where ;

$X(s,t)$ = Infinite dimensional state of the flexible system.

$f(s,t)$ = Finite dimensional control force (m dimensional)

$m(s)$ = The mass density of the mechanically flexible system, possibly a function of the spartial location

ξ = Nonnegative damping coefficient of the system, a function of the material and the method of construction.

A = Linear time invariant symmetric nonnegative differential operator with a square – root $A^{1/2}$ and domain $D(A)$, a dense separable Hilbert space. $H = L^2(\Omega)$ with $(.,.)$ denoting the usual inner product and $\|\cdot\|$, the associated norm.

Assumptions:

1. The spectrum of A contains only isolated eigenvalues, λ_k and the corresponding eigenvectors $\xi_k(s) \in D(A)$ $\ni$:

   $0 \le \lambda_1 \le \lambda_2 \le \lambda_3 \le . . . \le \lambda_k \le \lambda_{k+1}$ . . . . and

   $A\, \xi_k(s) = \lambda_k\, \xi_k(s)$ also: $A^{1/2}\xi_k(s) = \lambda_k^{1/2}\xi_k(s)$

2. The eigenvectors $\xi_k(s)$ form a basis for the Hilbert space H. This is satisfied because, A has compact resolvent.

Problem: It is required to find an optimal control of the form $K(s,t)\,X(s,t)$ so that the system is driven to and kept at the zero state with minimum effort $f^*(s,t)$. This is equivalent to the regulator problem.

Solution: Let:

$$\begin{bmatrix} X_1(s,t) \\ \dot{X}_1(s,t) \end{bmatrix} \triangleq [\, Z(s,t) \,] \triangleq \begin{bmatrix} X_1(s,t) \\ X_2(s,t) \end{bmatrix} \triangleq \begin{bmatrix} Z_1(s,t) \\ Z_2(s,t) \end{bmatrix} \text{; Therefore, since:}$$

$$\frac{\partial^2 X(s,t)}{\partial t^2} = -\frac{2\xi A^{1/2}}{m(s)} \frac{\partial X(s,t)}{\partial t} - \frac{A}{m(s)} X(s,t) + \frac{f(s,t)}{m(s)}$$

Define: $\dfrac{f(s,t)}{m(s)} = F(s,t)$; $A^{1/2} = P$; and $\dfrac{2\xi A^{1/2}}{m(s)} = D$: then,

$$\begin{bmatrix} \dot{Z}_1(s,t) \\ \dot{Z}_2(s,t) \end{bmatrix} = \begin{bmatrix} 0 & I \\ -A & -D \end{bmatrix} \begin{bmatrix} Z_1(s,t) \\ Z_2(s,t) \end{bmatrix} + \begin{bmatrix} 0 \\ 1 \end{bmatrix} F(s,t) \quad ; \quad \text{i.e.}$$

$$\dot{Z}(s,t) = A(s) \quad Z(s,t) + B\,F(s,t) \tag{*}$$

Note that (*) is of the general model of Eq. (1). The performance index can be expressed as:

$$I(.) = \frac{1}{2}\int_0^1 \| Z(s,t_f) \|^2 S_f \, ds + \frac{1}{2}\int_0^{t_f}\int_0^1 \| Z(s,t) \|^2 Q + \| F(s,t) \|^2 R \; ds\, dt$$

Applying the results of the main section, noting that $A_2 = A_1 = 0$, $A_0 = A(s)$, $E_0 = E_1 = B_0 = B_1 = 0$,
The optimal feedback controller is given by:

$$F^*(s,t) = -\int_0^1 R^{-1} B^T \lambda(s,t)\, ds$$

where $\lambda(s,t)$ is defined by Eq. (21). The parameter $K(r,s,t)$ is a solution of the following Riccati – like integro differential Eq.

$$\dot{K}(r,s,t) = -KA - A^T K + \int_0^1\int_0^1 K(r,s,t) B R^{-1}(r,s,t) B^T K(r,s,t)\, dr\, ds - Q(r,s,t)$$

$$F^*(s,t) = -\int_0^1\int_0^1 R^{-1}(r,s,t) B^T K(r,s,t)\, Z(s,t)\; dr\, ds$$

Example 2

Consider a long, thin rod heated in a multizone furnace. The temperature distribution along the rod could be modeled for each zone by the following equation.

$$\rho C_p \frac{\partial T(s',t')}{\partial t} = k \frac{\partial^2 T(s',t')}{\partial s'^2} + q(s',t')$$

Initial conditions: $T(s',0) = T_0$

Boundary conditions: $\frac{\partial T(0,t')}{\partial s'} = \frac{\partial T(1,t')}{\partial s'} = 0$

We assume that the heat loss at the rod ends is negligible; where:

$T(s',t')$ = Temperature distribution for each zone. The state variable for the D.P.S.

$q(s',t')$ = Heat flux for different zones of the heated rod

ρ = Density of the rod

C_p = Thermal capacity of the rod

k = Thermal conductivity of the rod

Problem: It is desired to keep the temperature distribution to a desired value (T_d) while the heat flux is kept to a specified quantity (q_d) for each zone of the heated rod.

Solution: Define : $t = \frac{t'k}{\rho C_p}$, $f'(s,t) = \frac{q(s',t')}{kT_0}$ and $s = s'$.

Note: The length of the rod is taken to be unity. Let the state of the system $= \frac{T(s',t')}{T_0} = X(s,t)$. If we further define:

$X(s,t) - X_d(s,t) = Z(s,t)$; and $f'(s,t) - f_d(s,t) = f(s,t)$; then we can repose the problem as follows:

Dynamics:

$$\frac{\partial Z(s,t)}{\partial t} = \frac{\partial^2 Z(s,t)}{\partial s^2} + f(s,t)$$

Boundary conditions

$$\frac{\partial Z(0,t)}{\partial s} = \frac{\partial Z(1,t)}{\partial s} = 0$$

Performance Index :

$$I(.) = \frac{1}{2}\int_0^1 \| Z(s,t_f) \|^2 S_f \, ds + \frac{1}{2}\int_0^{t_f}\int_0^1 \| Z(s,t_f) \|^2 Q + \| f(s,t)\|^2 R \, ds\, dt$$

Above is a regulator problem and the following identities could be made:

$S_f(r,s) = S_f\delta(r-s) : Q(r,s,t) = Q\delta(r-s) : R(r,s,t) = R\delta(r-s) :$

$R^{-1}(r,s,t) = R^{-1}\delta(r-s) : A_2 = 1, \ A_1 = A_0 = 0 :$

$B = 1, \ E_0 = B_0 = E_1 = B_1 = 0$ Hence, using the results of this

section, the optimal controller is:

$f^*(s,t) = -\int_0^1 R^{-1} K(r,s,t) \ Z(s,t) \, ds$; where $K(r,s,t)$ is the solution

of the following equation:

$$\dot{K}(r,s,t) = -K_{ss} - K_{rr} + \int_0^1\int_0^1 K(r,s,t) R^{-1} K(r,s,t) \, dr\, ds - Q\,\delta(r-s)$$

with the boundary conditions:

$$\frac{\partial K(r,0,t)}{\partial s} = \frac{\partial K(r,1,t)}{\partial s} = \frac{\partial K(0,s,t)}{\partial r} = \frac{\partial K(1,s,t)}{\partial r} = 0$$

and $K(r,s,t_f) = K_f\delta(r-s)$. Hence the infinite time controller is given by : $f^*(s,t) = -K_f \ Z(s,t_f)$

C. REDUCED ORDER (FINITE DIMENSIONAL) CONTROLLER SYNTHESIS FOR DISTRIBUTED PARAMETER SYSTEMS [LINEAR OPTIMAL REGULATOR PROBLEMS]

This sub-section addresses the development of the reduced order controller for the D.P.S. linear optimal regulator problem. Developing the reduced order controller (R.O.C.) based on the reduced order model dynamics of the D.P.S. will be referred to as the direct method . The indirect method will do an order reduction on the infinite dimensional controller to obtain a similar result. So far, we have not addresses the problem of a reduced order model dynamics for the D.P.S. The fundamental issues in the reduced order modelling for D.P.S. are well covered in the literature [1] . For completeness, we shall present an example of reduced order modelling for a class of D.P.S. since the discussion of Section C. 2 will be based on this result .of . In Section C. 3 we use the results of section B.2.d to obtain a similar result .

1. REDUCED ORDER MODELING FOR A CLASS OF DISTRIBUTED PARAMETER SYSTEMS

Consider a class of mechanically flexible systems governed by the following partial differential equation:

Dynamics

$$m(s)\frac{\partial^2 z(s,t)}{\partial t^2} = -\rho\frac{\partial^4 z(s,t)}{\partial s^4} + F(s,t) \ ; \ s \in (0,1)\, , t \geq 0 \tag{31}$$

Boundary Conditions

$$z(0,t) = 0 \tag{32}$$

$$\frac{\partial z(0,t)}{\partial s} = 0 \tag{33}$$

$$\frac{\partial^2 z(1,t)}{\partial s^2} = 0 \tag{34}$$

$$\frac{\partial^3 z(1,t)}{\partial s^3} = 0 \tag{35}$$

z(s,t) = the deflection of the flexible system at time t and space s.
m(s) = mass density of the system in unit of mass per unit length
ρ = flexural rigidity of the system (assumed uniform)
F(s,t) = the control input to the system at time t and space s

$$\text{Define: } z_1(s,t) = \frac{\partial z(s,t)}{\partial t} \; ; \quad z_2(s,t) = \frac{\partial^2 z(s,t)}{\partial s^2}; \quad Z(s,t) = \begin{bmatrix} z_1(s,t) \\ z_2(s,t) \end{bmatrix};$$

then :

$$\frac{\partial z_1(s,t)}{\partial t} = \frac{\partial^2 z(s,t)}{\partial t^2} = -\frac{\partial^2 z_2(s,t)}{\partial s^2} + f(s,t) = -\frac{\partial^4 z(s,t)}{\partial s^4} + f(s,t)$$

also :

$$\frac{\partial z_2(s,t)}{\partial t} = \frac{\partial^3 z(s,t)}{\partial t \partial s^2} = \frac{\partial^2 z_1(s,t)}{\partial s^2}; \text{ using these in Eqs. (31) – (35),}$$

we have :

$$m(s)\frac{\partial z_1(s,t)}{\partial t} = -\rho\frac{\partial^2 z_2(s,t)}{\partial s^2} + F(s,t) \tag{36}$$

$$m(s)\frac{\partial z_2(s,t)}{\partial t} = \rho\frac{\partial^2 z_1(s,t)}{\partial s^2} \tag{37}$$

hence ;

$$m(s)\begin{bmatrix} \dot{Z}_1(s,t) \\ \dot{Z}_2(s,t) \end{bmatrix} = \rho\begin{bmatrix} 0 & -\frac{\partial^2}{\partial s^2} \\ \frac{\partial^2}{\partial s^2} & 0 \end{bmatrix}\begin{bmatrix} z_1(s,t) \\ z_2(s,t) \end{bmatrix} + \begin{bmatrix} 1 \\ 0 \end{bmatrix} F(s,t) \tag{38}$$

$$\text{Defining: } \frac{\rho}{m(s)}\begin{bmatrix} 0 & -\frac{\partial^2}{\partial s^2} \\ \frac{\partial^2}{\partial s^2} & 0 \end{bmatrix} = A(s); \quad \frac{F(s,t)}{m(s)} = f(s,t) \tag{39}$$

$$\dot{Z}(s,t) = A(s)\, Z(s,t) + B\, f(s,t) \tag{40}$$

with the following boundary conditions (analog of Eqs. (32) – (35));

$$z_1(0,t) = 0 \tag{32a}$$

$$\frac{\partial z_1(0,t)}{\partial s} = 0 \tag{33a}$$

$$z_2(1,t) \quad = 0 \quad = \frac{\partial^2 z_1(1,t)}{\partial s^2} \tag{34a}$$

$$\frac{\partial z_2(1,t)}{\partial s} = 0 \quad = \frac{\partial^3 z_1(1,t)}{\partial s^3} \tag{35a}$$

Equation (40) is identical with Eq. (1) variable for variable . We are in search of an ordinary differential equation (ODE) that would approximate the partial differential equation (31 -35) . We use the finite difference method . If we partition the domain [0,1] into (N - 1) equal intervals of length Δs having N nodes, we describe the state of the system by giving the values of (z_1,z_2) at the nodes ($0=s_1$, s_2 ,s_3 , . . . $s_N=1$) labeled as (1, 2, 3, 4, 5, . . . N). Hence, the states are described by 2N variables . $[\, z_1(k,t) \;\; z_2(k,t)\,]^T$ for k = 1, 2, 3, . . . N as functions of time only . Now we assume the partitions are small enough to make the state values within them spatially constant. Using the central difference technique on the dynamics equation

$$\frac{\partial z_1(s,t)}{\partial t} = -\frac{\rho}{m(s)}\frac{\partial^2 z_2(s,t)}{\partial s^2} + f(s,t) = -\frac{\rho}{m(s)}\frac{\partial^4 z(s,t)}{\partial s^4} + f(s,t)$$

and

$$\frac{\partial z_2(s,t)}{\partial t} = \frac{\partial^2 z_1(s,t)}{\partial s^2}$$, we obtain a set of 2N ODEs as follows ;

$$\frac{dz_1(k,t)}{dt} = \frac{\rho}{m(k)}\frac{-[z_2(k+1,t)-2z_2(k,t)+z_2(k-1,t)]}{[\Delta s]^2} + f(k,t) \tag{41}$$

$$\frac{dz_2(k,t)}{dt} = \frac{\rho}{m(k)}\frac{[z_1(k+1,t)-2z_1(k,t)+z_1(k-1,t)]}{[\Delta s]^2} \tag{42}$$

for $k = 2,3,\ldots N-1$; This yields $2(N-2) = 2N-4$ equations .

To complete the set of 2N ordinary differential equations , the other 4 equations are obtained from the boundary conditions for $z_1(1,t), z_2(2,t)$, $z_1(N,t)$ and $z_2(N,t)$ using Eqs. (32) – (35). We have:

$$\frac{dz_1(1,t)}{dt} = 0 \tag{43}$$

$$\frac{dz_2(N,t)}{dt} = 0 \tag{44}$$

The boundary conditions for $z_2(1,t)$ and $z_1(N,t)$ would require using Eqs. (31), (34) and (35) ; a use of the Taylor series expansion of $z_2(.,t)$ and interpolations and extrapolations as follows :

Equation (36) at node N yields;

$$\frac{dz_1(N,t)}{dt} = -\frac{\rho}{m(s)} \left.\frac{\partial^2 z_2(s,t)}{\partial s^2}\right|_{s=N} + F(N,t) \tag{45}$$

We also know, by Taylor series, that:

$$z_2(N-1,t) = z_2(N,t) - \Delta s\left[\frac{\partial z_2(N,t)}{\partial s}\right] + \frac{[\Delta s]^2}{2}\left[\frac{\partial^2 z_2(N,t)}{\partial s^2}\right] + \ldots + \text{HOT}$$

(higher order terms) (46)

Now, using Eqs. (34) and (35), the first 2 terms on the right hand side of Eq. (46) vanish, and we have:

$$z_2(N-1,t) = \frac{[\Delta s]^2}{2}\left[\frac{\partial^2 z_2(N,t)}{\partial s^2}\right] + \ldots + \text{HOT} \tag{47}$$

i.e.

$$\frac{\partial^2 z_2(N,t)}{\partial s^2} = \frac{2}{[\Delta s]^2}\left[z_2(N-1,t)\right] + \ldots + \text{HOT} \tag{47a}$$

as $\Delta s \to 0$ the HOT vanish, and using Eq. (47a) in (45), we have:

$$\frac{dz_1(N,t)}{dt} = -\frac{\rho}{m(N)}\left[\frac{2}{[\Delta s]^2}\left[z_2(N-1,t)\right]\right] + F(N,t) \tag{48}$$

Invoking the Taylor series again, we have;

$$z_1(2,t) = z_1(1,t) + \Delta s\left[\frac{\partial z_1(1,t)}{\partial s}\right] + \frac{[\Delta s]^2}{2}\left[\frac{\partial^2 z_1(1,t)}{\partial s^2}\right] + \text{HOT} \tag{49}$$

Neglecting $\text{HOT}|_{\text{as } \Delta s \to 0}$, and using $\left[\frac{\partial z_2(s,t)}{\partial t} = \frac{\partial^2 z_1(s,t)}{\partial s^2}\right]$, we have:

$$\frac{\partial z_2(1,t)}{\partial t} = \frac{\partial^2 z_1(1,t)}{\partial s^2} = \frac{2}{[\Delta s]^2}\left[z_1(2,t) - z_1(1,t) - \Delta s\left[\frac{\partial z_1(1,t)}{\partial s}\right]\right]$$

Now, using Eqs. (32a) and (33a) we have:

$$\frac{\partial z_2(1,t)}{\partial t} = \frac{2}{[\Delta s]^2}\,[z_1(2,t)] \qquad (50)$$

Hence, the complete set of 2N ordinary differential equations are:

$$\frac{\partial z_1(1,t)}{\partial t} = 0 \qquad (51)$$

$$\frac{dz_1(k,t)}{dt} = \frac{\rho}{m(k)}\,\frac{-[z_2(k+1,t)-2z_2(k,t)+z_2(k-1,t)]}{[\Delta s]^2} + f(k,t) \qquad (41)$$

for $k = 2, 3, 4, \ldots N-1$;

$$\frac{dz_1(N,t)}{dt} = -\frac{\rho}{m(N)}\left[\frac{2}{[\Delta s]^2}[z_2(N-1,t)]\right] + F(N,t) \qquad (48)$$

$$\frac{dz_2(1,t)}{dt} = \frac{2}{[\Delta s]^2}\,[z_1(2,t)] \qquad (50)$$

$$\frac{dz_2(k,t)}{dt} = \frac{\rho}{m(k)}\,\frac{[z_1(k+1,t)-2z_1(k,t)+z_1(k-1,t)]}{[\Delta s]^2} \qquad (42)$$

for $k = 2, 3, 4, \ldots N-1$; and $\dfrac{dz_2(N,t)}{dt} = 0$ (52)

yielding a compact vector ordinary differential equation (53), where:

$$\dot{Z}(t) = A\,Z(t) + B\,u(t) \qquad (53)$$

$Z(t)$ = Finite dimensional state of the reduced order system and the dot (.) here implies total differential.

A = Constant element matrix representing the dynamics of the finite dimensional system.

B = Identically equal to the B (the control vector of the infinite dimensional system

$u(t)$ = The control vector, now a function of time only, identically equal to $f(t)$ in Eq. (41).

2. DIRECT METHOD

We will now synthesize the reduced order controller for the infinite dimensional system based on the approximate model - the reduced order model of Eq. (53) .

$$\dot{Z}(t) = \frac{dZ(t)}{dt} = AZ(t) + Bu(t) \; ; \qquad Z(t_0) = Z_0 \tag{53}$$

and the performance Index $J_L(.)$ (subscript 'L' denotes lumped), where:

$$J_L(.) = \frac{1}{2} \|Z(t_f)\|^2 S_f + \frac{1}{2} \int_0^{t_f} \|Z(t)\|^2 Q + \|u(t)\|^2 R \, dt \tag{54}$$

where all the parameters are as defined in subsection 1 above and t_f denotes the terminal time. The weighting matrices Q, R, and S_f are now constant element matrices, satisfying the following conditions;

(i) S_f and Q(t) are symmetric semipositive definite matrices.

(ii) R(t) is a symmetric positive definite matrix.

Note that the boundary terms in the performance index $J_D(.)$ have been absorbed in B u(t) vector via the R.O.M. process. Now, we go through exactly the same approach as we did in Section B for synthesizing the infinite dimensional controller. This time, we would be synthesizing the finite dimensional controller.

The Main and Only Hamiltonian:

$$H(.) = \frac{1}{2}\left\{ \|Z(t)\|^2 Q + \|u(t)\|^2 R \right\} + \lambda(t)\left\{ AZ(t) + Bu(t) \right\} \tag{55}$$

Using the minimum principle, the set of optimality conditions are;

System Dynamics:

$$\frac{\partial H(.)}{\partial \lambda(t)} = \dot{Z}(t) = AZ(t) + Bu(t) \tag{56}$$

Adjoint Equation

$$\frac{\partial \lambda(t)}{\partial t} = -\frac{\partial H(.)}{\partial Z(t)} = -QZ(t) - A^T \lambda(t) \tag{57}$$

Its Boundary Condition:

$$\lambda(t_f) = S_f Z(t) \tag{58}$$

Extremum Condition:

$$\frac{\partial H(.)}{\partial u(t)} = 0 = R\,u(t) + \lambda(t)B \tag{59}$$

Minimum Principle:

$$H(t,u^*(t),Z(t)) \le H(t,u(t),Z(t)) \tag{60}$$

The following provides the optimal control:

$$u^*(t) = -R^{-1} B^T \lambda(t) \tag{61}$$

where

$$\lambda(t) = P(t)\,Z(t) \tag{62}$$

and P(t) is the solution of the following finite dimensional Riccati differential equation:

$$\dot{P}(t) = -P(t)A - A^T P(t) + P(t)\,B^T R^{-1} B\,P(t) - Q \tag{63}$$

Hence by substitution:

$$u^*(t) = -R^{-1} B^T P(t)\,Z(t) \tag{64}$$

$$u^*(t) = K(t)\,Z(t) \tag{65}$$

where:

$$K(t) = -R^{-1} B^T P(t) \tag{66}$$

Eq. (66) is a feedback control law (matrix)

Infinite time (Steady state) Controller

As time $t \to \infty$, P(t) becomes a constant matrix, call it P. Then the steady state controller is given by:

$$u^*(t_f) = K_f\,Z(t) \tag{67}$$

where K_f is a constant element matrix. Eq. (67) is a constant linear state feedback controller.

Example 3.

Recall the finite dimensional model of Eq. (53), repeated below.

$$\dot{Z}(t) = \frac{dZ(t)}{dt} = A\,Z(t) + B\,u(t) \tag{***}$$

an Mth order linear time invariant lumped parameter system. The model has been the basis of the finite controller synthesis for the regulator problem of this section, hence a typical example.

3. INDIRECT METHOD

Recall that the infinite dimensional controller (finite time) is:

$$f^*(s,t) = \int_0^1 \int_0^1 R^{-1}(r,s,t)\, B^T K(r,s,t)\, Z(s,t)\ dr\, ds \tag{28}$$

where we noted that R(.), S(.) and Z(.) are all continuous functions of their arguments. Otherwise there exists the possibility of unboundedness, which we have excluded in the interest of practical applications and also

observations. Hence the integral in Eq. (28) can be approximated by finite or infinite summation as follows.

$$f^*(t,k) = \sum_{k=1}^{\infty} \sum_{k=1}^{\infty} R^{-1}(j,k,t)\ B^T K(j,k,t)\, Z(k,t) \tag{68}$$

$$\text{Define: } \left.\begin{aligned} f^*(t,k) &= f_k^*(t) \quad K(j,k,t) = K_{jk}(t) \\ R(j,k,t) &= R_{jk}(t) \quad Z(k,t) = Z_k(t) \end{aligned}\right\} \tag{69}$$

i.e. discretizing in the two spartial dimensions, Eq. (68) becomes:

$$f_k^*(t) = \sum_{k=1}^{\infty} \sum_{j=1}^{\infty} R_{jk}^{-1}(t)\, B^T K_{jk}(t)\, Z_k(t) \tag{70}$$

All variables in Eq. (70) are now functions of time only.

Assumptions:

1. If we let our spartial dimension be finite (i.e. $l < \infty$), and also note that in practice the spartial segments into which l is further divided is finite and > 0, then the number of such segments is finite and this places an upper bound on the number of summations. Call it N and M for the dimensions r and s.

2. Over the range of the spartial segments, the functions K(j,k,t), R(j,k,t) and Z(k,t) are only timevariant but are spartially constant. Hence we have:

$$f_k^*(t) = \sum_{k=1}^{N} \sum_{j=1}^{M} R_{jk}^{-1}(t)\, B^T K_{jk}(t)\, Z_k(t) \tag{71}$$

for $j=1,2,3,\ldots M;\ K=1,2,3,\ldots N$

$$f^*(t) = \left[f_1^*(t)\, f_2^*(t)\, f_3^*(t) \cdots f_N^*(t) \right]^T \tag{72}$$

Clearly for small M and N the number of computational operations is large for hand calculations. On the other hand, we have limited memory digital computer to help in the calculations. Hence the frequent need to approximate. For example for $M=10, N=8$ we have the following computations to do.

$3 \times 8 = 24$ matrix multiplications for each control element $f_i(t)$.

$10-1=9$ matrix additions. This implies:
$3 \times M \times N$ matrix multiplications and

$(M-1) \times N$ matrix additions for the entire control to be fully specified

For $M=10, N=8$: 240 matrix multiplications and

72 matrix additions are required. This almost prohibits hand computation. In the next subsection, a more systematic and computationally adaptive algorithm for real time computation of the control for computer control of the D.P.S. will be developed.

4 COMPARISON OF FINITE AND INFINITE DIMENSIONAL CONTROLLER FOR DISTRIBUTED PARAMETER SYSTEMS

(1) ***METHOD*** :

By the method used in synthesizing the controllers, the infinite dimensional controller (I.D.C.) is direct, and therefore more accurate, and more truely represents the controller for the D.P.S.

(2) ***STRUCTURE:***

In part the infinite dimensional controller is obtained by spatial integration. This explains the fact that the synthesis of the I.D.C. is spatial updated because of the spatial dependency of the D.P.S. . The finite dimensional controller (F.D.C.) assumes that the system parameter is constant in space which is the source of the difference between the I.D.C. and F.D.C. also , the source of the approximation of the F.D.C.

(3) ***IMPLEMENTATION:***

In practice the I.D.C. cannot be implemented in real time or on a limited memory digital computer. The F.D.C. has the advantage of being implemented in real time and on a limited memory digital computer .

III

OPTIMAL CONTROL IMPLEMENTATION ALGORITHMS

Most of the analytical and numerical techniques available for computing the optimal policy in finite dimensional control theory [e.g. , calculus of variations, Pontryagin minimum principle, Hamilton-Jacobi's theory, the gradient method etc.] are very applicable to the infinite dimensional control theory problems. However, what differs tremendously is the problem of dimension. While the advent of digital computers calls for the digital version of the optimal policy for both cases, the problem of dimension still lingers in the infinite dimensional problem.

The optimal linear regulator theory currently referred to as the linear quadratic gaussian (L. Q. G.) problem has become one of the most acceptable methods. This is so in part because the continuous -time dynamic control law needs be implemented in a digital version. Both the continuous and the discrete version of the problem provide us with a rigorous tool for developing the linear state-variable feedback control law. Never-the-less, occasions arise in practical problems demanding that the physical control law be implemented in a digital version. This is the case discussed in this chapter. Typical examples are : aircraft flight control applications with an on-board digital computer; military aircraft with an on-board digital computer on

active duty with the additional constraint of on-line control missions ; space-craft on a mission requiring on-line control problems to be solved.
The problem faced by these various control systems is to first find the digital control law algorithms that also achieve the continuous-design specifications. One approach in the context of the L.Q.G. is the optimal sampled-data regulator (O.S.R.) problems, followed by the implementation algorithms.

The first face of the problem is common to both the finite and infinite dimensional systems theory following the approach of Leondes and Salami [16]. The implementation phase creates a unique problem for the infinite dimensional systems (D.P.S.). This section extends the work of Leondes and Salami [16] to the infinite dimensional control problems and as may be envisaged, the extension is not trivial.

A. APPROACH

The O.S.R. problem is transformed into the discrete L.Q.G. (D.L.Q.G.) problem. The problem is then solved using the standard L.Q.G. solution algorithms. Originally the weighting matrices for the continuous O.S.R. . performance index are chosen to achieve the continuous design specifications. These are transformed into the weighting matrices for the discrete L.Q.G. problem computation. The transformation produces extra control-state cross terms in the digital performance index and off-diagonal entries in the weighting matrices for the other terms which have the effect of weighting the control and the state variables at points within the sampling intervals as well as the sampling instances. In summary, the O.S.R. approach provides a means whereby the designer's intuitive understanding of the continuous-time design problem, plus the L.Q.G. techniques aid in the determination of a digital feedback control laws which satisfies the continuous-time performance specifications.

Two methods are employed in achieving the algorithms for the computation of the weighting matrices for the D.P.S. model:

(a) The ROM for the D.P.S. as developed in Section II.A.1 (Eq. 53), i.e.

$$\frac{dZ(t)}{dt} = \dot{Z}(t) = A\,Z(t) + B\,u(t) \tag{53}$$

is used as the basis for developing the set of the weighting matrices required in implementing the O.S.R. methodology . This method is referred to as the direct method.

(b) The original model of the D.P.S. i.e.

$$\frac{\partial Z(s,t)}{\partial t} = \dot{Z}(s,t) = A(s)\,Z(s,t) + B\,f(s,t) \tag{1}$$

is used next as the basis for developing a similar set of weighting matrices for the same purpose . This is referred to as the indirect method.
The classes of D.P.S.. considered are the **hyperbolic** and the **parabolic** systems. Unbounded systems operators are encountered. Numerical technique is used for the representation of the corresponding semigroup in a form that has good convergence properties for the weighting matrices synthesis algorithms that followed. The methods of the algorithms in (a) are then applicable.

The major problem faced in using the O.S.R. approach is that the transformation equations defining the digital weighting matrices are complicated functions of the continuous system semigroup operator T(t) and other operators of the continuous formulation. In fact, the computation of these weighting matrices for systems with order higher than 3 are very discouraging both computationally and also to the O.S.R. methodology. A typical example is presented (computation carried out on an SR 52 TI programmable calculator) to demonstrate this point.

Fortunately, this problem is drastically reduced for the case of linear time-invariant systems that we shall be dealing with. More than that, the Analytic properties of the weighting matrices functions and the convergence properties of the series expansion are used to an advantage in deriving an infinite series representation for the same function. A digital computer implementation technique is finally recommended for the implementation of the algorithms on an on-line basis.

B. PROBLEM STATEMENT

Given the following control problem for an infinite dimensional controllable D.P.S. model (Ref. Section II . A), The essential part of the problem is repeated here for easy reference .

System Dynamics:

$$\dot{Z}(t,s) = A(s)\,Z(t,s) + B\,f(t,s) \;;\; \text{for} \;\; 0 \langle\, t_0 \,\langle\, T \tag{1}$$

Initial Conditions: $Z(0,s) = Z_0(s)$ (2)

Boundary Conditions:

$$\frac{\partial Z(t,s)}{\partial s} = -E_0\, Z(t,s) + B_0\, f_0(t) \;;\quad \text{for } s = 0 \tag{3a}$$

$$\frac{\partial Z(t,s)}{\partial s} = -E_1\, Z(t,s) + B_1\, f_1(t) \;;\quad \text{for } s = 1 \tag{3b}$$

Performance Index:

$$J_D(.) = \frac{1}{2}\int_0^1\int_0^1 \langle\, Z(t_f : r)\,, S_f(r,s)Z(t_f : s)\,\rangle\, dr\, ds \quad +$$

$$\frac{1}{2}\int_0^{t_f}\int_0^1\int_0^1 \langle\, Z(t : r)\,, Q(r,s)Z(t : s)\,\rangle + \langle\, f(t,s)\,, R(t{:}r,s)f(t,r)\,\rangle\, dr\, ds\, dt$$

$$+ \frac{1}{2}\int_0^{t_f} \langle\, f_0(t)\,, R_0(t)f_0(t)\,\rangle + \langle\, f_1(t)\,, R_1(t)\, f_1(t)\,\rangle\, dt \tag{4}$$

REMARK: These boundary conditions are relevant to the parabolic systems, but the results that follow can as well be applied to other classes of D.P.S., hyperbolic systems in particular.

1. PROBLEM

(a) Derive a digital version of Eq. (4) satisfying Eqs (1) (2) and (3)

(b) Develop a numerically attractive algorithm (simple to implement) for computing the set of discrete -time weighting matrices obtained in A. for implementing the linear optimal regulator problem on a limited memory digital computer :

(i) Based on the reduced order model of Eq. (53)
(ii) Based on the full order model of Eq. (1)

C. REDUCED ORDER WEIGHTING MATRICES SYNTHESIS [L. Q. G.] - DIRECT METHOD

The approach to solving the problem A. 1.(a) is what we refer to as the direct method. The reduced order model assumption transforms the above problem to the following :

Given the following finite dimensional system and the associated performance index (Eqs. (53) and (54))

$$\dot{Z}(t) = \frac{dZ(t)}{dt} = A\,Z(t) + B\,u(t) \;; \qquad Z(t_0) = Z_0 \tag{53}$$

and the performance index $J_L(.)$ (subscript " L " denotes lumped) where:

$$J_L(.) = \frac{1}{2}\|Z(t_f)\|^2 S_f + \frac{1}{2}\int_0^{t_f} \|Z(t)\|^2 Q + \|u(t)\|^2 R\,dt \tag{54}$$

with all the parameters as defined previously in Section II C.2 and also note that ;

$Z(t)$ = N dimensional state vector, with $Z(t) \in R^N$ the state space

$u(t)$ = M dimensional control vector, with $u(t) \in U$ -the control space

A = N x N Time-invariant matrix of the closed bounded linear time-invariant operator. The state matrix A , approximately describing the dynamics of the D. P. S.

B = N x M time-invariant control matrix.

Z_0 = state initial condition (73)

Instead of the continuous-time control u(t) , the proposition is to apply the discrete-time control u(k), for k = 0, 1, 2, . . . N - 1 to the system of Eq. (53) at regular intervals t_k, k = 0, 1, 2, . . . N - 1 where u(k) k = 0,1, 2, . . . N - 1 are assumed constant over the sub-intervals $t_k - t_{k-1}$, k = 0,1, 2, . . . N - 1 with t_0 = 0 (without any loss of generality since we are dealing with time-invariant system).

Let $t_k = k$ and $t_N = t_f$ (74)

The application of the control u(k) to the system of Eqs. (53) and (54) leads to the optimal sampled-data linear regulator problem (O. S. R.) and the problems posed above can now be tackled . The following technical facts would be of value in the sequel.

D. SOME FOUNDATION THEOREMS AND DEFINITIONS

Definition 1

Let B be a Banach space. A one parameter family of bounded operator **T**(t) : $0 < t < \infty$. from B into B is called a semigroup of bounded linear operator on B if :

(i) $T(0) = I$ (The identity operator on B) (75)

(ii) $T(t_1 + t_2) = T(t_1)\,T(t_2)$ for every $t_1, t_2 > 0$ (76)

Eq. (76) is the semigroup property. A semigroup of bounded linear operator T(t) is uniformly continuous if :

(iii) $\lim_{t \downarrow 0} \| T(t) - I \| = 0$ (77)

(iv) The Domain of the Linear Operator A defined by :

$$D(A) = \left\{ x \in B : \underset{t \downarrow 0}{\text{Lim.}} \left[\frac{T(t)x - x}{t} \right] \text{ exist} \right\} \tag{78}$$

$$Ax = \underset{t \downarrow 0}{\text{Lim.}} \left[\frac{T(t)x - x}{t} \right] = \left. \frac{d^{\pm} T(t)x}{dt} \right|_{t=0} \quad ; \forall \quad x \in D(A). \tag{79}$$

A is the infinitesimal generator of the Semi – group $T(t)$. The same family defined over the entire $\Re$ space $(-\infty < t < \infty)$ is called a group.

(i) $T(0) = I$ (The identity operator on B) (75)

(ii) $T(t_1 + t_2) = T(t_1)\,T(t_2)$ for every $t_1, t_2 > 0$ (76)

Eq. (76) is the semigroup property. A semigroup of bounded linear operator $T(t)$ is uniformly continuous if :

(iii) $\underset{t \downarrow 0}{\text{Lim.}} \, \| T(t) - I \| = 0$ (77)

(iv) The Domain of the Linear Operator A defined by :

$$D(A) = \left\{ x \in B : \underset{t \downarrow 0}{\text{Lim.}} \left[\frac{T(t)x - x}{t} \right] \text{ exist} \right\} \tag{78}$$

$$Ax = \underset{t \downarrow 0}{\text{Lim.}} \left[\frac{T(t)x - x}{t} \right] = \left. \frac{d^{\pm} T(t)x}{dt} \right|_{t=0} \quad ; \forall \quad x \in D(A). \tag{79}$$

A is the infinitesimal generator of the Semi – group $T(t)$. The same family defined over the entire $\Re$ space $(-\infty < t < \infty)$ is called a group.

Example 4.

Given an autonomous linear system :

$$\frac{dZ(t)}{dt} = AZ(t) \tag{80}$$

$$Z(0) = Z_0 \tag{81}$$

The state evolution of the system $Z(t)$ can be expressed as :

$$Z(t) = e^{A(t-0)} Z_0 = S(t)\, Z_0 \tag{82}$$

The operator $\mathbf{S}(t)$ weights the initial condition continuously in time to give the value of the state at any given future time t . The operator $\mathbf{S}(t)$ is the semigroup (denoted as $\mathbf{T}(t)$ above)

Theorem 1

A linear operator A is the infinitesimal generator of a uniformly continuous semigroup **T**(t) , if and only if A is a bounded linear operator.

Proof: **(Sufficiency)**

Let **A** be a bounded linear operator on B , and set :

$$T(t) = e^{At} = e^{tA} = \sum_{k=0}^{\infty} \frac{(tA)^k}{k!} \tag{83}$$

From the properties of the exponential series, we know that Eq. (83) conv −erges in the norm for $t > 0$ and therefore defines ($\forall$ t) a bounded linear operator T(t). Since $T(0) = I$; and

$$T(t_1 + t_2) = T(t_1)\, T(t_2) \tag{84}$$

Estimating the power series in Eqn. (83) gives: $T(t) = e^{At}$, and :

$$T(t) - I = e^{At} - T(0) = tAe^{At} \tag{85}$$

Taking norm in Eq. (85) yields:

$$\|T(t) - I\| \le t\,\|A\|\, e^{t\|A\|} \tag{86}$$

$$\left\|\frac{T(t) - I}{t} - A\right\| \le \left\|\, \|A\|\, e^{t\|A\|} - \|A\|\right\| = \|A\| \left\|e^{t\|A\|} - I\right\| = \|A\|\,\|T(t) - I\|$$

i.e.

$$\left\|\frac{T(t) - I}{t} - A\right\| \le \|A\|\,\|T(t) - I\| \tag{87}$$

This implies that **T(t)** is a uniformly continuous semigroup of bounded linear operators on B and that A is its infinitesimal generator.

Necessity Let **T(t)** be a uniformly continuous semigroup of bounded linear operators on **B**.

Fix $\rho \rangle 0$ small enough $\ni$; $\left\| I - \frac{1}{\rho} \int_0^\rho T(s)\, ds \right\| \langle 1$ (88)

$$\text{This} \Rightarrow \left[\frac{1}{\rho} \int_0^\rho T(s)\, ds \right]^{-1} \text{exists; i.e. } \frac{1}{\rho} \int_0^\rho T(s)\, ds \rangle 0 \tag{89}$$

$$\text{But}: \frac{1}{h}[T(h) - I] \int_0^\rho T(s)\, ds = \frac{1}{h}\left[\int_0^\rho T(s+h)\, ds - \int_0^\rho T(s)\, ds \right]$$

$$= \frac{1}{h}\left[\int_0^{\rho+h} T(s)\, ds - \int_0^h T(s)\, ds \right] \tag{90}$$

Hence :

$$\frac{1}{h}[T(h) - I] = \left[\frac{1}{h} \int_0^{\rho+h} T(s)\, ds - \frac{1}{h} \int_0^h T(s)\, ds \right]\left[\int_0^\rho T(s)\, ds \right]^{-1} \tag{91}$$

if $h \downarrow 0$ in Eq. (91), then $\frac{1}{h}[T(h) - I]$ converges in the norm and therefore converges strongly to the bounded operator:

$[T(\rho) - I]\left[\int_0^\rho T(s)\, ds \right]^{-1}$ which is the infinitesimal generator of the semigroup T(t). Following is a theorem on the uniqueness of T(t).

Theorem 2 **(Uniqueness)**

Suppose **T**(t) and **S**(t) are uniformly continuous semigroups of bounded linear operator A , then the following holds :

$$\lim_{t\downarrow 0} \frac{T(t) - I}{t} = A = \lim_{t\downarrow 0} \frac{S(t) - I}{t} \tag{92}$$

and :

$$T(t) = S(t) \quad \text{for} \quad t \geq 0 \tag{93}$$

The following corollary too will be of importance in our pursuit.

Corollary 1

Let **T**(t) be a uniformly continuous semigroup of bounded linear operators.

(i) $\exists$ a constant $k \ni : \| T(t) \| \leq e^{kt}$ (94)

(ii) There is a unique bounded linear operator $A \ni$:

$$T(t) = e^{tA} = e^{At}$$

(iii) The operator A in (ii) is the infinitesimal generator of T(t)

(iv) The operator : $t \rightarrow T(t)$ is differentiable in the norm and in particular :

$$\frac{dT(t)}{dt} = A\,T(t) = T(t)\,A \tag{95}$$

i.e. the analytic property of the semigroup T(t).

Proof

(i) , (iii) , and (iv) follow from (ii) . To prove (ii) , therefore, we note that the infinitesimal generator of **T**(t) is a bounded linear operator. Call it **A** Since **A** is also the infinitesimal generator of e^{tA} (defined by Eq. (83)) , then by Theorem 2: $\mathbf{T}(t) = e^{tA}$. Q.E.D.

Now we can resume the development of the algorithms.

E. ALGORITHMS FOR WEIGHTING MATRICES SYNTHESIS

The solution of Eq. (53) can be expressed as:

$$Z(t) = T(t,t_0)Z_0 + \int_{t_0}^{t_f} [\,T(t,\tau)\,B\,u(\tau)\,]\,d\tau \tag{96}$$

$$= T(t,0)Z_0 + \int_0^{t_f} [\,T(t,\tau)\,B\,u(\tau)\,]\,d\tau \tag{97}$$

for $t_0 = 0$: Using Eq. (97) in Eq. (54), we have:

$$J_L(.) = \tfrac{1}{2}\langle Z(t_f), Z(t_f)\rangle S + \tfrac{1}{2}\int_0^{t_f} \left\{ \left\langle T(t)Z_0 + \int_0^t T(t,\tau)\,B u(\tau)\,d\tau , \right.\right.$$

$$\left.\left. T(t)Z_0 + \int_0^t T(t,\tau)\,B\,u(\tau)\,d\tau \right\rangle Q + \langle u(t), u(t)\rangle \right\} dt \tag{98}$$

Here, we are dealing with continuous functions. Since u_k, $k = 0, 1, 2, \ldots N-1$ serves as the discrete control for stage-to-stage control problem and u_k is constant over the interval $t_k - t_{k-1}$, for the instance k the time is t_k and the outer integral can be replaced with the summation operation. Doing this and collecting terms, we have:

$$J_L(.) = \|Z(N)\|^2 S + \sum_{k=0}^{N-1} \|Z_k\|^2 P_k + \langle Z_k , u_k \rangle W_k + \|u_k\|^2 R_k \tag{99}$$

The factor 1/2 has been dropped without any loss of generality. Moreover,

$$P_k = \|T(t,t_k)\|^2 Q = \langle T(t,t_k), T(t,t_k) \rangle Q \tag{100}$$

$$W_k = 2 \langle T(t,t_k), Q \int_{t_k}^{t} T(t,t_k)\, G\, d\tau \rangle \tag{101}$$

$$R_k = R + \langle H(\tau,t_k), Q H(\tau,t_k) \rangle \tag{102}$$

$$H(t,t_k) = \int_{t_k}^{t} T(t,\tau)\, d\tau\, B = H^T(t,t_k) \tag{103}$$

If we further define $\Delta t_k \triangleq t_{k+1} - t_k$ (104)

then Eqs. (100) – (103) can be restated as :

$$P_k = P(\Delta t_k) = \int_0^{\Delta t_k} \langle T(\tau), Q T(\tau) \rangle\, d\tau \tag{105}$$

$$\tfrac{1}{2} W_k = \tfrac{1}{2} W(\Delta t_k) = \int_0^{\Delta t_k} \langle T(\tau), Q H(\tau,0) \rangle\, d\tau \qquad (106$$

and :

$$R_k = R(\Delta t_k) = \int_0^{\Delta t_k} \{ R + \langle H(\tau,0), Q H(\tau,0) \rangle \}\, d\tau \tag{107}$$

with:

$$H(t_{k+1}, t_k) = \int_0^{\Delta t_k} T(t,\tau)\, d\tau B = H^T(t_{k+1}, t_k) \tag{108}$$

Hence, the sampled-data linear regulator problem becomes the standard discrete linear quadratic regulator problem. Therefore choosing discrete time control to minimize the standard discrete linear quadratic problem, becomes the problem of minimizing Eq. (99). The problem of Eq. (99) in turn becomes that of evaluating the set of discrete weighting matrices (D.W.M.). Next we give the algorithms for evaluating these D.W.M. This job is the main problem of O.S.R. methodology.

1. EVALUATION OF $P(\Delta t_k)$

Recall that $\mathbf{T}(t) = \mathbf{e}^{At} = \mathbf{e}^{tA}$ where A is a bounded linear operator. Here in particular A is an N x N matrix of constant elements.

Define : $L(t) = \langle T(t), QT(t) \rangle$ (109)

then:

$$P(\Delta t_k) = \int_0^{\Delta t_k} L(\tau)\, d\tau \tag{110}$$

But T(t) is analytic in t, hence L(t) also is analytic in t, and so $P(\Delta t_k)$ is analytic. Denote the differential operator $\frac{d(.)}{dt}$ by D(.), and:

$$D^k(.) = \frac{d^k(.)}{dt^k} \tag{111}$$

Therefore, $D[\,L(t)\,] = A^T L(t) + L(t)A$ and :

$D^2[L(t)] = D\big[\,A^T L(t) + L(t)A\,\big] = A^T D[L(t)] + D[L(t)]A$. In general:

$$D^{k+1}[L(t)] = A^T D^k[L(t)] + D^k[L(t)]A \tag{112}$$

Using the analytic property of T(t) and the convergency property of Taylor series expansion of L(t), the following definitions hold.

$$L(t) = D^k[L(t)\,]\frac{t^k}{k!}\ ,\ k = 0, 1, 2, 3, \ldots \tag{113}$$

or

$$L(t) = C_p(k)\ \frac{t^k}{k!} \tag{114}$$

with :

$$C_p(k) = D^k[L(t)]_{t=0} \tag{115}$$

and:

$$M(k) = C_p(k)\frac{[\Delta t_k]^{k+1}}{(k+1)!} \tag{116}$$

with

$$C_p(0) = Q \tag{117}$$

$$M(0) = Q\Delta t_k \tag{118}$$

Moreover;

$$C_p(k+1) = A^T C_p(k) + C_p(k)A \tag{119}$$

$$M(k+1) = C_p(k+1)\frac{[\Delta t_k]^{k+2}}{(k+2)!} \tag{120}$$

$$= [\; A^T C_p(k) + C_p(k)A \;]\frac{[\Delta t_k]^{k+2}}{(k+2)!}$$

$$= A^T \frac{\Delta t_k}{k+2} M(k) + M(k)\frac{\Delta t_k}{k+2} A \tag{121}$$

for $k = 0,1,2,\ldots$

and

$$\left[\; P(\Delta t_k) = \sum_{k=0}^{k=\infty} M(k) \;\right] \tag{122}$$

Equation (122) is an infinite series. This series had to be truncated for practical purposes. Hence, we examine the convergence properties and the accuracy of the truncation procedure.

ERROR BOUNDS

Recall Eq. (119). Since $\|A^T\| = \|A\|$, then

$$\|C_p(k+1)\| \le \|A^T\|\,\|C_p(k)\| + \|C_p(k)\|\,\|A\| = 2\,\|A\|\,\|C_p(k)\| \tag{123}$$

Therefore; $$\|C_p(k+1)\| \le 2\,\|A\|\,\|C_p(k)\| \tag{124}$$

$$\left\| C_p(0) \right\| = \|Q\| \quad (\text{Eq. } (117)) \quad \text{and:} \quad \left\| C_p(k) \right\| \leq 2^k \|A\| \|Q\| \tag{125}$$

Therefore:

$$\left\| \sum_{k=m+1}^{\infty} \frac{C_p(k)}{(k+1)!} [\Delta t_k]^{k+1} \right\| \langle \|Q\| \left[\sum_{k=m+1}^{\infty} \frac{2^k \|A\|^k}{(k+1)!} \|\Delta t_k\|^{k+1} \right] \tag{126}$$

by Schwartz inequality

$$\left\| \sum_{k=m+1}^{\infty} \frac{C_p(k)}{(k+1)!} [\Delta t_k]^{k+1} \right\| \langle \left\{ 2^{m+1} \|A\|^{m+1} [\Delta t_k]^{m+2} \|Q\| \right\} X$$

$$\left\{ \sum_{K=0}^{\infty} \frac{2^k \|A\|^k}{(k+m+2)!} \|\Delta t_k\|^k \right\} \tag{127}$$

Hence denoting the term in $P(\Delta t_k)$ after the $(m+1)$th term by $E_p(m)$. $P(\Delta t_k)$ could be expressed as :

$$P(\Delta t_k) = \sum_{K=0}^{m} \frac{C_p(k)}{(k+1)!} [\Delta t_k]^{k+1} + E_p(m) \tag{128}$$

i.e.

$$E_p(m) \;\langle\; E(m) = 2^{m+1} \left[\frac{\|A\|^{m+1} \|Q\|}{(m+2)!} \exp[2\|A\| \, \Delta t_k] \right] \tag{129}$$

$$P(\Delta t_k) = \sum_{K=0}^{m} M(k) + E_p(m) \tag{130}$$

The function $E(m)$ is the upper bound limit on the error made in truncating the infinite series for $P(\Delta t_k)$ after m terms. As expressed in Eq. (129), the truncation error is a function of the desired accuracy. The accuracy, in turn, depends on the sampling interval (Δt_k), the operator matrix A, and the weghting matrix for the continuous performance index. Moreover, the convergence property of the series could be studied from the difference

$[E(m) - E(m+1)]$ or the ratio $\dfrac{E(m+1)}{E(m)}$.

2. EVALUATION OF $W(\Delta t_k)$

Recall that: $$\tfrac{1}{2}W_k = \tfrac{1}{2}W(\Delta t_k) = \int_0^{\Delta t_k} \langle T(\tau), Q\, H(\tau,0) \rangle \, d\tau \tag{131}$$

$$H(t_{k+1}, t_k) = \int_0^{\Delta t_k} T(t,\tau)\, d\tau\, G = H^T(t_{k+1}, t_k) \tag{132}$$

Now define: $$V(t) = \langle T(t), Q\, H(t,0) \rangle \tag{133}$$

then : $$\tfrac{1}{2}W(\Delta t_k) = \int_0^{\Delta t_k} V(t)\, dt \tag{134}$$

As we have seen previously V(t) is analytic in t, and using Taylor series expansion, Eq. (133) can be expressed as:

$$V(t) = \sum_{k=0}^{\infty} D^k V(t)\Big|_{t=0} \frac{t^k}{k!} \tag{135}$$

Step – by – step differentiation of Eq. (133) leads to Eq. (135). Further differentiation of Eq. (135) yields:

$$D^{k+1} V(t) = A^T D^k V(t) + D^k V(t) G \tag{136}$$

where : L(t) is as defined in Eq. (108). Further define:

$$C_w(k) = D^k V(t)\Big|_{t=0} \tag{137}$$

using Eqs. (136) and (133),

$$V(t) = \sum_{k=0}^{\infty} C_w(k) \frac{t^k}{k!}, \quad k = 0, 1, 2, 3, \ldots.. \tag{138}$$

where: $$C_w(0) = 0 \tag{139}$$

Recall that $C_p(0) = Q$ Hence, using Eqs. (138) and (134):

$$\tfrac{1}{2}W(\Delta t_k) = \int_0^{\Delta t_k} \sum_{k=0}^{\infty} C_w(k) \frac{t^k}{k!}\, dt \tag{140}$$

Therefore,

$$\tfrac{1}{2}\, W(\Delta t_k) = \sum_{k=0}^{\infty} C_w(k) \frac{(\Delta t_k)^{k+1}}{(k+1)!} \tag{141}$$

also define : $$D(k) = C_w(k) \frac{(\Delta t_k)^{k+1}}{(k+1)!} \tag{142}$$

then : $$\tfrac{1}{2}\, W(\Delta t_k) = \sum_{k=0}^{\infty} D(k) \tag{143}$$

Following up Eq. (142) recursively:

$$D(k+1) = A^T \frac{(\Delta t_k)\, D(k)}{(k+2)!} + \frac{(\Delta t_k)\, M(k)}{(k+2)!} \tag{144}$$

for $k = 0, 1, 2, 3, \ldots..$

where : $$M(k) = C_p(k) \frac{(\Delta t_k)^{k+1}}{(k+1)!} \tag{145}$$

$$D(0) = 0 \tag{146a}$$

$$M(0) = Q\,(\Delta t_k) \tag{146b}$$

Hence, $$\frac{M(k)}{D(k)} = \frac{C_p(k)}{C_w(k)} = \left. \frac{D^k\, L(t)}{D^k\, V(t)} \right|_{t=0} \tag{147}$$

Hence Eq. (144) becomes:

$$D(k+1) = A^T \frac{(\Delta t_k)\, D(k)}{(k+2)!} + G \frac{(\Delta t_k)\, C_p(k)\; D(k)}{(k+2)!\, C_w(k)} \tag{148}$$

since: $D(0) = 0$; $$W(\Delta t_k) = 2 \sum_{k=1}^{\infty} D(k) \tag{149}$$

Thus we again see that the state – control joint weighting matrix is an infinite sum of the series $D(k)$. As pointed out in the case of $P(\Delta t_k)$, evaluation of $W(\Delta t_k)$ has to be limited to finite sum of $D(k)$. As we did before, we will now pursue the error evaluation of the series sum limited to a finite n terms.

ERROR BOUNDS

Using Schwartz inequality, and assuming that A is non – singular:

$$\|C_w(k+1)\| \langle \|A\|\|C_w(k)\| + \|C_p(k)\|\|G\| \quad (150)$$

Using $C_p(k) = [2A]^k Q$ (151)

$$\|A\|^{-k}\|C_w(k+1)\| - \|A\|^{-k+1}\|C_w(k)\| \langle 2^k\|Q\|\|G\| \quad (152)$$

for $k = 0, 1, 2, 3, \ldots$

$$\text{L.H.S. of Eq. (152)} = \|A\|^{k+1}\left[\|C_w(k)\| - \|A\|\|C_w(0)\|\right] \quad (153)$$

R.H.S. of Eq. (152) is a geometric progression; first term = I common ratio = 2I, therefore, if S_k = sum of the first k terms

$$S_k = \frac{\|Q\|\|G\|\left[(2I)^k - I\right]}{[2I - I]} \quad (154)$$

substituting back in Eq. (152) and reducing terms leads to:

$$\|C_w(k)\| \langle \|A\|^{k-1}\|Q\|\|G\|\left[(2I)^k - I\right] \quad (155)$$

Hence $W(\Delta t_k)$ could be expressed as:

$$W(\Delta t_k) = 2\left[\sum_{k=0}^{n} \frac{C_w(k+1)\,(\Delta t_k)^{k+2}}{(k+2)!} + \sum_{k=n+1}^{\infty} \frac{C_w(k+1)\,(\Delta t_k)^{k+2}}{(k+2)!}\right] \quad (156)$$

Taking norm and using Schwartz inequality:

$$\tfrac{1}{2}\, W(\Delta t_k) = \sum_{k=1}^{n} D(k) + E_w(n) \quad (157)$$

where : $\|E_w(n)\| \langle E_w(n+1)\left\|\frac{G}{A}\right\|$ (158)

and

$$E_w(n+1) = 2^{n+2}\|A\|^{n+2}\left[\Delta t_k\right]^{n+2}\left[\|Q\|\left\|T(2\Delta t_k)\right\|\right] \quad (159)$$

The value of the error depends on n, and the desired accuracy could be used to determine n. This in turn stems from the sensitivity of the effect of the control on the state of the system controllability.

3. EVALUATION OF $R(\Delta t_k)$

Recall that:

$$R(\Delta t_k) = \int_0^{\Delta t_k} \left[R + H^T(t,0)\, Q\, H(t,0) \right] dt \tag{107}$$

restated as:

$$R(\Delta t_k) = \left[R_{cc} + R_{dd} \right] \tag{160}$$

where:

$$R_{cc} = \int_0^{\Delta t_k} R\, dt \tag{161}$$

$$R_{dd} = \int_0^{\Delta t_k} \left[H^T(t,0)\, Q\, H(t,0) \right] dt \tag{162}$$

$$\Gamma(t) \triangleq H^T(t,0)\, Q\, H(t,0) \tag{163}$$

and :

$$\frac{d\Gamma(t)}{dt} = G^T T^T(t)\, Q \int_0^t T(\tau)\, d\tau\, G + G^T \int_0^t T^T(\tau)\, d\tau\, Q\, T(t)\, G \tag{164}$$

using V(t) and repeated differentiation of Eq. (163), a recursive relation for $\Gamma(t)$ results as:

$$D^{k+1}\Gamma(t) = G^T D^k V(t) + D^k V(t) G \tag{165}$$

$$\text{for } k = 0, 1, 2, 3, \ldots$$

We also note that like L(t) and V(t); $\Gamma(t)$ is also analytic in t, which here denotes the sampling interval. Using Taylor series again:

$$\Gamma(t) = \sum_{k=0}^{\infty} D^k \Gamma(t)\big|_{t=0} \frac{t^k}{k!} \tag{166}$$

and define: $C_R(k) = D^k\, \Gamma(t)\big|_{t=0}$ (167)

From Eqs. (165) and (167):

$$C_R(k+1) = D^{k+1}\,\Gamma(t)\big|_{t=0} = G^T D^k V(t)\big|_{t=0} + D^k V(t) G\big|_{t=0} \tag{168}$$

Hence: $C_R(k+1) = G^T C_W(k) + C_W(k)G$ (169)

by appropriate substitution.

$$C_R(0) = G^T \int_0^0 T(t)\,dt\,Q \int_0^0 T(t)\,dt\,G = 0 \tag{170}$$

$$R_{dd}(\Delta t_k) = \int_0^{\Delta t_k} \Gamma(\tau)\,d\tau = \int_0^{\Delta t_k} \sum_{k=0}^{\infty} C_R(k)\frac{t^k}{k!}\,dt = \sum_{k=0}^{\infty} C_R(k)\frac{(\Delta t)^{k+1}}{(k+1)!} \tag{171}$$

Define $Y(k) = C_R(k)\dfrac{(\Delta t_k)^{k+1}}{(k+1)!}$ (172)

The recursive equation for Y(k) is:

$$Y(k+1) = C_R(k+1)\,\frac{(\Delta t_k)^{k+2}}{(k+2)!} \tag{173}$$

By substitution,

$$Y(k+1) = \left[G^T C_W(k) + C_W(k)G\right]\frac{(\Delta t_k)^{k+2}}{(k+2)!}, \tag{174}$$

for $k = 0, 1, 2, 3, \ldots$ and;

$$R_{dd}(\Delta t_k) = \sum_{k=0}^{\infty} Y(k) \tag{175}$$

Noting that: $C_R(0) = C_R(1) = 0$ (176)

$$R(\Delta t_k) = R_{cc} + \sum_{k=2}^{\infty} Y(k) \tag{177}$$

Again we can express $R(\Delta t_k)$ as a sum of infinite series.

ERROR BOUNDS

Similar to what has been done before, we express the error bounds as a function of the system's parameters. Let:

$$R(\Delta t_k) = R\,\Delta t_k + \sum_{k=2}^{s} Y(k) + \sum_{k=s+1}^{\infty} Y(k) \tag{178}$$

and if :

$$E_R(s) = \sum_{k=s+1}^{\infty} Y(k) \tag{179}$$

then :

$$R(\Delta t_k) = R\,\Delta t_k + \sum_{k=2}^{s} Y(k) + E_R(s) \tag{180}$$

where :

$$\| E_R(s) \| \langle E(s+2)\left[\left\| \frac{G}{A} \right\| \right]^2 \tag{181}$$

and for nonsingular A ,

$$R(\Delta t_k) = R_{cc} + \sum_{k=2}^{s} C_R(k) \frac{(\Delta t_k)^{k+1}}{(k+1)!} + E_R(s) \tag{182}$$

4. ESSENTIALS OF THE RESULTS

The main results of the section are :

$$P(\Delta t_k) = \sum_{k=0}^{m} M(k) + E_p(m) \tag{128}$$

where M(k) is given by Eq. (116), and

$$E_p(m) \langle E(m) = 2^{m+1} \| A \|^{m+1} \| Q \| e^{2\|A\|\Delta t_k} \tag{129}$$

$$\tfrac{1}{2} W(\Delta t_k) = \sum_{k=1}^{n} D(k) + E_W(n) \tag{157}$$

where D(k) is given by Eq. (142) ,

$$\left\| E_W(n) \right\| \langle \; E(n+1) \left\| \frac{G}{A} \right\| \tag{158}$$

$$\text{and};\; E(n+1) = 2^{n+2} \left\| A \right\|^{n+2} (\Delta t_k)^{n+2} \left\| Q \right\| \left\| T(2\Delta t_k) \right\| \tag{159}$$

$$R(\Delta t_k) = R_{cc} + \sum_{k=2}^{s} Y(k) + E_R(s) \tag{180}$$

where Y(k) is given by Eq. (172), and

$$\left\| E_R(s) \right\| \langle \; E(s+2) \left[\frac{G}{A} \right]^2 \tag{181}$$

5. COMPUTER IMPLEMENTATION

It has been established that the curse of dimensionality is one deterrent to the attraction of the O .S .R. methodology. Even for moderately sized systems, the series solution for $P(\Delta t_k)$, $W(\Delta t_k)$, and $R(\Delta t_k)$ leads to a large number of terms, not to talk of distributed parameter systems of infinite dimension. For such cases, computer storage locations and round-off errors become excessively prohibitive. This is one of the main objectives and problems of this work : To solve the problems of excessive need for computer storage locations or memory needs i.e. "the limited memory" requirements in the problem statement. In order to finally attain this objective, after going this far in terms of approximations and representations that produce limited terms algorithms, the system operator A has to be further preconditioned. One method of such preconditioning is hereby presented.

PRECONDITIONING THE SYSTEM OPERATOR-MATRIX A.

The method is directed at reconditioning of the independent variable Δt_k. The system operator A and sampling interval Δt_k are known. We find a parameter r such that :

$$\frac{1}{2^r}\,\Delta t_k \;<\; \frac{1}{\|A\|} \tag{182}$$

i.e.

$\Delta t_k\ \|A\| < 2^r$ one r exists that satisfies Eq. (182). Let:

$$\frac{1}{2^r}\,\Delta t_k \;=\; \in \tag{183}$$

Therefore, $\in < \frac{1}{\|A\|}$ i.e. $\in \|A\| < 1$ (184)

From which the eigenvalue of A is found within a unit circle in the complex plane. Hence:

$$\lim_{k\to\infty} (\in A)^k = 0 \tag{185}$$

A typical example will be given. This example was computed using the TI - SR52 programmable calculator. The example will also demonstrate both the use of the algorithm and its usefulness in overcoming the curse of dimensionality and the prohibitive nature of manual calculations for this type of problem.

6. EXAMPLE 5.

Assume, for the sake of a simple example to be computationally carried out with pencil and paper and a pocket calculator, that the following three-dimensional linear time-invariant system is an approximate model of a simple distributed parameter system.

$$\dot{Z}(t) = AZ(t) + Bu(t), \quad \text{with } Z(t_0) = Z_o \text{ where;}$$

$$A = \begin{bmatrix} 2 & 1 & 1 \\ 1 & 2 & 1 \\ 1 & 1 & 2 \end{bmatrix}, \quad B = \begin{bmatrix} 1 & 0 \\ 0 & 1 \\ 0 & 0 \end{bmatrix} \quad \text{with a cost function:}$$

$$J_L = \int_0^T \langle Z(t), Z(t) \rangle_Q + \langle u(t), u(t) \rangle_R \, dt \quad \text{where ;}$$

$$Q = \begin{bmatrix} 1 & 0 & 0 \\ 0 & 1 & 0 \\ 0 & 0 & 1 \end{bmatrix}, \quad R = \begin{bmatrix} 1 & 0 \\ 0 & 1 \end{bmatrix} \quad \text{with } t_0 = 0 \text{ and equal weighting}$$

on the state and control. Assume that a sampled-control input u_k is applied on the system, where the sampled/discrete signal u_k $k = 0, 1, 2, 3, \ldots N\text{-}1$ is constant over the subinterval t_k with $N \times t_k = T$. A sampled-data linear optimal regulator results. Suppose the sampling period is 0.5 second, compute the state, control, and joint state-control weighting matrices associated with the performance index.

$$J_N(.) = \sum_{k=0}^{N-1} Z^T(k)\,P(k)\,Z(k) + Z^T(k)\,W(k)\,u(k) + u^T(k)\,R(k)\,u(k)$$

COMPUTATION OF $P(\Delta)$.

Recall Eq. (128) where the series had been truncated after m terms. $E_p(m) < 7 \times 10^{-10}$. Sampled − time $\Delta t_k = \Delta = 0.5$ sec

$\frac{\Delta t_k}{2^r} < \frac{1}{\|A\|} = \frac{1}{2 x 2 x 2} = \frac{1}{8} \Rightarrow r = 2$, Hence $\Delta = \frac{1}{8}$ satisfying the requirement. Calculating $P(\Delta) = \sum_{k=0}^{m} C_p(k) \frac{(\Delta)^{k+1}}{(k+1)!}$

For $k = 0$; $P(\frac{1}{8}) = C_p(0) \times \frac{1}{8} \times \frac{1}{1!} = \frac{1}{8} C_p(0)$ (recall that $C_p(0) = Q$)

$P(\Delta) = \frac{Q}{8} = \begin{bmatrix} \frac{1}{8} & 0 & 0 \\ 0 & \frac{1}{8} & 0 \\ 0 & 0 & \frac{1}{8} \end{bmatrix}$; for $k = 1$; $P(\Delta) = C_p(1) \times \frac{(\frac{1}{8})^2}{2!} = \frac{C_p(1)}{128}$

Recall that;

$$C_p(1) = A^T\, T(\tau)\big|_{\tau=0} + T(\tau)\big|_{\tau=0} A + A^T Q + Q A \;\;(\text{since } C_p(0) = Q)$$

$$C_p(1) = 2\,A\,Q \;(\text{since A and Q are symmetric})$$

$$P(\tfrac{1}{8}) = \frac{2\,A\,Q}{128} = \begin{bmatrix} 0.03125 & 0.015625 & 0.015625 \\ 0.015625 & 0.03125 & 0.015625 \\ 0.015625 & 0.015625 & 0.03125 \end{bmatrix} = P(\tfrac{1}{8})\big|_{k=1}$$

similarly $C_p(2) = A^T C_p(1) + C_p(1) A = A^T (2AQ) + (2AQ) A$

$$= 2A^2Q + 2A^2Q = 4A^2Q \;(\text{A and Q symmetric})$$

$$4\,A^2Q = \begin{bmatrix} 24 & 20 & 20 \\ 20 & 24 & 20 \\ 20 & 20 & 24 \end{bmatrix} \quad \text{But} \quad P(\Delta)\big|_{K=2} = C_p(2)\,x\left[\tfrac{1}{8}\right]^3 x\,\frac{1}{3!} = \frac{4\,A^2Q}{512\,x\,64}$$

$$= \frac{1}{(10)^2}\begin{bmatrix} 0.78125 & 0.65104167 & 0.65104167 \\ 0.65104167 & 0.78125 & 0.65104167 \\ 0.65104167 & 0.65104167 & 0.78125 \end{bmatrix} = P(\tfrac{1}{8})\big|_{k=2} \quad \text{and}$$

for $k=3$: $C_p(3) = A^T\big[C_p(2)\big] + \big[C_p(2)\big]A = A^T\big[4A^2Q\big] + \big[4A^2Q\big]A$

$$\text{since } \frac{(\Delta)^4}{4!} = \frac{1}{64^2}\,x\,\frac{1}{24} \text{ and } A^3Q = \begin{bmatrix} 22 & 21 & 21 \\ 21 & 22 & 21 \\ 21 & 21 & 22 \end{bmatrix}, \text{ then } \frac{8\,A^3Q}{64^2 x\, 24} =$$

$$P(\tfrac{1}{8})\big|_{k=3} = \begin{bmatrix} 1.7903646 & 1.7089844 & 1.7089844 \\ 1.7089844 & 1.7903646 & 1.7089844 \\ 1.7089844 & 1.7089844 & 1.7903646 \end{bmatrix} x\,10^{-3} = P(\tfrac{1}{8})\big|_{k=3}$$

for $k=4$: $C_p(4) = 16A^4Q$: and $\dfrac{(\Delta)^5}{5!\,x\,64^2 x\,960}$ therefore ;

$$\frac{1}{64^2 x\,960}\begin{bmatrix} 86 & 85 & 85 \\ 85 & 86 & 85 \\ 85 & 85 & 86 \end{bmatrix} = P(\tfrac{1}{8})\big|_{k=4} = \frac{1}{10^4}\begin{bmatrix} 3.499349 & 3.458659 & 3.458659 \\ 3.458659 & 3.499349 & 3.458659 \\ 3.458659 & 3.458659 & 3.499349 \end{bmatrix}$$

$$= P(\tfrac{1}{8})\big|_{k=4} \text{ :and for } k=5;\ C_p(5)\,x\,\frac{(\Delta)^6}{6!} = \frac{32(\Delta)^5 Q}{64^2 x 64\ x\ 720}\begin{bmatrix} 342 & 341 & 341 \\ 341 & 342 & 341 \\ 341 & 341 & 342 \end{bmatrix}$$

$$= \begin{bmatrix} 5.79834 & 5.78139 & 5.78139 \\ 5.78139 & 5.79834 & 5.78139 \\ 5.78139 & 5.78139 & 5.79834 \end{bmatrix} x\,10^{-5} = P(\tfrac{1}{8})\big|_{k=5} : \text{ and for } k=6;$$

$$\frac{C_p(6)\,(\Delta)^7}{7!} = \begin{bmatrix} 1366 & 1365 & 1365 \\ 1365 & 1366 & 1365 \\ 1365 & 1365 & 1366 \end{bmatrix} x\,\frac{1}{64^2 x\,40320} =$$

$$\begin{bmatrix} 8.2712 & 8.2652 & 8.2652 \\ 8.2652 & 8.2712 & 8.2652 \\ 8.2652 & 8.2652 & 8.2712 \end{bmatrix} x\, 10^{-6} = \quad P(\tfrac{1}{8})\Big|_{k=6} : \text{ For } k=7\text{: } C_p(7)\, x \frac{(\Delta)^8}{8!} =$$

$$\frac{128\, A^7\, Q}{(128)^2\, x\, 40320} = \begin{bmatrix} 5462 & 5461 & 5461 \\ 5461 & 5462 & 5461 \\ 5461 & 5461 & 5462 \end{bmatrix} \frac{1}{32\, x\, 40320} =$$

$$\begin{bmatrix} 10.335 & 10.333 & 10.333 \\ 10.333 & 10.335 & 10.333 \\ 10.333 & 10.333 & 10.335 \end{bmatrix} x\, 10^{-7} = P(\tfrac{1}{8})\Big|_{k=7}\text{: Hence, } P(\tfrac{1}{8}) = \sum_{k=0}^{7} C_p(k) \frac{(\Delta)^{k+1}}{(k+1)!}$$

$$P(\tfrac{1}{8}) = \begin{bmatrix} 0.1662700875 & 0.0242573794 & 0.0242573794 \\ 0.0242573794 & 0.1662700875 & 0.0242573794 \\ 0.0242573794 & 0.0242573794 & 0.1662700875 \end{bmatrix}$$

CALCULATION OF $\frac{1}{2}W(\Delta)$.

Recall that

$$\frac{1}{2} W(\Delta t_k) = \sum_{k=1}^{\infty} C_w(k) \frac{(\Delta t_k)^{k+1}}{(k+1)!} = \sum_{k=1}^{n} C_w(k) \frac{(\Delta t_k)^{k+1}}{(k+1)!} + E_w(n)$$

All computational parameters remain as before:

$$\sum_{k=1}^{\infty} \frac{(\Delta)^{k+1}}{(k+1)!} = \frac{1}{8^2\, x\, 2!} + \frac{1}{8^3\, x\, 3!} + \frac{1}{8^4\, x\, 4!} + \ldots \frac{1}{8^n\, x\, n!} + \ldots$$

since $C_w(0) = 0$ and $C_w(k) = \left.\frac{d^k V(\tau)}{d\tau^k}\right|_{\tau=0}$ where:

$V(\tau) = e^{A\tau} Q \int_0^{\tau} e^{At} B\, dt$ so that, for $k = 1$: $C_w(1) = \left.\frac{dV(\tau)}{dt}\right|_{\tau=0} =$

$A^T V(\tau)\big|_{\tau=0} + T(\tau)\big|_{\tau=0} B$ where: $T(\tau) = e^{A'\tau} Q\, e^{A\tau}$; $C_w(1) = Q\,B$.

$$C_w(2) = \left.\frac{d^2 V(\tau)}{d\tau^2}\right|_{\tau=0} = A\left[A^T V(\tau)\big|_{\tau=0} + T(\tau)\big|_{\tau=0} B\right] + B\, x\, 2\, A\, T(\tau)\big|_{\tau=0}$$

$$= \left[A^2 V(\tau) + A\,B\,T(\tau) + 2\,A\,B\,T(\tau)\right]\Big|_{\tau=0} = \left[\, 3\,A\,B\,T(\tau)\right]\Big|_{\tau=0} = 3\,A\,B\,Q$$

i.e. $C_w(2) = 3\,A\,BQ$

Similarly: $C_w(3) = \left.\frac{d^3V(\tau)}{d\tau^3}\right|_{\tau=0} = \left.\frac{d[C_w(2)]}{d\tau}\right|_{\tau=0} = A^2BQ + 6\ A^2BQ$

$= 7\ A^2BQ$ i.e. $C_w(3) = 7\ A^2BQ$. Similar operations show that $C_w(4) = 15\ A^2BQ$. Following similar steps, we obtain the following coefficients for for $C_w(k)$: Let the coefficient of $C_w(k)$ be c(k), then:

$c(0) = 0$; $c(1) = 1$; $c(2) = 3$; $c(3) = 7$; $c(4) = 15$; $c(5) = 31$; $c(6) = 63$; $c(7) = 127$; $c(8) = 255$; etc.

Showing that the coefficients computation follows the algorithms:

$$c(k) = \sum_{j=1}^{k} {}^kC_j = \sum_{j}^{k} k\,\text{combination}\,j\ ;\ j = 1, 2, \ldots k\ ;\ k = 1, 2, \ \ldots \quad (*0*)$$

$$\text{Hence } C_w(k) = c(k)\,A^{k-1}BQ = \left[\sum_{j=1}^{k} {}^kC_j\right]\left[A^{k-1}BQ\right] \quad (*1*)$$

further giving the following few terms: $C_w(5) = 31\ A^4BQ$; $C_w(6) = 63\,A^5BQ$; $C_w(7) = 127\,A^6BQ$; $C_w(8) = 255\,A^7BQ$; etc. (*2*)

In relatively simpler language, we can note that defining $c(k) = 0$: for $k = 0$; the above coefficients can be obtained as follows;

$c(k) = c(k-1) + 1$ for $k = 1, 2, 3, \ldots n$ (*3*)

giving; $c(0) = 0$; $c(1) = 2 \times 0 + 1 = 1$; $c(2) = 2 \times 1 + 1 = 3$; $c(3) = 2 \times 3 + 1 = 7$; $c(4) = 2 \times 7 + 1 = 15$; $c(5) = 2 \times 15 + 1 = 31$; $c(6) = 2 \times 31 + 1 = 63$; $c(7) = 2 \times 63 + 1 = 127$; $c(8) = 2 \times 127 + 1 = 255$; etc. By direct computation:

$C_w(1) = \begin{bmatrix} 1 & 0 \\ 0 & 1 \\ 0 & 0 \end{bmatrix}$ since $Q = I$, the identity matrix. The set $C_w(k)$

could be expressed in terms of $C_w(2)$ by using (*2*) i.e.

$$C_w(2) = 3\,A\,B = \begin{bmatrix} 6 & 5 \\ 5 & 6 \\ 5 & 5 \end{bmatrix};\ C_w(k) = c(k)\,C_w(2)\,;\ k = 1, \ldots n \quad (*4*)$$

computer for the computation of the set of the discrete weighting matrices required in the similar regulator problem.

---

7. EXAMPLE 6.

Consider a R.O.M. of a D.P.S. governed by the vector O.D.E.;

$$\dot{Z}(t) = A\,Z(t) + G\,u(t) \quad \text{with } Z(0) = Z_0$$

It is assumed that the coefficient of the P.D.E. is unity. Moreover

$\Delta s = 0.1$ and A = a 10×10 tridiagonal matrix with $t_0 = 90$; $t_1 = -189$; $t_2 = 100$. These parameters completely specify the matrix A. $G = I$ = a 10×10 identity matrix.

Let the performace index be quadratic in state and control with: $Q = R = I$, also a 10×10 identity matrix. This means equal weightings on the state and control.

PROBLEM: The problem is to compute the discrete weighting matrices required in implementing the sampled–data optimal regulator problem.

SOLUTIONS This problem was simulated on the digital computer. The results are available in [21].

F. DISTRIBUTED PARAMETER SYSTEMS REDUCED ORDER WEIGHTING MATRICES SYNTHESIS [L. Q. G.] - INDIRECT METHOD.

Another method for synthesizing the set of discrete weighting matrices for implementing the D. P. S. optimal regulator problem will be presented in this section, using the full order model of the D. P. S. This we refer to as the indirect method. The approach would be similar but the model on which the algorithm is based would be different.

Recall the D.P.S. full order model: